Electroresponsive Molecular and Polymeric Systems

ELECTRORESPONSIVE MOLECULAR AND POLYMERIC SYSTEMS

A Series of Reference Books and Textbooks

EDITOR IN CHIEF

Terje A. Skotheim
Division of Materials Science
Department of Applied Science
Brookhaven National Laboratory
Upton, New York

EDITORIAL BOARD

Electroresponsive Molecular and Polymeric Systems

VOLUME 2

edited by

Terje A. Skotheim

BROOKHAVEN NATIONAL LABORATORY
UPTON, NEW YORK

CRC Press
Taylor & Francis Group
Boca Raton London New York

CRC Press is an imprint of the
Taylor & Francis Group, an **informa** business

First published 1991 by Marcel Dekker, Inc.

Published 2019 by CRC Press
Taylor & Francis Group
6000 Broken Sound Parkway NW, Suite 300
Boca Raton, FL 33487-2742

© 1991 by Taylor & Francis Group, LLC
CRC Press is an imprint of Taylor & Francis Group, an Informa business

First issued in paperback 2019

No claim to original U.S. Government works

ISBN 13: 978-0-367-45066-3 (pbk)
ISBN 13: 978-0-8247-8422-5 (hbk)

Visit the Taylor & Francis Web site at
http://www.taylorandfrancis.com

and the CRC Press Web site at
http://www.crcpress.com

Library of Congress Cataloging-in-Publication Data
(Revised for vol. 2)

Electroresponsive molecular and polymeric
    systems.

Includes bibliographies and indexes.
    1. Polymers--Electric properties.
I.   Skotheim, Terje A.
QD381.9.E38E395 1988   620.1'9204297   88-10855

About the Series

The science of electroresponsive polymers has become one of the most expansive and dynamic fields of research in recent years. Few other areas span to such an extent the whole spectrum of scientific endeavor, from synthetic organic chemistry to theoretical physics. The literature is therefore equally scattered, and any overview is obtained only with great difficulty. There is a need for a common forum to review the major developments of the field.

This series is devoted to discussing the central topics at a level that is useful to those actively working in the field as well as to those seeking entry into a new area. The areas we intend to cover include recent developments in electronically conducting conjugated polymers, ionically conducting polymers, redox polymers, piezo- and pyroelectric polymers, ferromagnetic systems, nonlinear optical systems, and developments relating to the prospects for achieving high-temperature superconductivity in organic materials. Other topics will be covered as they develop. Both molecular and polymeric systems will be covered.

Developments described in this series represent a dramatic expansion of the scope of polymer science, from being a field dealing primarily with structural and insulating materials to one that includes materials with a potentially endless variety of electrical and optical properties. With this leap we have truly come to the threshold of a new era, not only in polymer science, but in materials science itself. The prospect of employing the skills of synthetic chemists to tailor materials with electrical and optical properties as precise and specific as those of biological systems opens vistas we have not seen

Contributors

Larry R. Dalton, Ph.D. Professor, Department of Chemistry, University of Southern California, Los Angeles, California

A. F. Garito, Ph.D. Professor, Department of Physics, University of Pennsylvania, Philadelphia, Pennsylvania

J. R. Heflin Ph.D. Candidate in Physics, Department of Physics, University of Pennsylvania, Philadelphia, Pennsylvania

Malcolm R. McLean, Ph.D. Postdoctoral Research Associate, Department of Chemistry, University of Southern California, Los Angeles, California

Seizo Miyata, Ph.D. Professor, Department of Material Systems Engineering, Tokyo University of Agriculture and Technology, Tokyo, Japan

Hari Singh Nalwa, Ph.D* Lecturer, Department of Material Systems Engineering, Tokyo University of Agriculture and Technology, Tokyo, Japan

David W. Polis, Ph.D. Senior Postdoctoral Research Associate, Department of Chemistry, University of Southern California, Los Angeles, California

---

Current affiliation: Hitachi Research Laboratory, Hitachi, Ltd., Ibaraki, Japan

Martin Pomerantz, Ph.D.   Professor, Center for Advanced Polymer
  Research, Department of Chemistry, The University of Texas
  at Arlington, Arlington, Texas

John R. Reynolds, Ph.D.   Associate Professor, Center for Advanced
  Polymer Research, Department of Chemistry, The University
  of Texas at Arlington, Arlington, Texas

Kenneth D. Singer, Ph.D.*   Distinguished Member of Technical Staff,
  Optical Technologies Department, AT&T Bell Laboratories,
  Princeton, New Jersey

John E. Sohn, Ph.D.   Member of Technical Staff, AT&T Bell
  Laboratories, Princeton, New Jersey

Anuntasin Techagumpuch, Ph.D.†   Visiting Scientist, Material Systems
  Engineering, Faculty of Technology, Tokyo University of Agriculture and Technology, Tokyo, Japan

Luping Yu, Ph.D.   Department of Chemistry, University of Southern
  California, Los Angeles, California

Current affiliations:
*Department of Physics, Case Western Reserve University,
Cleveland, Ohio
†Associate Professor, Department of Physics, Faculty of Science,
Chulalongkorn University, Bangkok, Thailand

Contents of Other Volumes

Electroresponsive Molecular and Polymeric Systems

1

Electron Correlation Description of the Nonlinear Optical Properties of Lower Dimensional Conjugated Structures

J. R. Heflin and A. F. Garito / University of Pennsylvania, Philadelphia, Pennsylvania

I. INTRODUCTION

Conjugated π-electron organic and polymer structures are now well known to exhibit unusually large nonresonant macroscopic second-order $\chi_{ijk}^{(2)}(-\omega_3; \omega_1, \omega_2)$ and third-order $\chi_{ijkl}^{(3)}(-\omega_4; \omega_1, \omega_2, \omega_3)$ nonlinear optical susceptibilities [1-9], and their microscopic origin and mechanism can be successfully described by quantum field theory of many-electron systems in one and two dimensions [10-19]. In this description, as the spatial dimensionality of the many-electron system is effectively lowered, the motion among the many electrons becomes highly correlated [20]. In conjugated systems, the correlated motion of the π-electrons determines $\chi_{ijk}^{(2)}$ and $\chi_{ijkl}^{(3)}$, and, as we shall review, the many-body correlations that arise from electron-electron interactions must be explicitly taken into account in a thorough treatment of the nonlinear optical properties of these structures.

The spatial dimensionality of a many-electron system is effectively reduced in material structures having confinement length scales L of less than one to several tens of nanometers. The opposite limit occurs in regular three-dimensional bulk structures where L > 100 nm, and the electrons behave as a Fermi gas of weakly interacting particles. In this case, the electron motions are usually weakly correlated and well described by single-particle theory in the effective mass approximation.

Conjugated π-electron organic and polymer structures are natural lower-dimensional systems in which electron correlation effects are important. In the principal case of conjugated linear chains, the many-electron system is confined in two dimensions with L < 0.5 nm transverse to the linear chain axis, and thus the π-electrons are delocalized in their motion only in one dimension along the longitudinal chain axis. As a result, the repulsive Coulomb interactions between electrons are strong, and electron motion becomes highly correlated wherein the motion of any single electron depends on all the other remaining many electrons. The origin of the nonresonant $\chi_{ijkl}^{(3)}$ for these and related polymer structures then appears to reside in strong correlation behavior in virtual two-photon π-electron states [14–16].

In this review, we focus mainly on our recent studies [14–19] of the prototype one-dimensional case of conjugated linear chains known as polyenes. These are hydrocarbon chains in which each carbon site is covalently bonded to its two nearest-neighbor carbon atoms and a single hydrogen atom. The remaining valence electron of each carbon atom contributes to the delocalized π-electron system along the chain axis having an alternating single-bond/double-bond dimerized structure. The two most common structural conformations of polyene chains are known as the all-*trans* and *cis-transoid* conformations, hereafter referred to simply as *trans* and *cis*, respectively. An example of trans and cis conformations is illustrated in Fig. 1 for octatetraene (OT), the polyene chain with number of carbon sites N = 8.

The studies were performed from 1987 to 1989 using a CRAY X-MP/48 able to perform a 6 × 6 singly and doubly excited configuration calculation but requiring more than an hour of CPU time and 4 megawords of memory. To calculate longer chains required greater memory capacity beyond that of the CRAY X-MP, so approximations

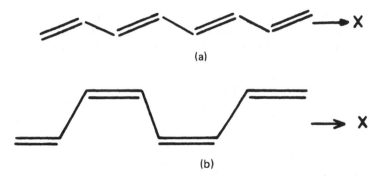

(a)

(b)

Figure 1. Schematic diagrams of the molecular structures for (a) all-*trans* and (b) *cis-transoid* polyenes.

were made by selecting important doubly excited configurations. In 1990, the CRAY X-MP was upgraded to a Y-MP with increased memory capacity of 32 megawords and three times faster speed, which together with software development enabled full 8 × 8 SDCI calculations in less than 2 minutes for calculations of both the real and imaginary parts of $\gamma_{ijkl}(-\omega_4; \omega_1, \omega_2, \omega_3)$. Recent results have revealed no significant changes in the origin and mechanism of $\gamma_{ijkl}(-\omega_4; \omega_1, \omega_2, \omega_3)$ in conjugated chains with the exception that the results for the longest chains lead to a smaller exponent in the power-law dependence of $\gamma_{ijkl}(-\omega_4; \omega_1, \omega_2, \omega_3)$ on chain length. The complete results from the full calculations are forthcoming.

This section reviews the formalism for the microscopic description of nonlinear optical responses in molecular structures with specific emphasis on the many-electron configuration interaction method as applied to low-dimensional conjugated chains. In Sections III and IV, we present the results for the microscopic third-order susceptibilities $\gamma_{ijkl}(-\omega_4; \omega_1, \omega_2, \omega_3)$ of trans and cis polyenes ranging in chain length from N = 4 to 16. After the calculated values are successfully compared with experimental measurements, the fundamental origin of the nonlinear optical response is described in detail, especially the discovery of the important role of virtual transitions to strongly correlated, high-lying, two-photon ^1A_g states. The results of these two sections have subsequently been independently confirmed by alternative theoretical approaches, such as diagrammatic valence-bond (DVB) theory [21] applied to the Pariser-Parr-Pople (PPP) Hamiltonian and multiply excited configuration interactions of INDO molecular orbital theory [22]. Prior calculations of the nonlinear optical properties of conjugated linear chains had employed free-electron [23], Huckel [24-26], and single particle theories [27-31]. All of these neglected electron correlation effects. In Section V, we consider the effect of lowered symmetry on $\gamma_{ijkl}(-\omega_4; \omega_1, \omega_2, \omega_3)$ by hetero-atomic substitution on a conjugated linear chain and find that an order-of-magnitude enhancement of the nonlinear optical response can be achieved.

The role of dimensionality is considered in Section VI by analysis of $\gamma_{ijkl}(-\omega_4; \omega_1, \omega_2, \omega_3)$ in quasi-two-dimensional conjugated cyclic structures. Increase of dimensionality reduces the isotropically averaged nonlinear susceptibility through reduction of the effective length over which π-electrons can respond to the applied optical electric field. Section VII concerns the microscopic origin of $\chi_{ijkl}^{(3)}(-\omega_4; \omega_1, \omega_2, \omega_3)$ in environmentally robust rigid rod polymers and concludes that the mechanism for nonlinear optical response in these materials is identical to that for the prototype polyene structures. Finally, closing remarks are made in Section VIII.

II. THEORETICAL FORMALISM

The macroscopic nonlinear optical properties of organic molecular and polymer structures in condensed states are best described starting from the individual responses of isolated molecular or polymer chain units [1]. This approach, which clarifies the origin of $\chi_{ijk}^{(2)}$ and $\chi_{ijkl}^{(3)}$, is, however, not applicable to other material classes, such as inorganic semiconductors, where it is difficult to identify analogous microscopic sources of nonlinear polarization. For the moment neglecting intermolecular interactions, if the nonlinear susceptibility of an isolated molecule is known, then $\chi_{ijk}^{(2)}$ or $\chi_{ijkl}^{(3)}$ of the macroscopic ensemble of molecules is determined by the orientational distribution function of the independent units. Local field factors must also be included to account for the effect of the dielectric environment on the electric field strength at the molecular site. The macroscopic frequency-dependent second- and third-order susceptibilities $\chi_{ijk}^{(2)}(-\omega_3; \omega_1, \omega_2)$ and $\chi_{ijkl}^{(3)}(-\omega_4; \omega_1, \omega_2, \omega_3)$ can then be expressed in terms of the molecular susceptibilities $\beta_{ijk}(-\omega_3; \omega_1, \omega_2)$ and $\gamma_{ijkl}(-\omega_4; \omega_1, \omega_2, \omega_3)$ as [32]

$$\chi_{ijk}^{(2)}(-\omega_3; \omega_1, \omega_2) =$$

$$N_u \sum_{s=1}^{c} R_{im'}{}^{s} R_{jn'}{}^{s} R_{ko'}{}^{s} f_{m'i'}{}^{\omega_3} \beta_{i'j'k'}{}^{s}(-\omega_3; \omega_1, \omega_2) f_{j'n'}{}^{\omega_1} f_{k'o'}{}^{\omega_2}$$

$$(1)$$

and

$$\chi_{ijkl}^{(3)}(-\omega_4; \omega_1, \omega_2, \omega_3) =$$

$$N_u \sum_{s=1}^{c} R_{im'}^s R_{jn'}^s R_{ko'}^s R_{lp'}^s f_{m'i'}^{\omega_4}$$

$$\times \gamma_{i'j'k'l'}^s(-\omega_4; \omega_1, \omega_2, \omega_3) f_{j'n'}^{\omega_1} f_{k'o'}^{\omega_2} f_{l'p'}^{\omega_3} \qquad (2)$$

where N_u is the number of unit cells per unit volume, the summation is over all molecules in the unit cell, R is a rotation matrix describing the orientation of each molecule in the unit cell, and f is the frequency-dependent local field factor. The description of the macroscopic nonlinear optical response is thus reduced to an understanding of the microscopic second- and third-order susceptibilities β_{ijk} and γ_{ijkl} and knowledge of the orientational distribution of the molecular units in the condensed phase. As special cases, isotropic gases and liquids reduce Eqs. (1) and (2) to simpler forms. For example, for an isotropic ensemble,

$$\chi_{1111}^{(3)}(-\omega_4; \omega_1, \omega_2, \omega_3) = N f^{\omega_4} f^{\omega_1} f^{\omega_2} f^{\omega_3} \gamma_g(-\omega_4; \omega_1, \omega_2, \omega_3) \qquad (3)$$

where N is the number density of molecules and γ_g is the isotropically averaged susceptibility defined by

$$\gamma_g = \frac{1}{5} \left[\sum_i \gamma_{iiii} + \frac{1}{3} \sum_{i \neq j} (\gamma_{iijj} + \gamma_{ijij} + \gamma_{ijji}) \right] \qquad (4)$$

where the indices i and j represent the Cartesian coordinates x, y, and z.

The above formalism is strictly appropriate only for gases such that the mean intermolecular distance is large enough for the interaction between molecules to be negligible. In condensed states, some modifications are required to include the effects of intermolecular interactions. In order to account for residual interactions with neighboring molecules, $\beta_{ijk}(-\omega_3; \omega_1, \omega_2)$ and $\gamma_{ijkl}(-\omega_4; \omega_1, \omega_2, \omega_3)$ can be "dressed" to extend their applicability beyond isolated molecular units. For instance, in dc-induced second-harmonic-generation (DCSHG) studies of dilute solutions [5,6], it was found that inclusion of dipole-mediated interactions between solvent and solute molecules provided excellent agreement of the theoretical values for an otherwise independent molecule with experimental measurements made in solution. In a more recent study [19,33] of phase-matched second harmonic generation in single crystals of 2-methyl-4-nitroaniline (MNA), the existence of closely oriented pairs of molecules in the unit cell crystal structure required treatment of an MNA-MNA pair as the fundamental source of second-order response. Because of the strong mutual effect of each molecule on the electronic structure of the other, calculation of $\beta_{ijk}(-2\omega;\omega,\omega)$ for an appropriately oriented pair of MNA molecules was necessary in order to obtain agreement with experiment. In any case, the fact that the intramolecular interaction energy is much stronger than the intermolecular interaction energy in organic molecular crystals, liquids, solutions, and polymer thin films means that the starting point for understanding the macroscopic nonlinear optical properties of these various condensed phases lies in an accurate description of the microscopic response from an isolated molecular unit.

The general theoretical expression for the components of the microscopic third-order susceptibility tensor $\gamma_{ijkl}(-\omega_4;\omega_1,\omega_2,\omega_3)$ is derived from time-dependent, quantum electrodynamic perturbation theory. In order to avoid secular divergences that would occur when any subset of the input frequencies sums to zero, one employs the Bogoliubov-Mitropolsky method of averages [34,35]. Owing to dispersive effects, $\gamma_{ijkl}(-\omega_4;\omega_1,\omega_2,\omega_3)$ is dependent on the input and output frequencies involved for each of the various possible nonlinear optical phenomena. The general expression for a third-order process is [35]

$$\gamma_{ijkl}(-\omega_4;\omega_1,\omega_2,\omega_3) = K(-\omega_4;\omega_1,\omega_2,\omega_3)\;\frac{e^4}{\hbar^3}\;I_{1,2,3}$$

$$\times\left[\sum_{n_1n_2n_3}'\left\{\frac{r_{gn_3}{}^i\;\bar{r}_{n_3n_2}{}^l\;\bar{r}_{n_2n_1}{}^k\;r_{n_1g}{}^j}{(\omega_{n_3g}-\omega_4)(\omega_{n_2g}-\omega_1-\omega_2)(\omega_{n_1g}-\omega_1)}\right.\right.$$

$$+\frac{r_{gn_3}{}^l\;\bar{r}_{n_3n_2}{}^i\;\bar{r}_{n_2n_1}{}^k\;r_{n_1g}{}^j}{(\omega_{n_3g}+\omega_3)(\omega_{n_2g}-\omega_1-\omega_2)(\omega_{n_1g}-\omega_1)}$$

$$+\frac{r_{gn_3}{}^j\;\bar{r}_{n_3n_2}{}^k\;\bar{r}_{n_2n_1}{}^i\;r_{n_1g}{}^l}{(\omega_{n_3g}+\omega_1)(\omega_{n_2g}+\omega_1+\omega_2)(\omega_{n_1g}-\omega_3)}$$

$$\left.+\frac{r_{gn_3}{}^j\;\bar{r}_{n_3n_2}{}^k\;\bar{r}_{n_2n_1}{}^l\;r_{n_1g}{}^i}{(\omega_{n_3g}+\omega_1)(\omega_{n_2g}+\omega_1+\omega_2)(\omega_{n_1g}+\omega_4)}\right\}$$

$$-\sum_{n_1n_2}'\left\{\frac{r_{gn_2}{}^i\;r_{n_2g}{}^l\;r_{gn_1}{}^k\;r_{n_1g}{}^j}{(\omega_{n_2g}-\omega_4)(\omega_{n_2g}-\omega_3)(\omega_{n_1g}-\omega_1)}\right.$$

$$+\frac{r_{gn_2}{}^i\;r_{n_2g}{}^l\;r_{gn_1}{}^k\;r_{n_1g}{}^j}{(\omega_{n_2g}-\omega_3)(\omega_{n_1g}+\omega_2)(\omega_{n_1g}-\omega_1)}$$

$$+ \frac{r_{gn_2}{}^l \; r_{n_2g}{}^i \; r_{gn_1}{}^j \; r_{n_1g}{}^k}{(\omega_{n_2g}+\omega_4)(\omega_{n_2g}+\omega_3)(\omega_{n_1g}+\omega_1)}$$

$$+ \left. \left. \frac{r_{gn_2}{}^l \; r_{n_2g}{}^i \; r_{gn_1}{}^j \; r_{n_1g}{}^k}{(\omega_{n_2g}+\omega_3)(\omega_{n_1g}-\omega_2)(\omega_{n_1g}+\omega_1)} \right\} \right] \tag{5}$$

where $r_{n_1 n_2}{}^i$ is the matrix element $<n_1|r^i|n_2>$ ($\bar{r}^i = r^i - <r^i>gg$), $\hbar\omega_{ng}$ is the excitation energy of state n, the prime on the summation indicates the ground state is omitted, $I_{1,2,3}$ denotes an average over all permutations of ω_1, ω_2, and ω_3 and their associated indices j, k, and l, and $K(-\omega_4;\omega_1,\omega_2,\omega_3)$ is a constant defined in Ref. 35. We will be concerned here with the case where all of the optical frequencies are above the molecular vibrational and rotational modes but below the electronic excitation energies, so the nonlinear optical response is strictly electronic in origin. In addition, for conjugated organic structures, γ is dominated by the delocalized π-electron contributions, which in general have both larger transition dipole moments and lower transition energies than the σ-electron excitations. Thus, with accurate description of both the excitation energies and transition moments of the π-electron manifold, one can calculate the frequency dependence of each of the different third-order nonlinear optical processes using Eq. (5).

The individual terms of Eq. (5) were directly evaluated from the singlet state excitation energies and transition dipole moments obtained by configuration interaction (CI) methods. All singly (SCI) and doubly (DCI) excited π-electron configurations are included in order to describe properly electron correlations and the resulting correlated π-electron excited states. (For example, the number of configurations for dodecahexaene with N = 12 are SCI 36 and DCI 666, and the total number of states is 703.) The CI π-electron basis sets were obtained by an all valence electron self-consistent field (SCF) molecular orbital (MO) method in the rigid lattice CNDO/S approximation. Although the calculation of the ground state includes all of the valence shell electrons for each atom in the molecule, we need only consider π-electron orbitals in configuration interaction theory, since low-lying excitations have predominantly $\pi \to \pi^*$ character, and it has also been generally established that for conjugated molecular systems the π-electron contribution to $\gamma_{ijkl}(-\omega_4;\omega_1,\omega_2,\omega_3)$ dominates that from σ-electrons.

Bond lengths and bond angles for the polyene molecular conformations are experimentally determined values, and, consequently, the bond alternation (carbon-carbon bond lengths of 1.34 and 1.46 Å) is treated directly. The coordinate x-axis is chosen along the chain

axis, and, of the y and z perpendicular axes, the z-axis is set
normal to the molecular plane. The hopping interaction between
all pairs of sites is included, and the electron-electron repulsion
is accounted for via the Ohno potential with the repulsion integral
between sites A and B given by

$$\gamma_{AB} = \frac{14.397 \text{ eV} \cdot \text{Å}}{\{[(28.794 \text{ eV} \cdot \text{Å})/(\gamma_{AA} + \gamma_{BB})]^2 + [R_{AB} \text{ Å}]^2\}^{1/2}} \qquad (6)$$

where γ_{AA} and γ_{BB} are empirical intra-atomic repulsion integrals
and R_{AB} is the interatomic distance.

   One measure of the success of the calculation method is provided
by the comparison between calculated and experimental transition
energies of the lowest-lying excited states. For a molecule belonging
to the C_{2h} symmetry group, such as an all-*trans* polyene, the π-
electronic states necessarily possess either A_g or B_u symmetry.
The optical absorption spectra of polyenes are dominated by the
lowest-lying ^1B_u excitation denoted by 1 ^1B_u. As an example, our
calculated values for the 1 ^1B_u excitation energies of *trans*-butadiene

Table 1. The Theoretical and Experimental Vertical Excitation
Energies for the Lowest-Lying One-Photon (1 ^1B_u) and Two-
Photon (2 ^1A_g) States of All-*trans* Polyenes

N (sites)	1 ^1B_u (eV)		2 ^1A_g (eV)	
	Exp.[a]	Theo.[b]	Exp.[a]	Theo.[b]
4	5.91	5.77	5.4[c,d]	5.31
6	4.93	4.94	4.0[c,e]	4.59
8	4.40	4.42	3.97	4.15
10	4.02	4.07	3.48	3.90
12	3.65	3.83	2.91	3.74

[a]From Ref. 22, p. 14.
[b]Following Schulten et al. (*J. Chem. Phys.*, 73:3927 (1976)), the
calculated energies are corrected by a fraction of the ground state
correlation energy to account for overcorrelation of the ground
state at the SDCI level of calculation.
[c]O-O excitation energies.
[d]From Ref. 23.
[e]From Ref. 24.

(BD) with number of carbon sites N = 4 and *trans*-hexatriene (HT) with N = 6, 5.77 and 4.94 eV, respectively, are in good agreement with the corresponding experimental vertical excitation energies of 5.91 and 4.93 eV [36]. Comparison is made to vertical rather than O-O transitions because the chain geometry is taken as frozen in the calculation. Also of fundamental importance to the study of polyenes is the one-photon forbidden ^1A_g state which has been found to lie below the 1 ^1B_u. These states are much more difficult to observe, particularly for the shortest chains, and are still the subject of intense experimental spectroscopic effort. The O-O transition energy of the 2 ^1A_g two-photon state of BD has recently been identified [37] at 5.4 eV and compares will to the calculated vertical energy of 5.31 eV. For HT, experimental data [38] are only available for the dimethyl-substituted chain in which the 2 ^1A_g state is observed at 4.0 ± 0.2 eV, as compared to the calculated value of 4.59 eV for HT. Similar satisfactory agreement between experimental and calculated transition energies for the 2 ^1A_g and 1 ^1B_u states is also found for the longer polyenes we have studied, as illustrated in Table 1.

III. $\gamma_{ijkl}(-\omega_4; \omega_1, \omega_2, \omega_3)$ OF ALL-*trans* LINEAR POLYENES

We will begin the discussion with the important example of HT, since the gas-phase value and sign of $\gamma_{ijkl}(-2\omega; \omega, \omega, 0)$ at 1.787 eV (λ = 0.694 μm) have been carefully determined, using gas-phase DCSHG [39]. The results for HT thus first serve as a comparison between theory and experiment, and, afterward, principal results for other length chains are discussed. Because the polyenes are members of the C_{2h} symmetry group, all of the π-electronic states must possess either A_g or B_u symmetry. Since the ground state is always ^1A_g, the ^1B_u excited states are one-photon allowed and two-photon forbidden; and the optical transition to the first ^1B_u excited state is the well-known dominant peak in the linear absorption spectrum. In contrast, the ^1A_g excited states are one-photon forbidden and directly observable optically only by two-photon spectroscopy. The parity selection rules are illustrated by the results for the lowest-lying states of HT in Table 2, which lists each x-component of the transition dipole moment $\mu_{g,n}^x$ between the excited n and the ground state g, and the x-component of the transition moment $\mu_{1^1B_u,n}^x$ between the excited state n and the 1 ^1B_u state. Further, it will be shown that $\gamma_{ijkl}(-\omega_4; \omega_1, \omega_2, \omega_3)$ is dominated by virtual transitions among states that have large values for these transition dipole moments.

Table 2. The Symmetries, Energies, and Selected Transition Dipole Moments of the Calculated Low-Lying States of *trans*-Hexatriene (N = 6)

State	Energy (eV)	$\mu_{g,n}{}^x$ (D)	$\mu_{1\ ^1B_u,n}{}^x$ (D)
$2\ ^1A_g$	4.59	0.0	2.0
$1\ ^1B_u$	4.94	6.6	0.0
$2\ ^1B_u$	5.22	0.3	0.0
$3\ ^1A_g$	6.69	0.0	2.0
$4\ ^1A_g$	6.80	0.0	1.2
$3\ ^1B_u$	7.55	0.6	0.0
$5\ ^1A_g$	7.97	0.0	11.0
$4\ ^1B_u$	8.07	1.0	0.0

The calculated values for the independent tensor components of the dc-induced second harmonic susceptibility $\gamma_{ijkl}(-2\omega,\omega,\omega,0)$ of HT at a nonresonant fundamental photon energy of 1.787 eV ($\lambda = 0.694$ μm) are $\gamma_{xxxx} = 49.7$, $\gamma_{xyyx} = 4.9$, $\gamma_{yxxy} = 3.5$, $\gamma_{xxyy} = 3.4$, $\gamma_{yyxx} = 2.6$, and $\gamma_{yyyy} = 1.2 \times 10^{-36}$ esu. Components of the form γ_{ijij} are necessarily equal to γ_{iijj} by symmetry of the second and third indices. All components vanish that involve the z-direction perpendicular to the molecular plane because of the antisymmetry of the π orbitals. The results demonstrate clearly that, as expected, the γ_{xxxx} component with all fields along the direction of conjugation is far larger than the other components.

In the gas-phase DCSHG experiments, since the molecules are isotropically oriented, the measured value for the third-order susceptibility necessarily involves orientational averaging over the different tensor components. The averaged gas-phase susceptibility $\gamma_g(-2\omega;\omega,\omega,0)$ is then related to the susceptibility tensor components according to Eq. (4). The experimentally obtained [39] value of γ_g for HT at 1.787 eV is $11.30 \pm 1.05 \times 10^{-36}$ esu, and the calculated value for γ_g for *trans*-HT from Eq. (4) is 11.5×10^{-36} esu. Here we note that the susceptibility thus calculated is solely the π-electron contribution. Although we anticipate the π-electron contribution to γ_g to be negligible for longer chains, it should have some contribution in the shorter-chain cases. After adding the σ-contribution estimated in Ref. 39 of 2.4×10^{-36} esu, we obtain the γ_g value of

13.9×10^{-36} esu, which is still in good correspondence with the experimental result. Experimental measurements were also performed on BD. Our calculated result for γ_g is 2.1×10^{-36} esu at 1.787 eV for the π-contribution to γ_g. After adding in the estimated value 1.5×10^{-36} esu for the σ-contribution, we obtain 3.6×10^{-36} esu for γ_g, which is also in good agreement with the experimental value [39] of $3.45 \pm 0.20 \times 10^{-36}$ esu at 1.787 eV.

As a further refinement in the comparison with experiment, we note that although the BD gas was believed to be more than 99% *trans*-BD, the HT may have contained as much as 40% of the cis conformation [39,40]. The values that we calculate for the independent tensor components of $\gamma_{ijkl}(-2\omega;\omega,\omega,0)$ of *cis*-HT at 1.787 eV are $\gamma_{xxxx} = 37.2$, $\gamma_{xyyx} = 5.4$, $\gamma_{yxxy} = 3.0$, $\gamma_{xxyy} = 4.7$, $\gamma_{yyxx} = 2.6$, and $\gamma_{yyyy} = 0.8 \times 10^{-36}$ esu. Our calculated π-electron contribution to $\gamma_g(-2\omega;\omega,\omega,0)$ for *cis*-HT is thus 9.1×10^{-36} esu. For a gas consisting of 60% trans and 40% cis conformations, the calculated π-electron contribution to $\gamma_g(-2\omega;\omega,\omega,0)$ becomes 10.5×10^{-36} esu. The addition of the estimated σ-electron contribution yields a total value for $\gamma_g(-2\omega;\omega,\omega,0)$ of 12.9×10^{-36} esu in even better agreement with the experimental value $11.30 \pm 1.05 \times 10^{-36}$ esu. We emphasize that in addition to the agreement on the magnitude of $\gamma_g(-2\omega;\omega,\omega,0)$, both the experiment and the calculation find the susceptibility to be positive in sign.

The third-order virtual excitation processes in the polyenes and their contribution to $\gamma_{ijkl}(-\omega_4;\omega_1,\omega_2,\omega_3)$ follow basic symmetry considerations. As evident from Eq. (5), all terms in γ_{ijkl} connect the ground state to itself through three virtual singlet intermediate states via four dipole moment operators. For centrosymmetric conjugated chains, electronic states have definite parity, and the one-photon transition moment vanishes between states of like parity. In a third-order process in the centrosymmetric polenes, states must be connected in the series $g \rightarrow {}^1B_u \rightarrow {}^1A_g \rightarrow {}^1B_u \rightarrow g$. Therefore, virtual transitions to both one-photon and two-photon states are necessarily involved. In the summation over intermediate states for HT, for example, there are two major terms which constitute 95% of γ_{xxxx}. In both of these terms, the ${}^1B_u$ state involved is the dominant low-lying one photon $1\ {}^1B_u$ π-electron excited state. In addition to its low energy, the importance of this state lies in the value of its transition dipole moment $\mu_{1\ {}^1B_u,g}$ with the ground state of 6.6 D being the largest of all those that involve the ground state.

One of the two significant terms results from the double summation of Eq. (5). This term has for the intermediate ${}^1A_g$ state the ground state itself and is denoted a type I term; but our finding of another, energetically high-lying, intermediate ${}^1A_g$ state is most important. Our calculations indicate that this state is the 5 ^1A_g state of HT lying at 7.97 eV. Since the 5 ^1A_g state has a transition moment with 1 ^1B_u of 11.0 D, it is much more significant than the first two-photon state 2 ^1A_g, which has a corresponding transition moment of only 2.0 D.

The numerators in the γ_{xxxx} component are positive definite in these two terms because both have the 1 ^1B_u as both the first and third intermediate states. Below the first resonance, each of the three factors in the denominator of the first term (which is the largest) of the triple summation of Eq. (5) is also positive when the second intermediate state is the 5 ^1A_g state. Thus, when the second state is the 5 ^1A_g state, the contributing term to γ_{xxxx} is positive in sign. This term is classified as a type II term. However, in the type I term where the second state is the ground state, although both the numerator and denominator are again positive definite, the double summation has an overall negative contribution to γ_{xxxx}. The two dominant terms in γ_{xxxx}, therefore, contribute with opposite sign. Since the positive, type II term for the 5 ^1A_g state is the larger term, the sign calculated for γ_{xxxx} is positive, and, in turn, the measured isotropic average susceptibility γ_g is also positive.

The 1 ^1B_u state is 96% comprised of a singly excited configuration of an electron from the highest occupied MO to the lowest unoccupied MO. The 2 ^1A_g and 5 ^1A_g states, on the other hand, are nearly 60% comprised of doubly excited configurations. The important distinction for γ_{xxxx} between these two highly electron correlated states is made most evident by the transition density matrix $\rho_{nn'}$ defined through the expression

$$<\mu_{nn'}> = -e \int r\rho_{nn'}(r) \, dr \qquad (7)$$

with

$$\rho_{nn'}(r_1) = \int \psi_n^*(r_1,r_2,\dots r_M) \psi_{n'}(r_1,r_2,\dots r_M) \, dr_2 \dots dr_M \qquad (8)$$

where M is the number of valence electrons included in the molecular wavefunction. Contour diagrams for $\rho_{nn'}$ of the ground, 2 ^1A_g, and 5 ^1A_g states with the 1 ^1B_u state are shown in Fig. 2, where solid and dashed lines correspond to increased and decreased charge density. The contour cut is taken 0.4 Å above the molecular plane since π orbitals vanish on the atoms. The contour diagram of $\rho_{2\,^1A_g,1\,^1B_u}$ shows the 2 $^1A_g \rightarrow 1\,^1B_u$ transition results in a charge redistribution concentrated at the center of the molecular structure, which yields a small transition moment of 2.0 D and, correspondingly, a small contribution to γ_{xxxx}. In sharp contrast, $\rho_{5\,^1A_g,1\,^1B_u}$ for the virtual transition between the 5 ^1A_g and 1 ^1B_u states produces

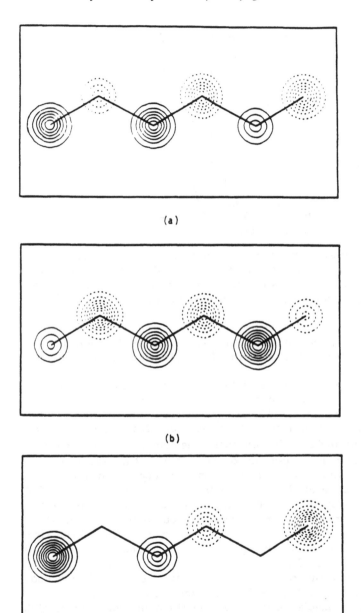

Figure 2. Transition density matrix contour diagrams of *trans*-HT for the (a) ground, (b) 2 ^1A_g, and (c) 5 ^1A_g states with the 1 ^1B_u state. The corresponding x-components of the transition dipole moments are 6.6, 2.0, and 11.0 D, respectively.

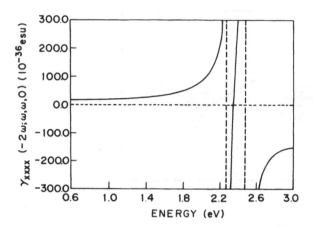

Figure 3. Calculated frequency dependence of $\gamma_{xxxx}(-2\omega;\omega,\omega,0)$
for *trans*-HT. The horizontal axis is the fundamental photon energy.
The first vertical dash locates the 2ω resonance to the $2\,^1A_g$ state,
and the second locates the 2ω resonance to the $1\,^1B_u$ state.

a large charge separation along the chain axis x-direction and an
associated large transition moment of 11.0 D which dominates the
contributing terms to γ_{xxxx}.

Figure 3 displays the calculated dispersion curve for the γ_{xxxx}
component of the DCSHG susceptibility of HT from 0.6 eV (λ =
2.07 μm) to 3.0 eV (λ = 0.41 μm). As can be seen in Eq. (5) for
$\gamma_{ijkl}(-2\omega;\omega,\omega,0)$, there can be 2ω resonances from both ^1B_u and
^1A_g states. The order in which these resonances appear in the
DCSHG dispersion is simply the order in which the states occur
energetically. Thus, the first singularity in Fig. 3, at 2.30 eV,
is the 2ω resonance of the $2\,^1A_g$ state, and the singularity at 2.47
eV is the 2ω resonance of the $1\,^1B_u$ state. Since these two states
are so close in energy, there is rapid variation in γ_{xxxx} in this
region. Of course, in real systems natural broadening of the electronic states will prevent divergence at the resonances and smooth
out this variation. In fact, since the $2\,^1A_g$ makes such a small
contribution below the resonances, when appropriately broadened,
it produces only a small peak in the dispersion of γ_{xxxx}.

As another example we consider the third harmonic susceptibility
$\gamma_{ijkl}(-3\omega;\omega,\omega,\omega)$ of *trans*-octatetraene (*trans*-OT) (Fig. 1a) consisting
of eight carbon sites (N = 8). The independent, nonvanishing tensor
components of $\gamma_{ijkl}(-3\omega;\omega,\omega,\omega)$ at a nonresonant fundamental photon
energy of 0.65 eV (λ = 1.907 μm) are γ_{xxxx} = 15.5, γ_{xyyx} = 0.6,
γ_{yxxy} = 0.5, and γ_{yyyy} = 0.2 × 10^{-36} esu. Again, all tensor com-

ponents involving the z-direction necessarily vanish because of
the symmetry constraints of the π-electron orbitals. We note here
that as a result of the two constitutive equations

$$p_i^{2\omega} = \gamma_{ijkl}(-2\omega;\omega,\omega,0)\, E_j^{\omega} E_k^{\omega} E_l^{0} \tag{9}$$

and

$$p_i^{3\omega} = \gamma_{ijkl}(-3\omega;\omega,\omega,\omega)\, E_j^{\omega} E_k^{\omega} E_l^{\omega} \tag{10}$$

the nominal value of $\gamma_{ijkl}(-2\omega;\omega,\omega,0)$ is six times larger than that
of $\gamma_{ijkl}(-3\omega;\omega,\omega,\omega)$ in the nonresonant frequency regime. This is
due to the three permutations of the dc field in $\gamma_{ijkl}(-2\omega;\omega,\omega,0)$
and to a factor of 2 difference in the definition of the Fourier components of the dc and oscillating fields. The factor of 6 is accounted
for by the constant $K(-\omega_4;\omega_1,\omega_2,\omega_3)$ in Eq. (5), which is equal to
3/2 for DCSHG and 1/4 for THG.

The symmetries, energies, and relevant transition dipole moments
of the 10 lowest calculated excited states of *trans*-OT are listed
in Table 3. All 153 of the calculated states are included in the
calculation of γ_{ijkl}, but Table 3 serves to illustrate the important
points. The columns $\mu_{n,g}^{x}$ and $\mu_{n,1\,^1B_u}^{x}$ refer to the x-components
of the transition dipole moment of each state with the ground state
and with the $1\,^1B_u$ state, respectively. The optical selection rules
are observed in the vanishing transition moments $\mu_{n,g}^{x}$ for all
the ^1A_g states and $\mu_{n,1\,^1B_u}^{x}$ for all the ^1B_u states. It is also seen
that the $1\,^1B_u$ state has by far the largest $\mu_{n,g}^{x}$, and the $6\,^1A_g$
state has the largest $\mu_{n,1\,^1B_u}^{x}$. As a result, these two are the
primary contributors to γ_{xxxx}, as will be described below.

It is instructive to consider the individual terms in the sum
over states perturbation expansion. Based on Eq. (5) and the symmetry selection rules described above, we reiterate that the π-electron
states in a third-order process must be connected in the series
$g \rightarrow {}^1B_u \rightarrow {}^1A_g \rightarrow {}^1B_u \rightarrow g$. For centrosymmetric structures, third-
order processes necessarily involve virtual transitions to both one-
photon and two photon states. For *trans*-OT, there are $(153)^3$ terms
involved in the summations of Eq. (5). However, again two of these
terms are an order of magnitude larger than all the others and
constitute 70% of γ_{xxxx}. The remaining terms to a large extent
cancel one another, resulting in a much smaller net contribution.
In both of the dominant terms, the only ^1B_u state involved is the
dominant low-lying one-photon $1\,^1B_u$ π-electron excited state. In
addition to its low excitation energy, this state is important because
its 7.8-D transition dipole moment with the ground state is more
than three times larger than any other ground state transition moment.

Table 3. The Symmetries, Energies, and Selected Transition Dipole Moments of the Calculated Low-Lying States of *trans*-Octatetraene (N = 8)

State	Energy (eV)	$\mu_{n,g}^{x}$ (D)	$\mu_{n,1\ ^{1}B_{u}}^{x}$ (D)
$2\ ^{1}A_{g}$	4.15	0.00	2.82
$1\ ^{1}B_{u}$	4.42	7.81	0.00
$2\ ^{1}B_{u}$	4.79	0.86	0.00
$3\ ^{1}A_{g}$	5.19	0.00	0.07
$4\ ^{1}A_{g}$	6.00	0.00	2.84
$5\ ^{1}A_{g}$	6.07	0.00	1.11
$3\ ^{1}B_{u}$	6.47	0.01	0.00
$4\ ^{1}B_{u}$	7.01	1.11	0.00
$6\ ^{1}A_{g}$	7.16	0.00	13.24
$5\ ^{1}B_{u}$	7.30	1.14	0.00

One of the two major terms is a type I term from the double summation of Eq. (5), with both of the intermediate states being the $1\ ^{1}B_{u}$. In the case of the double summation, the middle intermediate state is always the ground state. This term makes a negative contribution to γ_{xxxx} below resonance, since both the numerator and denominator are positive, but the double summation has an overall negative contribution. The other major term is a type II term from the triple summation with the $6\ ^{1}A_{g}$ state as the middle intermediate. This state, calculated at 7.2 eV, has a large transition moment with $1\ ^{1}B_{u}$ of 13.2 D. This term makes a positive contribution to γ_{xxxx} and is larger than the first, leading to an overall positive value for γ_{xxxx}. Importantly, the $6\ ^{1}A_{g}$ consists of 60% double-excited configurations, indicating it is highly correlated. SCI calculations obtain a negative value for γ_{xxxx} because they do not adequately describe this state and therefore omit its large contribution.

Figure 4 displays the calculated dispersion curve for $\gamma_{xxxx}(-3\omega; \omega,\omega,\omega)$ of *trans*-OT as a function of the input photon energy. The first resonance, located at 1.47 eV ($\lambda = 0.84\ \mu m$) and indicated by the vertical dash in the figure, is due to the 3ω resonance of the $1\ ^{1}B_{u}$ state. The second singularity, located at 2.08 eV ($\lambda = 0.6\ \mu m$), is from the 2ω resonance of the $2\ ^{1}A_{g}$ state. As seen in

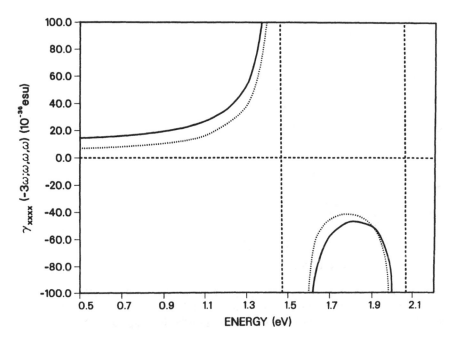

Figure 4. Calculated dispersion of $\gamma_{xxxx}(-3\omega;\omega,\omega,\omega)$ for *trans*-OT (solid curve). The vertical dashes locate the 3ω resonance to the $1\,^1B_u$ state and the 2ω resonance to the $2\,^1A_g$ state. The dotted curve shows the analogous calculated dispersion for *cis*-OT.

Eq. (5), the ^1B_u states will have both 3ω and ω resonances in third-harmonic generation, whereas the ^1A_g states will have only 2ω resonances. Other third-order processes have different resonance selection rules. For example, in DCSHG both the ^1B_u and ^1A_g states (or, more generally, both one-photon and two-photon states) have 2ω and ω resonances while there are no 3ω resonances. It is again noted that in real systems natural broadening of electronic states will prevent divergence at the resonances. This is accounted for theoretically by making the excitation energies in Eq. (5) complex with the inclusion of an imaginary damping term. Unfortunately, damping constants cannot be reliably calculated, and one must therefore use empirical values.

The general features of the origin of $\gamma_{ijkl}(-\omega_4;\omega_1,\omega_2,\omega_3)$ just described are common to all the chain lengths that we have studied. Importantly, there is a unifying general result that at fixed frequency, $\gamma_{xxx}(-\omega_4;\omega_1,\omega_2,\omega_3)$ for differing chain lengths exhibits a dramatic but well-defined increase with increased chain length. The

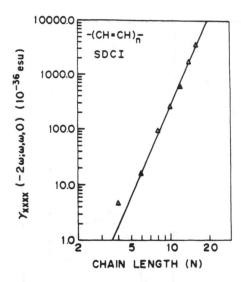

Figure 5. Log–log plot of $\gamma_{xxxx}(-2\omega;\omega,\omega,0)$ at 0.65 eV versus the number N of carbon atom sites in the polyene chain at the SDCI level of calculation. Linear fit corresponds to $\gamma_{xxxx} \propto N^{5.4}$.

values for $\gamma_{xxxx}(-2\omega;\omega,\omega,0)$ at a fundamental photon energy of 0.65 eV ($\lambda = 1.907$ µm) are plotted against the number of carbon atom sites N on a log–log scale in Fig. 5. The good linear fit indicates that γ_{xxxx} possesses a power-law dependence upon N with an exponent of 5.4 (the recent full calculations mentioned in the introduction yield 4.0) determined from the slope.

We have identified three length-dependent factors that lead to this rapid growth of $\gamma_{xxxx}(-\omega_4;\omega_1,\omega_2,\omega_3)$. First, the lowest optical excitation energy decreases proportionally to the inverse of the chain length as illustrated, for example, by the lowering from 5.9 eV in butadiene (N = 4) to 3.7 eV in the case of dodecahexaene (N = 12) [36]. Although we purposely chose a frequency which is far below the electronic resonances of all the chains we studied, it is clear from Eq. (5) that this factor will contribute to the growth of $\gamma_{xxxx}(-\omega_4;\omega_1,\omega_2,\omega_3)$. Second, the magnitudes of the transition dipole moments along the chain axis increase steadily with chain length. This is illustrated in Table 4 where the calculated transition dipole moments $\mu_{g,1\ ^1B_u}{}^x$ between the ground and $1\ ^1B_u$ states are listed as a function of chain length. Because such virtual excitations naturally produce a charge redistribution over a length comparable to that of the chain, longer chains have larger transition dipole moments and, correspondingly, larger γ_{xxxx} values. Third, although

Table 4.   The x-Component of the Transition Dipole Moment between the Ground and $1\,^1B_u$ States as a Function of the Number of Carbon Sites N For *trans* Polyenes

N (sites)	$\mu_{g,1\,^1B_u}{}^x$ (D)
4	5.2
6	6.6
8	7.8
10	8.8
12	9.7
14	11.1
16	12.2

we have pointed out that for the shorter chains, such as HT and OT, the nonresonant $\gamma_{xxxx}(-\omega_4; \omega_1, \omega_2, \omega_3)$ is dominated by virtual excitations involving primarily just three π-electron excited states, we have found that, for longer chains, an increasingly larger number of both ^1B_u and ^1A_g excited states play a significant role in γ_{xxxx} $(-\omega_4; \omega_1, \omega_2, \omega_3)$. Thus, in addition to lower excitation energies and larger transition dipole moments, there is a further increase in $\gamma_{xxxx}(-\omega_4; \omega_1, \omega_2, \omega_3)$ with increased chain length due to larger numbers of significant contributing terms.

The importance of electron correlations to $\gamma_{ijkl}(-\omega_4; \omega_1, \omega_2, \omega_3)$ of conjugated linear chains is further illustrated by results obtained from calculations at the SCI level that purposely omit doubly excited configurations (DCI) but are otherwise identical. As illustrated in Fig. 6, at this level of calculation, the values calculated for non-resonant $\gamma_{xxxx}(-2\omega; \omega, \omega, 0)$ are negative in sign for all of the polyene chains, which is contrary to the experimental results. This disagreement occurs because the SCI calculation improperly describes electron correlation which we have found to be of primary importance, such as in the illustrative cases of the $5\,^1A_g$ state of HT and $6\,^1A_g$ state of OT. Instead, at the SCI level, $\gamma_{xxxx}(-\omega_4; \omega_1, \omega_2, \omega_3)$ is predicted to be dominated solely by the virtual excitation process that involves only the ground and $1\,^1B_u$ states. A further consequence of incomplete account of correlation is the prediction of a smaller increase in $\gamma_{xxxx}(-\omega_4; \omega_1, \omega_2, \omega_3)$ with increased chain length. The power-law dependence of $\gamma_{xxxx}(-2\omega; \omega, \omega, 0)$ on N at the SCI level has an exponent of 3.9, compared with 5.4 for the full SDCI calculation.

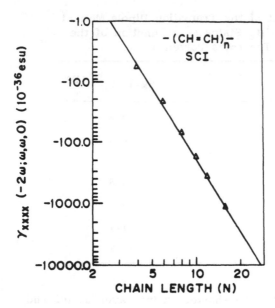

Figure 6. Log-log plot of $\gamma_{xxxx}(-2\omega;\omega,\omega,0)$ at 0.65 eV versus the number N of carbon atom sites from a calculation at the SCI level. The calculated values for γ_{xxxx} are negative, and the slope corresponds to $\gamma_{xxxx} \propto N^{3.9}$.

IV. EFFECT OF CONFORMATION ON $\gamma_{ijkl}(-\omega_4;\omega_1,\omega_2,\omega_3)$

We now turn to a discussion of our calculations of the cis polyenes. Although the trans form is the lowest-energy equilibrium conformation, commercially available supplies of HT(N = 6), for example, have been found to possess as much as 40% of the cis isomer [40]. If the two conformations are found to have substantially different values of $\gamma_{xxxx}(-\omega_4;\omega_1,\omega_2,\omega_3)$, experimental measurements must be very careful to identify the composition of the samples.

No *cis-transoid* form of BD (N = 4) is possible, so we will make comparison between our trans and cis calculations for chains from N = 6 to 16 sites. The first point to note is that not all of these chains belong to the same symmetry group. The cis chains with an odd number of short bonds (N = 6, 10,14, ...) are not centro-symmetric and possess C_{2v} symmetry, whereas the chains with an even number of short bonds (N = 8, 12, 16, ...) belong to the C_{2h} group along with the trans conformations. The states of the C_{2v} group are identified as A_1, A_2, B_1, and B_2 in contrast to the familiar A_g and B_u of C_{2h}. For $\pi \to \pi^*$ excitations the states must be either A_1 or B_2, the ground state being always $^1A_1$.

The corresponding one-photon dipole-allowed selection rules are also somewhat different for the two symmetry groups. For C_{2h}, only the ^1B_u states are one-photon allowed transitions from the $1\,^1A_g$ ground state for electric fields polarized along both the x- and y-axes. The y-axis is here defined as the second axis in the molecular plane. However, for the cis chains in the C_{2v} symmetry group, where the y-axis becomes the axis of C_2 twofold rotational symmetry, the $^1B_2$ states are one-photon allowed for fields polarized along x, and the excited $^1A_1$ states are the one-photon allowed transitions for polarizations along the y-axis. This implies that all of the singlet $\pi \rightarrow \pi^*$ excitations are one-photon allowed transitions for the chains possessing this symmetry. But it should be noted that because these chains are quasi-one-dimensional, the x-components of the transitions dipole moments will, in general, far exceed the y-components, and the oscillator strengths of the $^1B_2$ states will therefore dominate those of the $^1A_1$ states. We will therefore consider the $^1A_1$ and $^1B_2$ states to be analogous to the ^1A_g and ^1B_u states, respectively.

For the range of chain lengths that we have considered, we find that the transition energies of the $1\,^1B_u$ and $2\,^1A_g$ states of the cis conformations are slightly red-shifted from the values for the trans conformations by 0.02 to 0.10 eV, with the shift constantly increasing with increasing chain length. These results are in agreement with Soos and Ramasesha's findings for the PPP model [41]. Furthermore, an experimental gas-phase absorption study of the cis and trans forms of HT [40] found transition energies of 4.919 and 4.935 eV, respectively, which is in very good correspondence with our calculated transition energies of 4.92 and 4.94 eV.

The calculated values of the dominant component of the non-resonant third-order susceptibility $\gamma_{xxxx}(-3\omega;\omega,\omega,\omega)$ for the cis conformations are smaller than the values of $\gamma_{xxxx}(-3\omega;\omega,\omega,\omega)$ for the trans conformations of an equal number of sites for all of the chains considered, and the percentage difference between the trans and cis values monotonically increases with increasing chain length. The independent tensor components of $\gamma_{ijkl}(-3\omega;\omega,\omega,\omega)$ at 0.65 eV for cis-OT (N = 8) are calculated to be $\gamma_{xxxx} = 7.7$, $\gamma_{xxyy} = 0.7$, $\gamma_{yyxx} = 0.6$, and $\gamma_{yyyy} = 0.1 \times 10^{-36}$ esu. Thus it can be seen that although γ_{xxxx} for the cis is only about half the value for the trans, it is still this component with all fields aligned in the chain axis direction which dominates the susceptibility.

Figure 4 shows the calculated value of $\gamma_{xxxx}(-3\omega;\omega,\omega,\omega)$ for cis-OT as a function of the input photon energy. This dispersion curve is essentially the same as that calculated for *trans*-OT with the exception that the low-energy, nonresonant value for the cis is smaller than that for the trans, as mentioned earlier. Since the calculated transition energies for the cis are only slightly red-shifted, the resonant behavior of the two conformations is very similar. The first resonance, located at 1.46 eV, is the 3ω resonance of the $1\,^1B_u$ state. The second singularity, located at 2.05 eV, is the 2ω resonance of the $2\,^1A_g$.

The microscopic origin of $\gamma_{ijkl}(-3\omega;\omega,\omega,\omega)$ for the cis polyenes is in direct analogy with that for the trans conformations. For the specific case of *cis*-OT, there are again two terms in Eq. (5) which constitute 89% of the total value for $\gamma_{xxxx}(-3\omega;\omega,\omega,\omega)$ at 0.65 eV. In both of these terms, the B_u state is the $1\,^1B_u$ state with $\mu_{g,1\,^1B_u}{}^x = 7.9$ D. This is the largest of all transition moments involving the ground state. The important ^1A_g states are the ground state and the $6\,^1A_g$ state. Although the $6\,^1A_g$ state is calculated to have a high transition energy of 7.0 eV, this is again counteracted by the very large transition dipole moment with the $1\,^1B_u$ state of 12.0 D.

Contour diagrams for $\rho_{nn'}$ of the ground and $6\,^1A_g$ states with the $1\,^1B_u$ state for both the cis and trans forms of OT are compared in Fig. 7. The (a) cis and (b) trans virtual $g \rightarrow 1\,^1B_u$ transitions result in a somewhat modulated redistribution of charge with transition moments of 7.9 and 7.8 D, respectively. For the $1\,^1B_u \rightarrow 6\,^1A_g$ virtual transition, however, there is a resultant highly separated charge redistribution in the (c) cis and (d) trans conformations. The corresponding transition moments are 12.0 and 13.2 D for the cis and trans cases, respectively.

The variation in $\gamma_{xxxx}(-3\omega;\omega,\omega,\omega)$ with the size of the chain is also very similar in the cis and trans cases. Figure 8 is a log-log plot comparison of the calculated values for $\gamma_{xxxx}(-3\omega,\omega,\omega,\omega)$ at 0.65 eV against the number N of carbon atom sites in the cis and trans polyenes. There is a good linear fit to the calculated points, which indicates that γ_{xxxx} for the cis chains possesses a power-law dependence on N with an exponent of 4.7 (3.8 in the full calculations) determined from the slope. The same three length-dependent factors that lead to the power-law dependence of $\gamma_{xxxx}(-\omega_4;\omega_1,\omega_2,\omega_3)$ for the trans polyenes are also responsible for the rapid growth in the cis case.

Although there is an excellent qualitative analogy between the descriptions of $\gamma_{xxxx}(-3\omega;\omega,\omega,\omega)$ for the cis and trans polyenes, there are two important quantitative distinctions. First, for chains with equal numbers of sites, the value of $\gamma_{xxxx}(-3\omega;\omega,\omega,\omega)$ at a fixed frequency is in all cases calculated to be smaller for the cis chain than the trans chain. Second, there is a lesser rate of growth for γ_{xxxx} with increasing N for the cis. The power-law exponent in the cis case is 4.7, versus 5.4 for the trans.

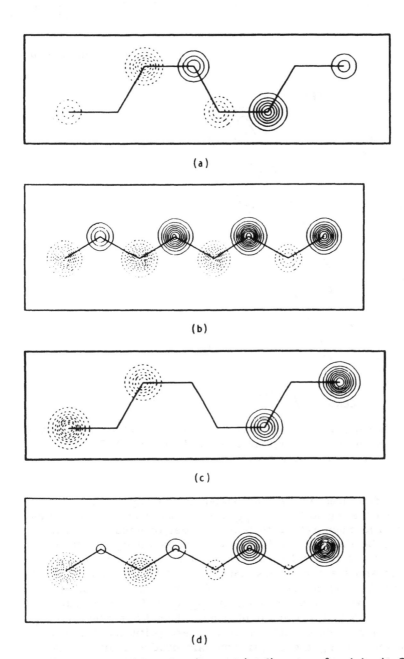

(a)

(b)

(c)

(d)

Figure 7. Transition density matrix diagrams for (a) *cis*-OT and
(b) *trans*-OT ground states with the $1\ ^1B_u$ state and (c) *cis*-OT
and (d) *trans*-OT $6\ ^1A_g$ states with the $1\ ^1B_u$ state. In both cases,
these are the most important virtual transitions for $\gamma_{xxxx}(-\omega_4;$
$\omega_1,\omega_2,\omega_3)$. The corresponding x-components of the transition dipole
moments are 7.9, 7.8, 12.0, and 13.2 D, respectively.

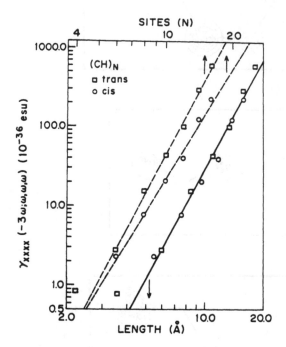

Figure 8. Log-log plot of $\gamma_{xxxx}(-3\omega;\omega,\omega,\omega)$ at 0.65 eV versus the number of carbon sites N (upper axis) and the length L (lower axis) for *cis* and *trans* polyenes.

These differences are well accounted for in Fig. 8, in which we also plot on a log-log scale $\gamma_{xxxx}(-3\omega;\omega,\omega,\omega)$ against the actual length L of the chain rather than the number of sites. We have defined L to be the distance in the x-direction (along the chain axis) between the two end carbon sites. The calculated values for both the cis and trans polyenes are very well fit by a single line. This plot unifies the calculated values for the two conformations and yields a power-law dependence of γ_{xxxx} on L with an exponent of 4.6 ± 0.2 (3.5 in the full calculations). The differences in γ_{xxxx} $(-3\omega;\omega,\omega,\omega)$ for an equal number of sites are primarily due to the shorter distance along the conjugation axis produced by the cis geometry. For equal numbers of carbon sites, the cis conformation is always shorter than the trans conformation. Figure 8 implies that the values of $\gamma_{xxxx}(-3\omega;\omega,\omega,\omega)$ will be the same for cis and trans polyene chains which have equal lengths along the chain axis rather than equal numbers of carbon sites. We conclude that although the trans third-order susceptibilities are significantly larger than those for the cis conformation of an equal number of carbons,

$\gamma_{xxxx}(-3\omega;\omega,\omega,\omega)$ is in fact much more sensitive to the physical
length of the chain than to the conformation.

Finally, we can make a rough estimate of the implications of
these calculations for polymers by extrapolating to longer chains
the power-law dependence that we have observed. A typical value
of the nonresonant macroscopic third-order susceptibility $\chi^{(3)}(-3\omega;$
$\omega,\omega,\omega)$ observed for polymers is 10^{-10} esu [9,42]. For an (nearly)
isotropic distribution of chains considered as independent sources
of nonlinear response with a single dominant tensor component
$\gamma_{xxxx}(-3\omega;\omega,\omega,\omega)$, we have

$$\chi_{1111}^{(3)} = \frac{1}{5}\, N(f^{\omega})^3 f^{3\omega}\, \gamma_{xxxx} \tag{11}$$

where N is the number density of chains and f^{ω} is the Lorentz-
Lorenz local field factor. Using typical values of $N = 10^{20}$ molecules/
cm^3 and 1.8 for the refractive index, we derive a γ_{xxxx} of roughly
2×10^{-31} esu. An extrapolation of our power-law dependence of
γ_{xxxx} on N for the *trans*-polyenes yields a value of N ≈ 50 carbon
sites or a length of approximately 60 Å. Since many of these polymers
consist of much longer chains, we infer that γ_{xxxx} must deviate
from the power-law dependence and begin to saturate at some length
shorter than 60 Å. If we estimate the corresponding values for a
cis chain of 50 sites using a similar extrapolation, we find that
the cis has $\gamma_{xxxx}(-3\omega;\omega,\omega,\omega)$ and $\chi^{(3)}$ approximately seven times
smaller than the trans chain of the same number of sites. Of course,
a chain in the cis conformation with a length of 60 Å is extrapolated
to have the same value of γ_{xxxx} as the same length chain in the
trans conformation since Fig. 8 indicates that the two conformations
have the same dependence of γ_{xxxx} on the actual length of the
chain. It is then concluded that large values of γ, and correspondingly $\chi^{(3)}$, may only require chains of intermediate length of order
100 Å with little to be gained by increasing the chain length beyond
this limit.

V. EFFECT OF LOWERED SYMMETRY
   ON $\gamma_{ijkl}(-\omega_4;\omega_1,\omega_2,\omega_3)$

In previous sections, we have considered the third-order nonlinear
optical properties of centrosymmetric linear chains and have reviewed
the important role of definite parity selection rules in the third-
order virtual excitation processes. We will demonstrate in this section
that lowering the symmetry to a noncentrosymmetric structure can
act as a mechanism for the enhancement of nonresonant $\gamma_{ijkl}(-\omega_4;$

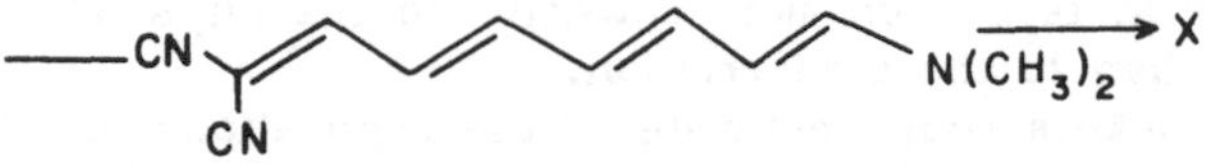

Figure 9. Schematic diagram of the molecular structure of 1,1-dicyano,8-N,N-dimethylamino-1,3,5,7-octatetraene (NOT).

ω_1, ω_2, ω_3). As seen in Eq. (5), the Bogoliubov-Mitropolsky formalism admits new types of virtual excitation processes otherwise forbidden under centrosymmetric conditions.

The symmetry is lowered by heteroatomic substitution on the linear chain. A principal noncentrosymmetric analog to OT is 1,1-dicyano-8-N,N-dimethylamino-1,3,5,7-octatetraene (NOT) having a dicyano acceptor group on one end and a dimethylamino donor group on the other, as shown in Fig. 9. Comparison with the earlier detailed discussion of unsubstituted OT will allow direct understanding of the effect of lowered symmetry on $\gamma_{ijkl}(-\omega_4; \omega_1, \omega_2, \omega_3)$.

The calculation of the electronic states and nonlinear optical properties of NOT involved all single- and double-excited configurations of the six occupied and six unoccupied π-electron molecular orbitals. This leads to 703 configurations in the CI matrix, which is then diagonalized to produce 703 singlet π-electron states. The complete calculation including computation of all transition dipole moments and evaluation of Eq. (5) for γ_{ijkl} required 5.5 CPU hours on a CRAY-X/MP. The calculated excitation energies and oscillator strengths of NOT are given in Table 5. The dominant excitation is that from the lowest energy π-electron singlet excited state located at 3.03 eV. This is 1.4 eV lower than the energy of the dominant one-photon 1 ^1B_u state of OT. There is a secondary peak predicted in the optical absorption spectrum at 3.69 eV which actually corresponds to the 2 ^1A_g state of OT. Because of the lowered symmetry of NOT, there are no one-photon selection rules as there are in OT. Instead, all of the π-electron states of NOT possess A' symmetry, and all are allowed one-photon excitations from the ground state. Thus, in addition to the lowering in energy of the analog to the 1 ^1B_u state, the symmetry lowering has two interesting effects on linear optical properties. The analog of the 2 ^1A_g state becomes a one-photon allowed transition, which turns out to have a sizable oscillator strength, and the ordering of the analogs of the 2 ^1A_g and 1 ^1B_u states is inverted. We wish to emphasize, however, that although the existence of the 2 ^1A_g below the 1 ^1B_u provided the first definitive evidence of the importance of electron correlation in polyenes, the inverted order in the substituted chain is not due to any less correlation. The 3 $^1A'$ state of NOT, which is the 2 ^1A_g analog, is still composed of 48% double-excited configurations.

Table 5. The Symmetries, Energies and Oscillator Strengths of the Calculated Low-Lying States of NOT

State	Energy (eV)	Oscillator strength
2 ^{1}A'	3.03	0.88
3 ^{1}A'	3.69	0.26
4 ^{1}A'	4.21	0.00
5 ^{1}A'	4.55	0.02
6 ^{1}A'	4.87	0.01
7 ^{1}A'	4.87	0.01
8 ^{1}A'	5.35	0.00
9 ^{1}A'	5.88	0.00
10 ^{1}A'	5.98	0.00
11 ^{1}A'	6.21	0.05

The double-excited configurations, of course, are those that represent the many-electron nature of the excited state and provide the electron correlation in the CI formalism.

The principle symmetry constraint in the case of centrosymmetric structures that the intermediate states must alternate between one-photon states and two-photon states is lifted upon symmetry lowering. Matrix elements of the form $\langle n|r^i|n\rangle$ are no longer symmetry-forbidden and can have an important role in γ_{ijkl}. Diagonal transitions of this form are best illustrated in the difference density matrix $\Delta\rho_n$, where

$$\Delta\rho_n(r) = \rho_n(r) - \rho_g(r) \qquad (12)$$

and

$$\langle\Delta\mu_n\rangle = -e \int r\, \Delta\rho_n(r)\, dr \qquad (13)$$

The function $\rho_n(r)$ is given by Eq. (8) with n equal to n'. The contour diagram of $\Delta\rho_2\, {}^1{}_{A'}(r)$ is shown in Fig. 10, where the solid and dashed lines correspond to increased and decreased electron density, respectively. There is a large redistribution of electron density along the dipolar x-axis, leading to a large dipole moment

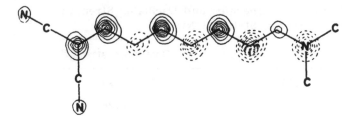

Figure 10. Difference density matrix contour diagram for the $2\,^1A'$ state with the ground state of NOT.

difference $\Delta\mu_2\,_{1A'}{}^x$ of 12.6 D. The sign for $\Delta\mu_2\,_{1A'}{}^x$ is seen to be positive as electron density is decreased in the region of the electron donor and increased in the region of the electron acceptor group upon excitation. This is consistent with the experimentally observed shift to lower energies of the first optical absorption peak in increasingly polar solvents. The magnitude of $\Delta\mu_2\,_{1A'}{}^x$ is relatively large and leads to important terms in γ_{ijkl} that involve the matrix element $<2\,^1A'|x|2\,^1A'>$. There are no analogous terms in γ_{ijkl} in the case of centrosymmetric linear chains since the dipole moments of the ground state and all excited states are zero by symmetry. Contour diagrams for $\rho_{nn'}$ between the $2\,^1A'$ state and the $1\,^1A'$, $3\,^1A'$, and $7\,^1A'$ states are similar to the OT diagrams between the $1\,^1B_u$ state and the $1\,^1A_g$, $2\,^1A_g$, and $6\,^1A_g$ states, but these transitions are dominated in NOT by the unusually large terms involving $<\Delta\mu_2\,_{1A'}>$.

For the third-harmonic susceptibility $\gamma_{ijkl}(-3\omega;\omega,\omega,\omega)$, it is found once again, that the γ_{xxxx} component is by far the largest. At 0.65 eV, the independent tensor components are $\gamma_{xxxx} = 173.2$, $\gamma_{xxyy} = 1.6$, $\gamma_{yyxx} = 1.0$, and $\gamma_{yyyy} = 0.1 \times 10^{-36}$ esu. The calculated dispersion curve of $\gamma_{xxxx}(-3\omega;\omega,\omega,\omega)$ in Fig. 11 smoothly increases to the first resonance occurring at 1.0 eV, which is the 3ω resonance of the $2\,^1A'$ state. Because of the lowered symmetry of NOT, the 3ω and 2ω resonance selection rules for the centrosymmetric polyenes are no longer applicable. Thus, every excited state has allowed 3ω, 2ω, and ω resonances, and the dispersive behavior of $\gamma_{xxxx}(-3\omega;\omega,\omega,\omega)$ exhibits all of these many resonances at frequencies beyond the first resonance.

As described earlier, for centrosymmetric structures there are two important types of virtual excitation processes that dominate $\gamma_{xxxx}(-\omega_4;\omega_1,\omega_2,\omega_3)$. We have found for the noncentrosymmetric chains a third type of process is allowed and, in fact, makes a larger contribution to γ_{xxxx} than the other two. These three types of virtual excitation processes in NOT are illustrated in Fig. 12.

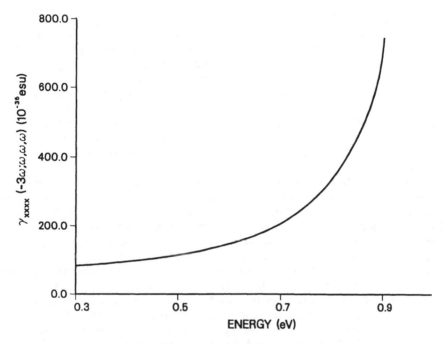

Figure 11.   Dispersion of $\gamma_{xxxx}(-3\omega;\omega,\omega,\omega)$ for NOT.

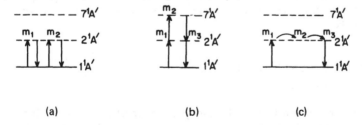

(a)   (b)   (c)

Figure 12.   Schematic of the three important types of contributing terms to γ_{xxxx} for NOT: (a) type I, (b) type II, and (c) type III.

Types I and II are analogous to the dominant processes for centro-symmetric OT. For NOT, the $2\,^{1}A'$ state plays the role of the $1\,^{1}B_{u}$ state of OT because it has the largest transition dipole moment (8.6 D) with the ground state. The type I term is a result of the double summation in the Bogoliubov-Mitropolsky formalism and has the ground state as the middle intermediate state. The largest term of this type is the one with $2\,^{1}A'$ as the first and last intermediate state because of its large transition moment with the ground state. The type I term illustrated in Fig. 12 makes a negative contribution to γ_{xxxx} because, although the numerator and denominator are both positive, the double sum makes a negative contribution to γ_{xxxx}. In the type II process, there is a high-lying middle intermediate state that has a large transition moment with $2\,^1A'$. While for OT this state was the $6\,^1A_g$, both the $7\,^1A'$ and the $11\,^1A'$ states of NOT occupy this role. The type II terms make a positive contribution to γ_{xxxx} because the numerator is effectively the square of two matrix elements and the denominator is positive when below all resonances.

Most importantly, however, for noncentrosymmetric structures, there is a new type of process which is allowed, denoted as type III. For NOT, this is the dominant type of term contributing to γ_{xxxx}. Type III terms involve a diagonal matrix element and are therefore forbidden in centrosymmetric structures which cannot possess a permanent dipole moment. The important quantity in this term is the dipole moment difference between an excited state and the ground state. For the $2\,^1A'$ state, the value of 12.6 D leads to a very large term in the triple sum in which all three intermediate states are the $2\,^1A'$. Since the numerator and denominator are both positive, the contribution of this term to γ_{xxxx} is positive. The lowered symmetry of NOT, as compared to OT, produces a new type of virtual excitation process which dominates γ_{xxxx} and causes the value of γ_{xxxx} to be an order of magnitude larger for NOT compared to OT. For example, the calculated nonresonant values of $\gamma_{xxxx}(-3\omega; \omega,\omega,\omega)$ at 0.65 eV for NOT and OT are 173×10^{-36} and 15.5×10^{-36} esu, respectively.

VI. EFFECT OF DIMENSIONALITY
ON $\gamma_{ijkl}(-\omega_4;\omega_1,\omega_2,\omega_3)$

In this section, we extend the microscopic description of γ_{ijkl} for one dimension to two dimensions and consider a specific example within the class of conjugated cyclic structures known as annulenes. As a major case, we consider the planar structure of cyclooctatetraene (COT), the cyclic analog to OT with $N = 8$, illustrated schematically in Fig. 13. Although the geometrically relaxed ground state of COT is known to be a nonplanar, bent structure, we will examine only the planar structure of COT in this section. The purpose is to explore the effect of dimensionality on the microscopic features of $\gamma_{ijkl}(-\omega_4;\omega_1,\omega_2,\omega_3)$. Because the inclusion of the geometrically relaxed distortions of the physically observed COT structure unnecessarily complicates comparison of cyclic structure results with

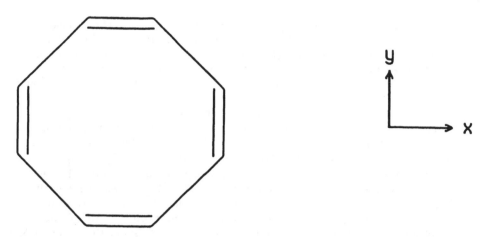

Figure 13. Schematic diagram of the molecular structure of planar cyclooctatetraene (COT).

those described above for linear chains, we consider only the planar structure. Analogies and contrasts between linear and cyclic structures are clearest when the only distinction is the increase in dimensionality from one to two so that one need not be concerned with decoupling dimensionality from nonplanarity effects.

For linear chains, since the x-component of the transition dipole moments is much larger than the y- and z-components, the isotropically averaged susceptibility γ_g is effectively determined by the γ_{xxxx} component while all components involving transverse fields make negligible contributions. This reduces Eq. (4) to

$$\gamma_g \approx \frac{1}{5}\, \gamma_{xxxx} \tag{14}$$

But for cyclic structures, since the in-plane x- and y-directions are equivalent, γ_{xxxx} and γ_{yyyy} should be of equal magnitude. Furthermore, components such as γ_{xxyy} will also be significant since they involve these two directions as well. The corresponding terms of Eq. (4) that cannot be neglected are

$$\gamma_g \approx \frac{1}{5}\, [\gamma_{xxxx} + \gamma_{yyyy} + \frac{1}{3}\, (\gamma_{xxyy} + \gamma_{xyxy} + \gamma_{xyyx}$$

$$+ \gamma_{yyxx} + \gamma_{yxyx} + \gamma_{yxxy})] \tag{15}$$

It seems, therefore, that one might be able to enhance γ_g by moving from linear to cyclic conjugated structures and opening pathways for

new components of the γ_{ijkl} tensor to contribute. We shall demonstrate, however, a most striking, opposite finding. We show that,
because of the relevant length scales involved in the two problems,
the conjugated cyclic structure will necessarily have a smaller γ_g
than the corresponding conjugated linear chain with an equal number
of carbon sites.

Our planar model of COT is a member of the dihedral D_{4h}
symmetry group which is non-Abelian and, hence, has two-dimensional
irreducible representations denoted as E classes. The allowed state
symmetries for π-electron excitations are A_{1g}, A_{2g}, B_{1g}, B_{2g}, and
E_u. Of these, only states of ^1E_u symmetry are one-photon allowed
excitations from the $^1A_{1g}$ ground state. The ^1E_u states are doubly
degenerate with the two representations related by a $\pi/2$ rotation
about the z-axis perpendicular to the molecular plane. All of the
remaining symmetries listed above describe nondegenerate, two-photon
states. A typical feature of conjugated cyclic molecules, including
phthalocyanines and porphyrins, is the existence of a relatively
low-frequency absorption in the visible or near ultraviolet and a
higher-frequency absorption deeper in the ultraviolet [43]. This
feature appears in our model COT with the weak low-frequency
$1\ ^1E_u$ state at 4.4 eV and the much stronger higher-frequency
$2\ ^1E_u$ state at 6.5 eV. In the case of phthalocyanines, however,
the low-frequency band is always stronger.

In Table 6 we list the symmetries, energies, and relevant transition dipole moments for the eight lowest calculated excited states of
COT. Again, a total of 153 states are calculated and used in the
evaluation of $\gamma_{ijkl}(-\omega_4;\omega_1,\omega_2,\omega_3)$. The third and fourth columns
list the x- and y-components, respectively, of the transition moment
between a given state and the ground state, while the column labeled
$\mu_{n,2\ ^1E_u}{}^x$ lists the x-component of the transition moments of that
state with the two degenerate representations of the $2\ ^1E_u$ state;
the y-components are given in the $\mu_{n,2\ ^1E_u}{}^y$ column. Although
the x- and y-directions are equivalent in COT, it is seen in the
table that these components of the transition moments are not always
equal, and this is a direct result of the double degeneracy of the
E_u states. Transition moments involving any degenerate pair of ^1E_u
states with a two-photon state are clearly related by a $\pi/2$ rotation
or x $\to$ y, y $\to$ -x. Thus, the appearance of negative signs in some
of the transition moments merely reflects the choice of basis in
Hilbert space and has no absolute physical meaning. By choosing
an appropriate basis for the degenerate pair, the magnitudes of
the x- and y-components can be made equal, although this is not
necessary as illustrated in Table 6. All physically observable quantities are, however, still symmetric in the x- and y-directions.

Table 6.　The Symmetries, Energies, and Selected Transition Dipole
Moments of the Calculated Low-Lying States of Cyclooctatetraene

State	Energy (eV)	$\mu_{n,g}{}^x$ (D)	$\mu_{n,g}{}^y$ (D)	$\mu_{n,2\,^1E_u}{}^x$ (D)	$\mu_{n,2\,^1E_u}{}^y$ (D)
$1\,^1A_{2g}$	2.24	0.00	0.00	-1.60, 2.40	-2.40, -1.60
$2\,^1A_{1g}$	3.19	0.00	0.00	0.14, 0.09	-0.09, 0.14
$1\,^1E_u$	4.41	0.05	-0.03	0.00, 0.00	0.00, 0.00
$1\,^1E_u$	4.41	0.03	0.05	0.00, 0.00	0.00, 0.00
$1\,^1B_{2g}$	5.21	0.00	0.00	-0.33, 0.50	0.50, 0.33
$1\,^1B_{1g}$	5.94	0.00	0.00	1.03, 0.68	0.68, -1.03
$2\,^1E_u$	6.48	4.58	-3.04	0.00, 0.00	0.00, 0.00
$2\,^1E_u$	6.48	3.04	4.58	0.00, 0.00	0.00, 0.00

The first one-photon doubly degenerate $1\,^1E_u$ state has very
small transition moments with the ground state and a correspondingly
small oscillator strength. The $2\,^1E_u$ doubly degenerate state, on the
other hand, has a maximum transition moment of 4.6 D in this xy
basis and is a large-oscillator-strength one-photon state in analogy
to the $1\,^1B_u$ state of OT. Thus, the transition moments of the two-
photon states with the $2\,^1E_u$ state determine their importance of
$\gamma_{ijkl}(-\omega_4; \omega_1, \omega_2, \omega_3)$ and are also listed in Table 6 for the lowest-
lying states.

Analysis of the various terms in the summations of Eq. (5)
for COT reveals much similarity to the linear chain analog. As in
OT, there is one dominant type I term in the double summation
which is due to the state with the largest oscillator strength, in
this case the $2\,^1E_u$ state. The dominant type II contributions from
the triple sum also all involve the $2\,^1E_u$ state, but rather than just
one term there are several significant contributions for COT. Each
one is smaller than the dominant negative term from the $2\,^1E_u$ state,
but together they again lead to positive values for the nonresonant
tensor components γ_{ijkl}. The significant contribution from several
terms involving different intermediate two-photon states is similar
to what was found for the linear polyenes longer than OT.

In Fig. 14, the transition density matrix $\rho_{nn'}$ is shown for the
ground state with the two representations of the $2\,^1E_u$ state. The
$\pi/2$ rotational relationship of the two representations of $2\,^1E_u$ is
also clear here as in Table 6. The charge redistribution for this

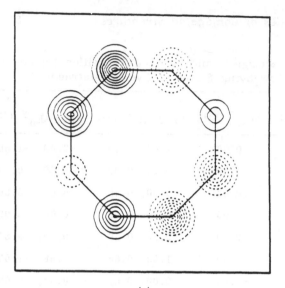

(a)

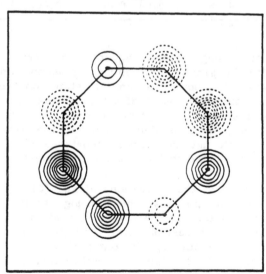

(b)

Figure 14. Transition density matrix diagrams for the ground state
of COT with each of the two representations of the doubly degenerate
$2\ ^1E_u$ state. the $\pi/2$ rotational relationship between the two representations is clear. In this basis for the $2\ ^1E_u$ degenerate pair, the
x- and y-components of the transition dipole moments are 4.6 and
-3.0 D for (a) and 3.0 and 4.6 D for (b), respectively. Other
choices for the $2\ ^1E_u$ basis will result in different values for the
individual components of the transition dipole moments, but the magnitude of the transition dipole moment vector must remain invariant.

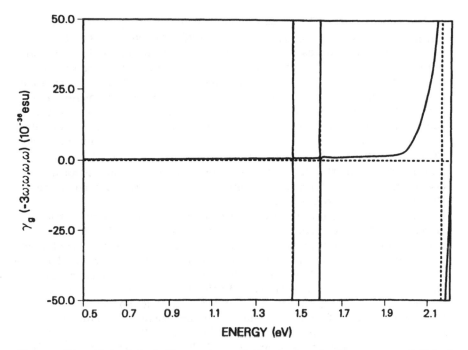

Figure 15. Calculated dispersion of the isotropically averaged third-order susceptibility $\gamma_g(-3\omega;\omega,\omega,\omega)$ for COT.

transition is fairly modulated as it is for the linear chains. In the particular basis shown for the representation of the doubly degenerate $2\,^1E_u$ state, the transition moments along the x- and y-directions are comparable as seen in the figure.

The calculated dispersion of the isotropic third-order susceptibility $\gamma_g(-3\omega;\omega,\omega,\omega)$ is given in Fig. 15. The nonresonant value is quite small compared to the linear chains. Very sharp resonances occur at 1.47 eV due to a 3ω resonance to $1\,^1E_u$ and 1.60 eV due to a 2ω resonance with $2\,^1A_{1g}$. Since both of these states have very small transition moments in Table 6, they only make significant contributions directly on resonance. Calculations including imaginary damping terms to account for the finite width of the electronic excitations show that these resonances become completely washed out and the dispersion remains flat in this part of the spectrum. Experimental dispersion measurements, therefore, likely would not observe significant resonant enhancement until the 3ω resonance to the $2\,^1E_u$ state which occurs at 2.16 eV.

The calculated values of $\gamma_{ijkl}(-3\omega;\omega,\omega,\omega)$ for COT at the nonresonant fundamental photon energy of 0.65 eV are $\gamma_{xxxx} = 0.75$

and $\gamma_{xxyy} = 0.21 \times 10^{-36}$ esu. Because of the D_{4h} symmetry, $\gamma_{xxxx} = \gamma_{yyyy}$ and γ_{xxyy} is equal to all components that involve two x-component fields and two y-component fields. This equivalence between the x- and y-directions is in contrast to the linear polyene case, where the γ_{xxxx} component dominates all others. Furthermore, the γ_{xxxx} component of OT (15.5×10^{-36} esu) is far larger than any of the components of COT. However, since COT has significant components in both the x- and y-directions, it is more reasonable to compare values of the isotropically averaged susceptibility. For COT, $\gamma_g = 0.38 \times 10^{-36}$ esu as compared to 3.4×10^{-36} esu for *trans*-OT. As in the case of comparing values for the trans and cis conformations of linear polyenes, here also we must consider the actual length scale involved in the problem.

The important length for nonlinear optical responses is the largest length over which charge can be separated due to the presence of an optical electric field. For the linear polyenes, that length is the distance along the conjugation axis separating the end carbons. For COT, the equivalent length is between two points on either end of a diameter of the ring. Additionally, since π-electron motion is constrained along the carbon lattice, the relevant length is one-half of the circumference of the molecular structure. This distance is 5.3 Å for COT. For comparison, for *trans*-HT (N = 6) $\gamma_g = 0.75 \times 10^{-36}$ esu at $\hbar\omega = 0.65$ eV and the end-to-end length L is 6 Å. Thus, for these two very different structures where the relevant lengths are nearly equal, we find that the calculated values of γ_g are also comparable. In the case of cyclic structures, the length which determines the magnitude of γ_g appears to be one-half of the circumference; this is a general finding in calculations that we have performed for other sizes of rings.

VII. $\gamma_{ijkl}(-\omega_4;\omega_1,\omega_2,\omega_3)$ of RIGID-ROD POLYMERS

The preceding sections describe the origin of the large, nonresonant $\gamma_{ijkl}(-\omega_4;\omega_1,\omega_2,\omega_3)$ for the prototype systems for conjugated linear polymers in terms of the virtual transitions of the π-electronic states in which the many-electron nature of the excitations is of fundamental importance. Delocalization of the π-electrons along the axis of the linear chain allows a large redistribution of charge density in response to the optical field. It is because of this mechanism for $\chi^{(3)}(-\omega_4;\omega_1,\omega_2,\omega_3)$ that the nonresonant nonlinear optical response for conjugated polymers is also ultrafast. For some materials, thermal and/or molecular reorientational effects can make significant contributions to third-order optical processes. Since these mechanisms involve translational or rotational motion of molecules, these effects have relatively slow response times ($\sim 10^{-6}$ s for thermal and $\sim 10^{-12}$ s for reorientational). In another example, the large nonlinearities observed in inorganic materials such as GaAs are due to resonant electronic processes. These suffer from being narrow band, since optical pumping resonant with the exciton lines is required; but more importantly, since these processes involve the real absorption of photons, they also can give at best only picosecond recovery times. On the other hand, the virtual excitation of π-electrons responsible for the nonresonant $\chi^{(3)}$ of conjugated organic and polymeric materials occurs on the femtosecond (10^{-15} sec) time scale. This prospect for optical devices with switching speeds orders of magnitude faster than those of current semiconductor electronic devices is the source of the great technological interest in developing polymers as nonlinear optical materials.

In this section, we consider the origin of the nonlinear optical response in the more complex but environmentally robust rigid-rod polymer structures. In addition to possessing a large, ultrafast $\chi^{(3)}$, materials for nonlinear optical devices must be easily processable and remain stable under typical working conditions. The rigid-rod polymers PBO (poly-*p*-phenylene benzobisoxazole) and PBT (poly-*p*-phenylene benzobisthiazole), as an example, yield high-quality thin films and may be quite suitable for optical waveguide architectures. Values of $\chi^{(3)}$ larger than 10^{-11} esu have been observed [44,45] for these materials with a response time shorter than picoseconds. Here we discuss the origin of $\gamma_{ijkl}(-\omega_4; \omega_1, \omega_2, \omega_3)$ for the repeat unit structure of PBO, known as *t*-BOZ (*trans*-benzobisoxazole) and shown in Fig. 16, based on the same theoretical method employed in the previous sections. Comparison of the theoretical results with our third-harmonic generation measurements demonstrates that the origin of $\chi^{(3)}$ in rigid-rod polymers is identical to that described above for the prototype polyenic conjugated structures.

The coordinates used for the rigid ionic lattice of *t*-BOZ were taken from x-ray diffraction studies of the crystal structures [46], and standard values were employed for the atomic CNDO/S parameters (Table 7). The *t*-BOZ molecule is fully conjugated, and thus the

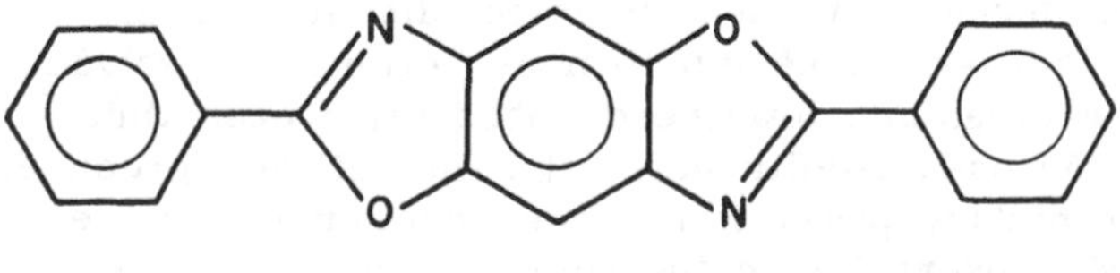

Figure 16. Rigid rod structure of *trans*-benzobisoxazole (*t*-BOZ). For *trans*-benzobisthiazole (*t*-BTZ) replace O with S.

Table 7. Input Parameters for ζ: Slater Exponents, β: Resonance Integrals, γ: Coulomb Repulsion Integrals, and I: Valence State Ionization Energies for Hydrogen, Carbon, Nitrogen, and Oxygen

Parameter	H	C	N	O
ζ (Å^{-1})	2.30	3.78	3.03	4.06
β_s (eV)	10.0	20.0	23.0	26.8
β_p (eV)	—	17.0	18.0	21.8
γ (eV)	12.8	10.6	12.4	13.1
I_s (eV)	13.6	21.3	27.5	35.5
I_p (eV)	—	11.5	14.3	17.9

ground state includes 13 occupied π-electron molecular orbitals. The capability to perform a complete π-electron CI calculation for this structure and then compute the complete set of transition dipole moments between all the resulting excited states is beyond current computational power. The linear optical absorption spectrum, how-ever, does not require a complete calculation since the dominant transitions are typically composed of excitations from the first few highest occupied to first few lowest unoccupied π-electron molecular orbitals. The correspondence between the calculated and experimental absorption spectra thus provides proof of the suitability of the theoretical method.

The predicted optical absorption spectrum [with 0.5-eV FWHM (full width at half-maximum) peaks] from an SDCI calculation includ-ing the three highest occupied and three lowest unoccupied MOs is compared to the gas-phase and solution-phase spectra of the *t*-BOZ compound in Fig. 17 with the three spectra normalized to a common peak height. The experimental gas-phase spectrum, which was obtained by sublimation of the compound at 280°C, shows a dominant peak at 3.8 eV (326 nm), in good agreement with the calculated value for an isolated molecule at 4.06 eV (306 nm), but the signal was too weak to resolve any further structure. The spectrum for *t*-BOZ dissolved in dioxane is solvatochromatically red-shifted due to induced dipole interactions between the solvent and the solute. The dominant, low-energy absorption peak thus occurs at 3.6 eV (350 nm). Additionally, the theoretically predicted 5.5-eV (225-nm) secondary peak, which was unobtainable in the gas phase, is clearly observable in the solution spectrum. The existence of fine structure in the peak at 350 nm (rather than just

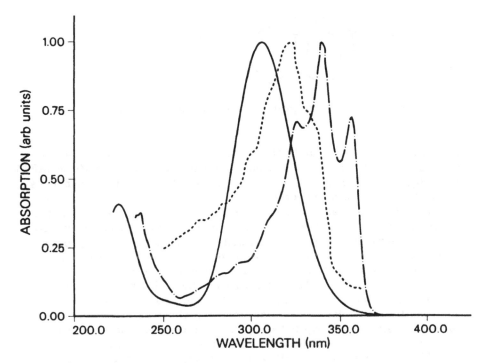

Figure 17.   Experimental and calculated optical absorption spectra of *t*-BOZ; solid line, theoretical; dashed line, experimental gas phase; chain-dot line, experimental solution phase in dioxane.

a single peak) is due to vibrational excitations which are not included in the calculation and unimportant to the nonresonant $\gamma_{ijkl}(-\omega_4; \omega_1, \omega_2, \omega_3)$.

   The dominance of the optical absorption spectrum by the lowest-energy optically allowed electronic transition which occurs in the near-uv is a general feature of conjugated linear chains and is an example of the remarkable similarity of *t*-BOZ to simpler structures, such as the polyenes. As will be shown later, this similarity holds for the third-order optical properties as well. The analogy between the *t*-BOZ structure and polyenes allows us to simplify the computational problem by considering a modified version of *t*-BOZ (Fig. 18) in which each of the end phenyl groups is replaced by a single, conjugated carbon-carbon bond. Since the modified *t*-BOZ structure is only 3 Å shorter than full *t*-BOZ, it should have nearly the same $\gamma_{ijkl}(-\omega_4; \omega_1, \omega_2, \omega_3)$. The advantage of considering the modified compound is that its seven double bonds now pose a tractable problem for current computing technology. Since

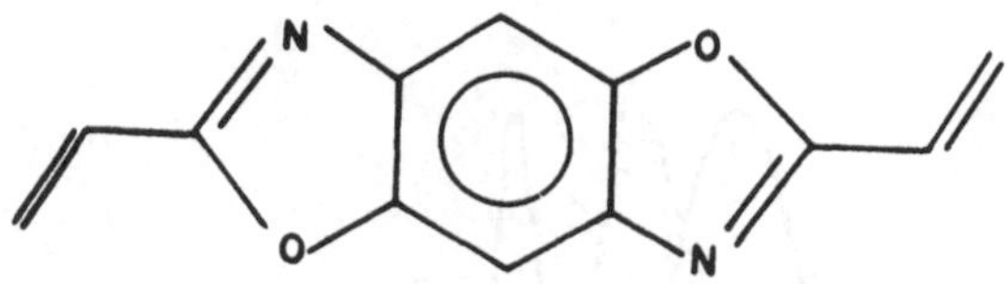

Figure 18.   Schematic diagram of the modified structure of *t*-BOZ.

the calculation of the nonlinear optical response function requires
an accurate description of the entire π-electron excited state manifold
(see Eq. (5)), it requires a more complete CI calculation than does
the optical absorption spectrum.

The calculated absorption spectrum for modified *t*-BOZ provides
clear evidence that the π-electronic states are not strongly altered
by the substitution for the end phenyls. The lowest-energy peak
occurs as a pair of states at 3.69 and 3.97 eV in comparison to
the 4.06-eV peak in full *t*-BOZ. The secondary peak at 5.52 eV
becomes a doublet at 5.06 and 5.54 eV. The primary difference
in the absorption spectra calculated for the two structures is that,
whereas the secondary peak was only one-third as strong as the
primary peak in the original compound, it is almost as strong as
the primary in the calculation for the modified structure.

The modified form of *t*-BOZ has tensor components of $\gamma_{ijkl}(-3\omega;$
$\omega,\omega,\omega)$ of $\gamma_{xxxx} = 68.7$, $\gamma_{xxyy} = 0.2$, $\gamma_{yyxx} = 0.1$, and $\gamma_{yyyy} =$
1.3×10^{-36} esu from a calculation at 0.65 eV (1.907 μm). Since
the length L between the end carbons is 10.7 Å for modified *t*-BOZ,
it can be seen from Fig. 8 that this value for γ_{xxxx} is nearly equal
to that of a polyene of the same length. The complicated heteroatomic
structure of *t*-BOZ does not appear to have a drastic effect on
$\gamma_{ijkl}(-\omega_4;\omega_1,\omega_2,\omega_3)$. The resultant value for the isotropically averaged
susceptibility γ_g determined from Eq. (4) is 14.0×10^{-36} esu.

Since *t*-BOZ and the polyene series both are members of the
C_{2h} symmetry group, the same selection rules for one-photon transi-
tions and $\gamma_{ijkl}(-\omega_4;\omega_1,\omega_2,\omega_3)$ described earlier apply here as well.
The virtual excitation series $g \rightarrow 1\,^1B_u \rightarrow x\,^1A_g \rightarrow 1\,^1B_u \rightarrow g$ is the
most important one for *t*-BOZ just as it is for the polyenes. How-
ever, there are not just two dominant intermediate ^1A_g states for
the case of *t*-BOZ. All of the terms of the above form, where x ^1A_g
is any of the calculated ^1A_g states, together make up 70% of the
total value for γ_{xxxx}. (The remaining 30% comes from terms in which
^1B_u states other than 1 ^1B_u are involved.) The existence of large
terms involving several different intermediate ^1A_g states in *t*-BOZ,
as opposed to just the ground and 6 ^1A_g states for OT, is a common
feature of the longer polyenes, also, and was one of the factors
responsible for the rapid growth of γ_{xxxx} with increased chain length.

Transition density matrix contour diagrams for three of the important virtual transitions of the modified t-BOZ structure are displayed in Fig. 19. The ground → $1\,^1B_u$ diagram shows the same modulated charge redistribution as was observed for the polyenes in Figs. 2 and 7. The x-component of the transition dipole moment in this case is 6.5 D. The first excited two-photon state, $2\,^1A_g$, also has a fairly large transition moment with the $1\,^1B_u$ of 5.2 D. It is interesting to note that, in contrast to the polyenes, the $2\,^1A_g$ state is calculated 0.9 eV above the $1\,^1B_u$ state in t-BOZ. Finally, the $1\,^1B_u \rightarrow 8\,^1A_g$ transition, with a 7.2-D transition moment, has the same large transfer of electron density across the chain as was found in the $1\,^1B_u \rightarrow 5\,^1A_g$ transition of HT and $1\,^1B_u \rightarrow 6\,^1A_g$ transition of OT.

Figure 20 shows the calculated dispersion of $\gamma_{xxxx}(-3\omega;\omega,\omega,\omega)$ for the modified form of t-BOZ. The first resonance, at 1.23 eV ($\lambda = 1.01\ \mu m$), is due to the 3ω resonance of the $1\,^1B_u$ state. Since the $2\,^1B_u$ state lies nearby, its 3ω resonance also contributes. The second resolvable resonance is the 3ω resonance of the $3\,^1B_u$ state. The 2ω resonance of the $2\,^1A_g$ state does not occur, in this case, until 2.30 eV ($\lambda = 0.54\ \mu m$) because it lies so far above the $1\,^1B_u$.

The results presented in this section demonstrate that the mechanism of $\gamma_{ijkl}(-\omega_4;\omega_1,\omega_2,\omega_3)$, and thus $\chi^{(3)}(-\omega_4;\omega_1,\omega_2,\omega_3)$, is the same for the repeat unit of the rigid-rod PBO polymer as it is for the polyene chain. The results for the magnitude and sign of the nonresonant $\gamma_{ijkl}(-\omega_4;\omega_1,\omega_2,\omega_3)$, the important virtual excitation processes, and the transition density matrix diagrams appear to be generalizable to a large class of conjugated linear chains. For this class of structures, the large, ultrafast nonresonant $\chi^{(3)}$ is a result of virtual transitions among strongly correlated π-electron states.

We have performed [18] a liquid-phase third-harmonic generation experiment in order to confirm the calculation presented above. Because the melting point of the t-BOZ compound was beyond the range of our heating equipment, we performed the experiment on t-BTZ liquid, which melts at ~300°C. Since the t-BOZ and t-BTZ compounds differ only in the replacement of the two oxygens with sulfurs, it is not expected that there will be a significant difference in the values for $\chi^{(3)}(-3\omega;\omega,\omega,\omega)$. The wedge maker fringe technique was employed, and a fundamental photon energy of 0.65 eV ($\lambda = 1.907\ \mu m$) was used to ensure that both the fundamental and third-harmonic light were far from electronic resonances.

For liquid t-BTZ we measure a value for the liquid third-order susceptibility $\chi_L^{(3)}(-3\omega;\omega,\omega,\omega)$ at 1.907 μm (0.65 eV) of 0.16 × 10^{-12} esu. The macroscopic susceptibility for an isotropic liquid $\chi_L^{(3)}$ is related to the isotropic liquid molecular susceptibility γ_L by

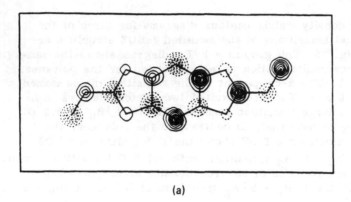

(a)

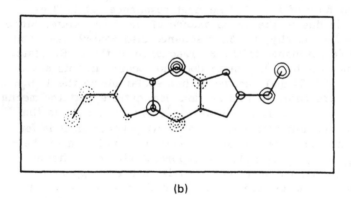

(b)

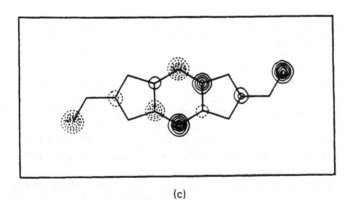

(c)

Figure 19. Transition density matrix contour diagrams for (a) ground, (b) $2\,^1A_g$, and (c) $8\,^1A_g$ states of modified t-BOZ with the $1\,^1B_u$ state.

$$\chi_L^{(3)}(-3\omega;\omega,\omega,\omega) = N\left(\frac{2+n_\omega^2}{3}\right)^3\left(\frac{2+n_{3\omega}^2}{3}\right)\gamma_L(-3\omega;\omega,\omega,\omega)$$

(16)

where N is the number density of molecules and the refractive index terms are Lorentz–Lorenz local field factors. The isotropic liquid molecular susceptibility γ_L requires precisely the same orientational averaging as that for an isotropic gas given in Eq. (4). Approximating the number density as $N = 2.5 \times 10^{21}$ cm^{-3} based on the *t*-BTZ crystal density [46] of 1.42 g/cm^3, we obtain $\gamma_L(-3\omega;\omega,\omega,\omega)$ = 10.9×10^{-36} esu for *t*-BTZ at 1.907 μm. The agreement with the calculated value 14.0×10^{-36} esu given above for the modified *t*-BOZ structure is quite satisfactory, considering the slight difference in the molecular structures and that the calculation is for an isolated (gas) molecule. It is thus shown that the theoretical analysis of the preceding sections is also the appropriate description for $\gamma_{ijkl}(-\omega_4;\omega_1,\omega_2,\omega_3)$ of the repeat unit structures of conjugated rigid-rod polymers.

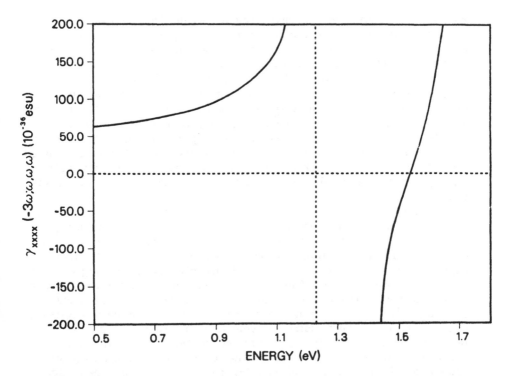

Figure 20.   Calculated dispersion of $\gamma_{xxxx}(-3\omega;\omega,\omega,\omega)$ for modified *t*-BOZ. The vertical dash locates the 3ω resonance to the $1\,^1B_u$ state.

The $\chi^{(3)}$ of polymer PBT can be estimated in the approximation of a densely packed, linked array of monomer units. For an isotropic film of PBT with a density [47] of 1.69 g/cm^3 and an estimated refractive index of 1.8, one obtains $\chi^{(3)} = 0.3 \times 10^{-12}$ esu. If one takes as the active optical group two repeat units, then the power-law dependence of Fig. 8 yields values of γ and $\chi^{(3)}$ 24 times larger than that of the monomer. The resulting estimated polymer $\chi^{(3)}$ is then 7×10^{-12} esu in good agreement with the measurements [44,45] of partially ordered thin films of PBT. We thus conclude that the effective source of nonlinear optical polarization in PBT is at most two or three repeat unit structures, and that the phase coherence of the electronic wavefunctions and induced polarization does not extend beyond this limit. We expect that dimers or trimers of PBT should possess nearly as large a $\chi^{(3)}$ as has been observed for the polymer and that further increase of the chain length will not increase the nonlinear optical response.

VIII. CONCLUSION

Electron correlation effects markedly determine the virtual electronic excitation processes and nonlinear optical properties of the low-dimensional structures of conjugated linear and cyclic chains. Microscopic understanding of the origin of the large, ultrafast nonresonant $\chi_{ijk}^{(2)}(-\omega_3;\omega_1,\omega_2)$ and $\chi_{ijkl}^{(3)}(-\omega_4;\omega_1,\omega_2,\omega_3)$ of these materials is obtained through many-electron calculations of the molecular susceptibilities $\beta_{ijk}(-\omega_3;\omega_1,\omega_2)$ and $\gamma_{ijkl}(-\omega_4;\omega_1,\omega_2,\omega_3)$ and direct comparison with available experimental results. Details of the mechanism for $\gamma_{ijkl}(-\omega_4;\omega_1,\omega_2,\omega_3)$ of one-dimensional conjugated linear chains have been explained in terms of virtual transitions among highly correlated π-electron states, and contour diagrams of transition density matrices $\rho_{nn'}$ provided direct illustration of the most significant virtual transitions.

Theoretical results for $\gamma_{ijkl}(-2\omega;\omega,\omega,0)$ in two short linear polyene structures have been demonstrated to be in agreement with independent gas-phase experimental measurements. The origin of $\chi_{ijkl}^{(3)}(-\omega_4;\omega_1,\omega_2,\omega_3)$ in these conjugated structures has been shown to involve important contributions from energetically high-lying, strongly correlated two-photon states. The symmetry selection rule for third-order processes that virtual transitions must be connected in the series $g \rightarrow {}^1B_u \rightarrow {}^1A_g \rightarrow {}^1B_u \rightarrow g$ has been demonstrated to govern $\gamma_{ijkl}(-\omega_4;\omega_1,\omega_2,\omega_3)$ in centrosymmetric polyenes. Two dominant types of terms have been identified and denoted as types I and II, and the type II term was shown to be the larger of the two and positive, thus determining the sign of $\gamma_{ijkl}(-\omega_4;\omega_1,\omega_2,\omega_3)$. The dependences of $\gamma_{ijkl}(-\omega_4;\omega_1,\omega_2,\omega_3)$ on chain length and structural conformation have been examined, and conformation was found to be significant only inasmuch as it affects the actual physical length of the chain. Results for the all-*trans* and *cis-transoid* conformations of polyenes are unified by a power-law dependence of the dominant tensor component $\gamma_{xxxx}(-3\omega;\omega,\omega,\omega)$ on the chain length L.

For linear polyene chains in which the centrosymmetry is removed by heteroatomic substitution on the chain ends, the nonresonant value of $\gamma_{xxxx}(-\omega_4;\omega_1,\omega_2,\omega_3)$ is observed to increase by at least an order of magnitude. Noncentrosymmetry leads to new types of virtual excitation processes involving diagonal transition moments, and the dominant terms of this sort have been denoted as type III terms. The important quantity in this type of term, which greatly enhances $\gamma_{xxxx}(-\omega_4;\omega_1,\omega_2,\omega_3)$, is the dipole moment difference between a principal excited state and the ground state.

A theoretical study of $\gamma_{ijkl}(-\omega_4;\omega_1,\omega_2,\omega_3)$ in a two-dimensional conjugated cyclic structure allowed examination of the role of dimensionality in the nonlinear optical properties of conjugated systems. The origin of $\gamma_{ijkl}(-\omega_4;\omega_1,\omega_2,\omega_3)$ in this case was found to be very similar to that for the one-dimensional case, but the magnitudes of the various components of γ_{ijkl} are all smaller than the γ_{xxxx} component of the one-dimensional chain with an equal number of carbons. More important, the isotropically averaged third-order susceptibility γ_g is also much less for the two-dimensional structure as compared to the one-dimensional chain. Our analysis shows that this is a result of a reduction in the effective length over which π-electrons can respond to an applied optical electric field. For two-dimensional cyclic structures, this length is one-half of the circumference of the ring as compared to the full end-to-end chain length in one-dimensional structures. As a general rule, two-dimensional conjugated cyclic structures are therefore expected to have smaller nonresonant $\chi_{ijkl}^{(3)}$ values than their one-dimensional analogs.

Analysis of calculations for rigid-rod polymer model compounds indicates that the origin of $\chi_{ijkl}^{(3)}(-\omega_4;\omega_1,\omega_2,\omega_3)$ in these more complicated but environmentally robust materials is entirely analogous to that discovered for the polyene structures. The transition density matrix diagrams of the primary states in this case bear remarkable resemblance to those of the polyenes. The symmetry restrictions, dispersion, and magnitude of $\gamma_{ijkl}(-\omega_4;\omega_1,\omega_2,\omega_3)$ in the rigid-rod polymers also are in correspondence with the polyene results. An experimental third-harmonic generation measurement in the model rigid-rod polymer compound *t*-BTZ proved the theoretical approach to be equally valid in these structures. The results presented in detail in this review appear to be generalizable to the complete class of oligomeric and polymeric conjugated organic materials.

This research was generously supported by the U.S. Air Force Office of Scientific Research and the U.S. Defense Advanced Research Projects Agency Grant No. F49620-85-C-0105 and National Science Foundation/Materials Research Laboratories Program Grant No. DMR-85-19059. The calculations were performed on the CRAY X-MP of the Pittsburgh Supercomputing Center. We gratefully acknowledge the contributions of Drs. K. Y. Wong and O. Zamani-Khamiri and Mr. Qi-Lu Zhou.

REFERENCES

1. See, for example, D. J. Williams (ed.), *Nonlinear Optical Proper-ties of Organic and Polymeric Materials*, ACS Symposium Series 233, American Chemical Society, Washington, D.C., 1983; A. J. Heeger, D. Ulrich, and J. Orenstein (eds.), *Nonlinear Optical Properties of Polymers*, Mater. Res. Soc. Proc. *109*, Pittsburgh, PA, 1988; G. Khanarian (ed.), *Nonlinear Optical Properties of Organic Materials*, SPIE PROC. *971*, 1989; R. A. Hann and D. Bloor, *Organic Materials for Nonlinear Optics*, Royal Soc. of Chem., London, 1989; J. Messier et al. (eds.), *Nonlinear Optical Effects in Organic Polymers*, NATO ASI Ser. E, Vol. *162*, Kluwer Academic, Boston, 1989.
2. S. K. Kurtz and T. T. Perry, *J. Appl. Phys.*, *39*:3798 (1968).
3. B. F. Levine and C. G. Bethea, *J. Chem. Phys.*, *63*:2666 (1975).
4. J. L. Oudar and D. S. Chemla, *J. Chem. Phys.*, *66*:2664 (1977).
5. K. D. Singer and A. F. Garito, *J. Chem. Phys.*, *75*:3572 (1981).
6. C. C. Teng and A. F. Garito, *Phys. Rev. Lett.* *50*:350 (1983); *Phys. Rev. B28*:6766 (1983).
7. J. P. Herrmann, D. Ricard, and J. Ducuing, *Appl. Phys. Lett.*, *23*:178 (1973).
8. J. P. Herrmann and J. Ducuing, *J. Appl. Phys.*, *45*:5100 (1974).
9. C. Sauteret et al., *Phys. Rev. Lett.*, *36*:956 (1976).
10. S. J. Lalama and A. F. Garito, *Phys. Rev.*, *A20*:1179 (1979).
11. A. F. Garito, C. C. Teng, K. Y. Wong, and O. Zamani-Khamiri, *Mol. Cryst. Liq. Cryst.*, *106*:219 (1984).
12. A. F. Garito, Y. M. Cai, H. T. Man, and O. Zamani-Khamiri, *Crystallographically Ordered Polymers*, ACS Symposium Series 337, D. J. Sandman, ed., American Chemical Society, Washington, D.C., 1987, Chap. 14.
13. A. F. Garito, K. Y. Wong, and O. Zamani-Khamiri, *Nonlinear Optical and Electroactive Polymers*, D. Ulrich and P. Prasad, eds. Plenum, New York, 1987.
14. J. R. Heflin, K. Y. Wong, O. Zamani-Khamiri, and A. F. Garito, *J. Opt. Soc. Am. B4*:136 (1987).

15. J. R. Heflin, K. Y. Wong, O. Zamani-Khamiri, and A. F. Garito, *Phys. Rev. B38*:1573 (1988).
16. A. F. Garito, J. R. Heflin, K. Y. Wong, and O. Zamani-Khamiri, *Nonlinear Optical Properties of Polymers*, A. J. Heeger, D. Ulrich, and J. Orenstein, eds., Mater. Res. Soc. Proc. 109, Pittsburgh, PA, 1988, pp. 91-102; *Mol. Cryst. Liq. Cryst. 160*:37 (1988); *Proc. SPIE, 825*:56 (1988).
17. J. W. Wu, J. R. Heflin, R. A. Norwood, K. Y. Wong, O. Zamani-Khamiri, A. F. Garito, P. Kalyanaraman, and J. Sounik, *J. Opt. Soc. Am. B6*:707 (1989).
18. A. F. Garito, J. R. Heflin, K. Y. Wong, and O. Zamani-Khamiri, *Organic Materials for Nonlinear Optics*, R. A. Hann and D. Bloor, eds. Royal Soc. of Chem., London, 1989, pp. 16-27; *Proc. SPIE, 971*:9 (1989); *Photoresponsive Materials*, S. Tazuke, ed., Mat. Res. Soc. Proc. IMAM-12, Pittsburgh, PA, 1989, pp. 3-20.
19. C. Grossman, J. R. Heflin, K. Y. Wong, O. Zamani-Khamiri, and A. F. Garito, *Nonlinear Optical Effects in Organic Polymers*, J. Messier et al., eds., NATO ASI Ser. E, Vol. 162, Kluwer Academic, Boston, 1989, pp. 61-78.
20. See, for example, J. Solyom, *Adv. in Phys.*, *28*:201 (1979) and references therein.
21. Z. G. Soos and S. Ramasesha, *Chem. Phys. Lett.*, *153*:171 (1988); *J. Chem. Phys.*, *90*:1067 (1989).
22. B. M. Pierce, *J. Chem. Phys.*, *91*:791 (1989).
23. K. C. Rustagi and J. Ducuing, *Opt. Comm.*, *10*:258 (1974).
24. H. Hameka, *J. Chem. Phys.*, *67*:2935 (1977).
25. G. P. Agrawal, C. Cojan, and C. Flytzanis, *Phys. Rev.*, *B17*:776 (1978).
26. D. N. Beratan, J. N. Onuchic, and J. W. Perry, *J. Phys. Chem.*, *91*:2696 (1987).
27. M. G. Papadopolous, J. Waite, and C. A. Nicolaides, *J. Chem. Phys.*, *77*:2527 (1982).
28. G. J. B. Hurst, M. Dupuis, and E. Clementi, *J. Chem. Phys. 89*:385 (1988).
29. C. P. de Melo and R. Silbey, *J. Chem. Phys.*, *88*:2567 (1988).
30. J. L. Bredas, *Proc. SPIE, 971*:42 (1989).
31. P. Chopra, L. Carlacci, H. F. King, and P. N. Prasad, *J. Phys. Chem.*, *93*:7120 (1989).
32. A. F. Garito and K. Y. Wong, *Polymer J.*, *19*:51 (1987).
33. C. H. Grossman, Ph.D. Thesis, University of Pennsylvania, 1987.
34. N. N. Bogoliubov and Y. A. Mitropolsky, *Asymptotic Methods in the Theory of Nonlinear Oscillations*, Gordon and Breach, 1961 (translated from Russian).

35. B. J. Orr and J. W. Ward, *Mol. Phys.*, *20*:513 (1971). In contrast to Eq. (5), Orr and Ward chose to include the factor $K(\omega_4;\omega_1,\omega_2,\omega_3)$ in the constitutive equation for the microscopic polarization rather than in $\gamma_{ijkl}(-\omega_4;\omega_1,\omega_2,\omega_3)$.
36. See, for example, B. S. Hudson, B. E. Kohler, and K. Schulten, *Excited States*, Vol. 6, E. C. Lim, ed., Academic Press, New York, 1982, p. 1 and references therein.
37. R. R. Chadwick, D. P. Gerrity, and B. S. Hudson, *Chem. Phys. Lett.*, *115*:24 (1985).
38. B. S. Hudson and B. E. Kohler, *Synth. Metals*, 9:241 (1984).
39. J. F. Ward and D. S. Elliott, *J. Chem. Phys.*, 69:5438 (1978). We have converted Ward and Elliott's $\chi^{(3)}$ to our notation by $\gamma_g(-2\omega;\omega,\omega,0) = \frac{3}{2} \chi^{(3)}$.
40. R. M. Gavin, Jr., S. Risemberg, and S. A. Rice, *J. Chem. Phys.*, *58*:3160 (1973).
41. Z. G. Soos and S. Ramesesha, *Phys. Rev.*, *B29*:5410 (1984).
42. F. Kajzar and J. Messier, *Polym. J.*, *19*:275 (1987).
43. R. P. Linstead, *J. Chem. Soc.*, 2873 (1953).
44. A. F. Garito and C. C. Teng, *Nonlinear Optics*, P. Yeh, ed., SPIE, San Diego, 1986.
45. D. N. Rao, J. Swiatkiewicz, P. Chopra, S. K. Ghoshal, and P. N. Prasad, *Appl. Phys. Lett.*, *48*:1187 (1986).
46. M. W. Wellman, W. W. Adams, D. R. Wiff, and A. V. Fratini, Air Force Technical Report AFML-TR-79-4184, 1980.
47. S. G. Wierschke, Air Force Technical Report AFWAL-TR-88-4201, 1988.

2

Organic Materials for Second-Order Nonlinear Optical Devices

Kenneth D. Singer* and John E. Sohn / AT&T Bell Laboratories, Princeton, New Jersey

I. INTRODUCTION

The advent of the laser has driven the field of nonlinear optics in two important ways. First, the high-intensity electric field associated with laser light has allowed the observation of the small optical nonlinearities in materials; and second, lasers have permitted optical information transmission at very high speed which will eventually require active photonic devices based on various nonlinear optical phenomena. Thus, these two applications of the laser have stimulated much work both from scientific and technological viewpoints. Much of this work has been materials related—that is, the properties of materials that give rise to the nonlinearities and the methods to exploit these properties to produce useful devices. These studies

---

Current affiliation: Case Western Reserve University, Cleveland, Ohio.

have spanned the range of condensed matter: atomic and molecular gases and liquids, metals, semiconductors, and insulators. Many of these studies and initial applications centered on single-component materials, such as pure inorganic and organic crystals; however, a great deal of current interest and the hopes for many future applications is focused on composite materials, such as multiple-quantum-well semiconductors, semiconductor-insulator composites, metal-containing composites, multicomponent Langmuir-Blodgett films, and guest-host polymers [1,2].

Interest in organic materials arose from the large optical non-linearities observed in organic molecular crystals [3]. These large nonlinearities have been traced to efficient charge reorganization between the electronic states in the π-electron system of their molecular constituents. Much of the early work in the field was focused on the origin of the second-order nonlinear optical properties in organic molecules and the crystals they comprise. This work on molecular crystals has culminated in the first practical organic nonlinear optical devices. These devices have resulted from the "molecular engineering" made possible by the nearly endless synthetic possibility of organic molecules coupled with the understanding both on the molecular level and on the relationship between the molecular arrangement in crystals and the operation of nonlinear optical devices. Recently, it has been recognized that, in addition to synthetic flexibility, a wide variety of bulk fabrication techniques are available for organic and polymeric materials as well. Although work continues on molecular crystals, much current interest is also focused on other, more flexible fabrication approaches, including pure and guest-host Langmuir-Blodgett films, liquid crystal polymers, and amorphous polymers [4]. Renewed interest in the second-order nonlinear optical properties of molecules has developed due to the new requirements of these systems and to a new interest in possible electrooptic applications [5]. The first steps along the way to practical devices in these materials have been taken. A great deal of current activity is also directed at the basic understanding of third-order nonlinearities. Third-order properties will be covered elsewhere in this volume.

We begin by introducing the basic concepts of nonlinear optics as well as the phenomena that arise from second-order optical non-linearities. The basic mechanisms of molecular and bulk nonlinearities in organic and polymeric materials will then be described, and the application of this understanding to the development of materials appropriate for device use will be discussed. This is followed by a section describing possible second-order nonlinear devices, with particular attention being paid to the material requirements of devices as well as our outlook relating to material issues for devices.

II. NONLINEAR OPTICAL PHENOMENA

The arrangement of electrical charge in matter can be described in a multipole expansion. For most practical purposes, the interaction of that charge distribution with external fields is given by the lowest multipole in charge neutral matter, namely the electric dipole per unit volume, $\underline{P}(\underline{r},t)$. This vector quantity describing the distribution will itself depend on the presence of external fields. The precise functional dependence on the field is unknown, but since externally applied fields are usually much weaker than the atomic fields binding the charges, an expansion of $\underline{P}(\underline{r},t)$ in powers of the applied field will suffice. In fact, before the advent of the laser, optical fields were so weak that the electric polarization was linear in those fields, and terms higher order in the field could only be observed at microwave frequency or lower, since electrical sources were the only intense ones available.

Formally, the field expansion of the dipole per unit volume is given by

$$\underline{P}(\underline{r},t) = \underline{P}^{(1)}(\underline{r},t) + \underline{P}^{(2)}(\underline{r},t) + \underline{P}^{(3)}(\underline{r},t) + \cdots \tag{1}$$

where the terms of nth order in the field are

$$\underline{P}^{(n)}(\underline{r},t) = \int_{-\infty}^{+\infty} d\tau_1 \int_{-\infty}^{+\infty} d\tau_2 \cdots \int_{-\infty}^{+\infty} d\tau_n \int d\underline{r}_1 \int d\underline{r}_2 \cdots \int d\underline{r}_n$$

$$\times \underline{R}^{(n)}(\underline{r},\underline{r}_1\tau_1 \cdots \underline{r}_n\tau_n)$$

$$\times E(\underline{r}_1,t-\tau_1)E(\underline{r}_2,t-\tau_2) \cdots E(\underline{r}_n,t-\tau_n) \tag{2}$$

where $\underline{R}^{(n)}(\underline{r},t,\underline{r}_1\tau_1 \cdots \underline{r}_n\tau_n)$ is the nth-order real, causal, and time-invariant response function depending on the time intervals τ_n referred to t. A more complete discussion of the polarization is given elsewhere [6-8].

The response function, in general, can be nonlocal in both time and space, allowing the polarization to grow as the integral implies. It is found that the instantaneous and local polarization is small; however, significant effects can be observed in configurations where the polarization is allowed to build up in a coherent way. The polarization response can grow coherently if the interaction occurs over a long time, such as when a material resonance occurs and energy is stored over the period of the relaxation time determined by the Q of the resonance. This corresponds to an integral over time in Eq. (2). Quantum confined excitons in semiconductors are an example where these effects have been used to great advantage. The response also builds spatially over the

volume that the light propagates during the relaxation time. As
we will see below, a coherent growth of polarization can also be
attained for an instantaneous response when the interaction adds
over long lengths of propagation. This effect has been utilized
in crystals for second-harmonic generation. These schemes for
obtaining large (and useful) polarizations, have their price. In
the case of long-time interactions, the speed of the device is reduced,
and if one is to recover the advantageous high-speed expected for
photonic devices, a high degree of parallelism is required. For
utilizing a long length to build the nonlinear polarization, the requirements for low loss and spatial homogeneity demand very high quality
materials. Since organics are most suited for nonresonant operation,
much of this chapter will deal with the work aimed at producing
such high-quality materials.

Since laser sources are, for many practical purposes, monochromatic, and since the nonlinear response for the devices we
will describe arise from local interactions, it is convenient to cast
Eq. (2) in terms of the frequency and wave vector variables, ω
and $\underline{k}$:

$$\underline{P}(\underline{k},-\omega) = \underline{P}^{(1)}(\underline{k},-\omega) + \underline{P}^{(2)}(\underline{k},-\omega) + \underline{P}^{(3)}(\underline{k},-\omega) + \cdots \tag{3}$$

where

$$\underline{P}^{(n)}(\underline{k},-\omega) = \underline{X}^{(n)}(-\omega;\omega_1 \cdots \omega_n)\underline{E}(\underline{k}_1,\omega_1) \cdots \underline{E}(\underline{k}_n,\omega_n) \tag{4}$$

and where $\underline{X}^{(n)}(-\omega;\omega_1 \cdots \omega_n)$ is the susceptibility of nth order,
the negative sign denotes an emitted photon, and tensor multiplica-
tion is implied. The quantity $\underline{X}^{(n)}(-\omega;\omega_1 \cdots \omega_n)$ is related to the
response function $R(\underline{r},\underline{r}_1\tau_1 \cdots)$ through the appropriate Fourier
transform, as are the appropriate field quantities. The locality
of the response yields no $\underline{k}$ dependence of the susceptibility, and
this $\underline{k}$-independent susceptibility is known as the dipole susceptibility.
For the second-order processes of interest here,

$$P_i(\underline{k},-\omega;\omega_1,\omega_2) = \chi_{\underline{ijk}}^{(2)}(-\omega;\omega_1,\omega_2)E_j(\underline{k}_1,\omega_1)E_k(\underline{k}_2,\omega_2) \tag{5}$$

where the electric fields are given by the Fourier transform of
the time-dependent fields:

$$E_j(\underline{k}_n,t) = \frac{1}{2}[E_j^0(\underline{k}_n,\omega_1)\exp\{-i(\omega_1 t)\}$$

$$+ E_j^0(\underline{k}_n,\omega_2)\exp\{-i(\omega_2 t)\} + c,c] \tag{6}$$

and the susceptibility, $\chi_{\underline{ijk}}^{(2)}(-\omega;\omega_1\omega_2)$ provides the connection between the material property and the optical phenomenon.

It is readily seen from Eq. (5) that $\chi_{\underline{ijk}}^{(2)}(-\omega;\omega_1,\omega_2)$ is a third-rank polar tensor which connects its product with two polar vectors (the electric fields) with the polar $\underline{P}(\underline{k},-\omega)$. When the inversion operation is applied to the polar vectors, then $\underline{E}(-\underline{r}_n) \rightarrow -\underline{E}(\underline{r}_n)$ and $\underline{P}(-\underline{r}) \rightarrow -\underline{P}(\underline{r})$. Equation (5) is consistent under this operation only if $\chi_{\underline{ijk}}^{(2)}(-\underline{r}) \rightarrow -\chi_{\underline{ijk}}^{(2)}(\underline{r})$. Thus if the material is invariant under inversion (centrosymmetric), $\chi_{\underline{ijk}}^{(2)} \equiv 0$. The requirement that the material be noncentrosymmetric in order to exhibit second-order nonlinear optical properties constrains the possible materials for use in second-order nonlinear optical devices. Of the 32 crystal point groups, 21 are noncentrosymmetric. In general, there are 27 possible components, but the symmetries of the point groups generally reduce the number of independent components [9].

The number of independent tensor components can be further reduced by noting other symmetries that the nonlinear optical tensor possesses. In loss-free media, energy conservation principles and the reality of the response function can be applied to yield the symmetry

$$\chi_{\underline{ijk}}^{(2)}(-\omega;\omega_1,\omega_2) = \chi_{\underline{jik}}^{(2)}(-\omega_1;\omega,-\omega_2) = \chi_{\underline{kij}}^{(2)}(-\omega_2;\omega,-\omega_1) \qquad (7)$$

In the case of small dispersion, one obtains the Kleinman symmetries [10], where

$$\chi_{\underline{ijk}}^{(2)}(-\omega;\omega_1,\omega_2) = \chi_{\underline{jik}}^{(2)}(-\omega;\omega_1,\omega_2) = \chi_{\underline{kij}}^{(2)}(-\omega;\omega_1,\omega_2) \qquad (8)$$

For example, when $\omega_1 = \omega_2$ and Kleinman symmetry applies, there is only one component to the second-order susceptibility tensor in the point group $\bar{4}2m$.

Additional degeneracies may be present due to the frequencies involved, which lead to additional symmetries in the polarization. For second-harmonic generation, the fields at ω_1 and ω_2 are degenerate in Eq. 6) so that $\chi_{\underline{ijk}}^{(2)}(-2\omega;\omega,\omega)$ is symmetric in j and k, and a second-harmonic coefficient can be defined as

$$d_{il}(-2\omega;\omega,\omega) = \frac{1}{2}\chi_{\underline{ijk}}^{(2)}(-2\omega;\omega,\omega) \qquad (9)$$

where the polarization is now given by

$$P_i^{(2)}(2\omega) = 2 \sum_{l=1}^{6} d_{il}(-2\omega;\omega,\omega) \left[1 - \frac{1}{2} \delta_{jk} \right] E_j(\omega)E_k(\omega) \qquad (10)$$

The index l compresses the last two indices due to their degeneracy so that l = 1, 2, 3, 4, 5, and 6 correspond to xx, yy, zz, zy = yz, zx = xz, and xy = yx, respectively, and δ_{ij} is the Kronecker delta. Figure 1 depicts the nonzero components of the second-harmonic coefficient as defined in Eq. 10) belonging to these point groups. Isotropic materials such as liquids and glasses are centrosymmetric, but orientationally ordered materials, which will be discussed below, belong to the point group ∞mm, which has the same symmetries as 6mm in Fig. 1. A similar degeneracy in j and k arises in the case of optical rectification, i.e., $d_{il}(0;\omega,\omega)$. In most electrooptic measurements the detector cannot distinguish between the modulation sidebands, so both the sum and difference frequency are detected. In this case, the measured polarization is the sum of the sum- and difference-generation processes involving the fields in Eq. (6); therefore, the polarization is given by

$$P_i^{(2)}(\omega) = 2\chi_{\underline{ijk}}^{(2)}(-\omega;\omega,0)E_j(\omega)E_k(0) \qquad (11)$$

The linear electrooptic coefficient r_{ij} has been traditionally defined with respect to the optical indicatrix or the index ellipsoid [11]:

$$\frac{x^2}{n_x^2} + \frac{y^2}{n_y^2} + \frac{z^2}{n_z^2} = 1 \qquad (12)$$

When an electric field is applied, the indicatrix will change in proportion to the field; thus,

$$\Delta\left[\frac{1}{n^2} \right]_i = \sum_{j=1}^{3} r_{ij}E_j \qquad (13)$$

where j runs over 1 = x, 2 = y, 3 = z. The index i runs from 1 to 6 as defined above, due to the degeneracy of the fields at ω. When the sum over sum- and difference-generation processes is taken into account, the relationship between the electrooptic coefficient tensor defined in Eq. (13) is related to the nonlinear optical susceptibility through

$$-n_{\underline{u}}^2(\omega)n_{\underline{v}}^2(0)r_{uvw}(-\omega;\omega,0) = 2\chi_{\underline{uvw}}^{(2)}(-\omega;\omega,0) \qquad (14)$$

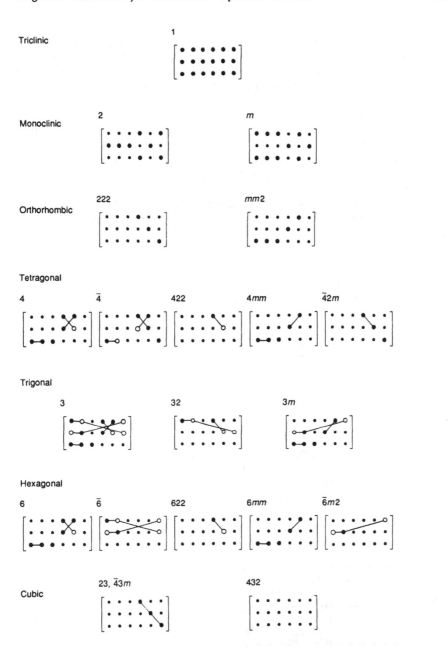

Figure 1. Second-order nonlinear optical tensor components for the 21 noncentrosymmetric point groups. • = Zero component; • = non-zero component; •——• = equal components; and •——○ = equal components, but opposite sign. (Adapted from Ref. 16.)

where indices of refraction are taken at the indicated frequencies. The linear electrooptic and optical rectification susceptibilities are related by the relation given in Eq. (7). Further, as will be shown below, the second-harmonic and electrooptic coefficients can be related to each other if the material dispersion is appropriately taken into account.

There are only two physically distinct second-order nonlinear optical processes which arise from the two sets of quantum electro-dynamical interactions, examples of which are depicted in the Feyn-man diagrams in Figs. 2 and 3 [12,13]. These two interactions, namely, sum and difference frequency generation are the two that conserve energy in the sense described by the Manley-Rowe relations; i.e., the number of photons of *each* incident field is the same and equal to the number of photons of *each* outgoing field. Thus, ω =

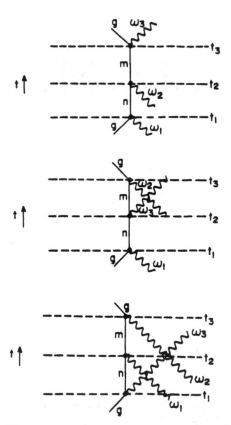

Figure 2.   Quantum electrodynamical interaction diagrams for sum-frequency generation.

$\omega_1 \pm \omega_2$. The energy conservation principles described here refer to the lossless regime, which will be the regime that the devices described below operate. When losses are present, all the above phenomena are still observable; however, Raman processes, where the material adds or subtracts energy so that $\omega \neq \omega_1 \pm \omega_2$ does not hold, also occur. Damping processes must also be considered in resonant operation.

The three frequencies in $\chi_{ijk}^{(2)}(-\omega; \omega_1, \omega_2)$ determine the nonlinear optical phenomenon observed due to the material processes contributing to the susceptibility, as given in Figs. 2 and 3. In Fig. 2, we see two incoming photons combining into one at the sum frequency. The diagrams in Fig. 3 represent the ways to generate the difference frequency in interaction with the quantum mechanical states of the material while conserving energy. Thus, a single photon breaks up into two lower-energy photons. Since

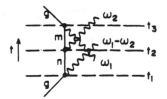

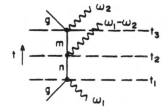

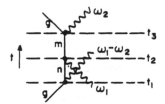

Figure 3.   Quantum electrodynamical interaction diagrams for difference-frequency generation.

the cross section for such a process is distributed throughout the
range of frequencies from zero to ω_1, the efficiency at a given
wavelength would be nonzero but very small. However, in difference-
frequency generation experiments, two incident frequencies, one
at ω_1 and ω_2, are present; thus the photons at ω_2 stimulate the
difference-frequency generation process, increasing the cross section.
This is illustrated by considering the interpretation of Fig. 3 in
the fully quantized picture where the susceptibility contains matrix
elements containing permutations of the free-photon creation and
annihilation operators, $a_{\omega_1}^\dagger a_{\omega_2}^\dagger a_{\omega_1 - \omega_2}^\dagger$, which when operating on
the initial free-photon state generates an eigenvalue proportional
to $\sqrt{n_{\omega_2} + 1}$. Thus, the cross section is proportional to the number
of photons in the initial state. In difference-frequency generation,
the beam at ω_2 is not depleted but amplified, in contrast to the
situation in sum-frequency generation, where both of the input
frequencies are depleted. This feature of difference-frequency
generation can also be derived from the classical point of view from
the Manley-Rowe relations. This effect is well known in the phase-
matched case of difference-frequency generation (parametric
amplification). Optical rectification is also a special case of difference-
frequency generation, where $\omega_1 = \omega_2$; parametric amplification is
the special case of phase-matched difference-frequency generation;
and parametric oscillation is the special case of parametric amplification where the effective interaction length is increased by placing
the nonlinear optical material in a Fabry-Perot cavity.

The second-order nonlinear optical susceptibility can be calculated from a complete set of the quantum mechanical states of the
system by using perturbation theory. The interaction diagrams
depicted in Figs. 2 and 3 can be used to generate the perturbation
expansion. The complete set of six terms in the sum for both processes are obtained from the diagrams shown and by interchanging
the photons at ω_1 and ω_2 in the diagrams. The diagrams are evaluated
by performing the time-ordered product integration, starting at
earliest time (bottom of diagram) and reading of the propagators
and transition moments in turn and using the energy conservation
condition. The terms are summed over all states, and the singularities
are renormalized by using the method of averages [13,14]. For
sum-frequency generation the susceptibility is given by

$$\chi_{\underline{ijk}}^{(2)}(\omega_1 + \omega_2; \omega_1, \omega_2)$$

$$= \frac{e^3}{2\hbar^2} \times \sum_{nm}{}' \left\{ \frac{\langle g| \; x_i \; |m\rangle \langle m| \; \bar{x}_k \; |n\rangle \langle n| \; x_j \; |g\rangle}{(\Omega_{ng} - \omega_1)(\Omega_{mg} - \omega_1 - \omega_2)} \right.$$

$$+ \frac{\langle g| \; x_i \; |m\rangle \langle m| \; \bar{x}_j \; |n\rangle \langle n| \; x_k \; |g\rangle}{(\Omega_{ng}-\omega_2)(\Omega_{mg}-\omega_2-\omega_1)}$$

$$+ \frac{\langle g| \; x_k \; |m\rangle \langle m| \; \bar{x}_i \; |n\rangle \langle n| \; x_j \; |g\rangle}{(\Omega_{ng}-\omega_1)(\Omega_{mg}^{\gamma *}+\omega_2)}$$

$$+ \frac{\langle g| \; x_j \; |m\rangle \langle m| \; \bar{x}_i \; |n\rangle \langle n| \; x_k \; |g\rangle}{(\Omega_{ng}-\omega_2)(\Omega_{mg}^{\gamma *}+\omega_1)}$$

$$+ \frac{\langle g| \; x_k \; |m\rangle \langle m| \; \bar{x}_j \; |n\rangle \langle n| \; x_i \; |g\rangle}{(\Omega_{ng}^{*}+\omega_1+\omega_2)(\Omega_{mg}^{\gamma}+\omega_2)}$$

$$\left. + \frac{\langle g| \; x_j \; |m\rangle \langle m| \; \bar{x}_k \; |n\rangle \langle n| \; x_i \; |g\rangle}{(\Omega_{ng}^{*}+\omega_1+\omega_2)(\Omega_{mg}^{\gamma}+\omega_1)} \right\} \qquad (15)$$

where the $\langle m| \; \bar{x}_u \; |n\rangle = \langle m| \; x_u \; |n\rangle - \langle g| \; x_u \; |g\rangle$ and the prime on the summation denotes the exclusion of the ground state from the sum. The states $|n\rangle$ are eigenstates of the unperturbed Hamiltonian, and the frequency factors Ω_{mn} are given by $\Omega_{mn} + i\Gamma_{mn}$, where ω_{mn} are the transition energies and Γ_{mn} are phenomenological damping terms. The frequency factor Ω_{mg} is given by $\Omega_{mg}^{\gamma} \equiv \omega_{mg} + i(\Gamma_{mn} - \Gamma_{ng})$.

The matrix elements for difference-frequency generation can be read from Fig. 3, again performing the operations as above, yielding

$$\chi_{ijk}^{(2)} (\omega_1-\omega_2; \omega_1, \omega_2)$$

$$= \frac{e^3}{2\hbar^2} \times {\sum_{nm}}' \left\{ \frac{\langle g| \; x_k \; |m\rangle \langle m| \; \bar{x}_i \; |n\rangle \langle n| \; x_j \; |g\rangle}{(\Omega_{ng}-\omega_1)(\Omega_{mg}-\omega_2)} \right.$$

$$+ \frac{\langle g| \; x_j \; |m\rangle \langle m| \; \bar{x}_i \; |n\rangle \langle n| \; x_k \; |g\rangle}{(\Omega_{ng}-\omega_2)(\Omega_{mg}-\omega_1)}$$

$$+ \frac{\langle g| \; x_k \; |m\rangle \langle m| \; \bar{x}_j \; |n\rangle \langle n| \; x_i \; |g\rangle}{(\Omega_{ng}+\omega_1-\omega_2)(\Omega_{mg}^{\gamma}-\omega_2)}$$

$$+ \frac{\langle g| \; x_j \; |m\rangle \langle m| \; \bar{x}_k \; |n\rangle \langle n| \; x_i \; |g\rangle}{(\Omega_{ng}-\omega_1+\omega_2)(\Omega_{mg}^{\gamma}-\omega_1)}$$

$$+ \frac{\langle g| \; x_j \; |m\rangle \langle m| \; \bar{x}_k \; |n\rangle \langle n| \; x_i \; |g\rangle}{(\Omega_{ng}+\omega_1-\omega_2)(\Omega_{mg}^{*}+\omega_1)}$$

$$\left. + \frac{\langle g| \; x_k \; |m\rangle \langle m| \; \bar{x}_j \; |n\rangle \langle n| \; x_i \; |g\rangle}{(\Omega_{ng}-\omega_1+\omega_2)(\Omega_{mg}^{*}+\omega_2)} \right\} \qquad (16)$$

The optical rectification susceptibility is obtained, again, by taking
$\omega_1 = \omega_2$ and multiplying by $1/2$. The electrooptic susceptibility is
obtained by summing Eqs. (15) and (16) with $\omega_1 = 0$. The damping
terms correspond to spontaneous losses due to coupling to the radiation field, and it is assumed that the populations of any excited
states at resonance is small. The expressions in Eqs. (15) and (16)
are valid at zero temperature. It is usual to carry out the perturbation at finite temperature, where the losses are due to random thermal
perturbations, in the density matrix formalism. However, for purposes
here, the expressions above are adequate. The second harmonic
susceptibility is obtained by taking the sum with $\omega_1 = \omega_2$ and multiplying by $1/2$ so as not to double count the degenerate fields.

The susceptibilities can, in principle, be calculated when a
complete set of states and energies of the system are known [3,4].
Such calculations have been carried out for organic molecules using
a variety of techniques. It is interesting, however, to examine
the frequency dependence of the susceptibility, for instance the
sum susceptibility of Eq. (15) [6,15]. The energy denominators
in Eqs. (15) and (16) depend both on the energy of the photons
and the characteristics energies of the material system. Unlike the
linear susceptibility, nonlinear susceptibilities contain "hybrid"
terms where one intermediate state is electronic and one vibrational;
thus these effects are not necessarily totally separable. The susceptibility, then, will contain "pure" electronic and vibrational terms as
well as the hybrid terms. The nonlinear optical phenomena which
we have described above involve either all optical frequencies or
a combination of optical and low frequencies. We discuss each in
turn.

The relative contribution of electrons and phonons can be estimated by examining the expression Eqs. (15) and (16). In phenomena
involving only optical frequencies, the energy denominators belonging
to terms due to electronic excitations will be roughly a few electron
volts or less. For phonon states, the characteristic energies are
of the order of tenths of electron volts, while the optical energy
is generally greater than 1 electron volt. Thus, the denominator
is dominated by the energy of the photon. The energy denominators
for both electronic and phonon processes is of the order of an
electron volt. However, for the linear susceptibility, for each process
there are two terms with the photon energy having opposite sign
[14]. Thus, in the case of phonon processes, the two terms will
nearly cancel, so that, regardless of the matrix elements, the phonons
will contribute little to the linear susceptibility. For the nonlinear
susceptibilities, this pairwise cancellation does not occur, and, in
principle, the phonons could make a substantial contribution to all
optical frequency susceptibilities, such as second-harmonic generation.

In practice, however, the optical phonons will not contribute much to the all-optical frequency susceptibilities. This is due to the small size of the phonon matrix elements compared to the electronic matrix elements. It should be noted that the acoustic phonons will make no contribution since the dipole approximation leading to Eqs. (15) and (16) does not apply; i.e., a nonlocal response is required for acoustic phonons. In that case the fact that the sound velocity is much smaller than the velocity of light leads to the intersection of the phonon and photon dispersion curve being restricted to the origin.

The relative size of the phonon and electronic matrix elements can be estimated by considering the classical anharmonic oscillators in the Born-Oppenheimer approximation. In that approximation the two interactions, phonon-electron and phonon-phonon, are considered separately. The equation of motion for displacement from equilibrium is given by [16],

$$\ddot{X} + \gamma\dot{X} + \omega_0^2 X + DX^2 = \frac{eE_0}{2m}(e^{i\omega t} + e^{-i\omega t}) \tag{17}$$

where γ is the damping, D the nonlinearity, e the electronic charge, and m the electron or ion mass for the electronic or phonon interaction, respectively. The second-harmonic coefficient obtained from this equation of motion is given by

$$d^{(2\omega)} = \frac{-DNe^3}{2m^2[(\omega_0^2 - \omega^2) + i\omega\gamma]^2[\omega_0^2 - 4\omega^2 + 2i\omega\gamma]} \tag{18}$$

where N is the number of charges involved. It is evident, that the phonon terms will indeed be much smaller than the electronic terms since the ratio of the electronic to phonon mass is of the order 5×10^{-4}. The "hybrid" terms will be intermediate in magnitude between the "pure" electronic and vibrational terms, and, depending on the frequencies and the exact magnitudes may make small but measurable contributions to the all-optical frequency process.

The situation where a low-frequency field is involved is generally more complicated. In this case, the characteristic material frequency dominates the energy denominator, so that low-frequency excitations can lead to large terms in the susceptibility even with small matrix elements. All terms both pure and hybrid can contribute.

Noncrystalline polymeric materials will also have contributions from orientational motion of dipoles contained in the polymer. These can have characteristic frequencies in the radio to high-microwave regime. As in the optical phonon modes, these will mostly contribute to processes involving a low-frequency electric field.

The relative contribution of the various processes to second-order effects can be estimated by measuring the dispersion of the susceptibility. This method was first utilized on potassium dihydrogen phosphate (KDP) [15], and involves the measurement of the optical frequency dependence of both the electrooptic and second-harmonic coefficients. The two coefficients can be related using a two-state model taking into account local field factors. The optical dispersion eliminated pure vibrational terms from both coefficients. It was found that pure electronic terms accounted for almost all of the second harmonic coefficient as expected from the above arguments. At least half of the electrooptic coefficient was found to arise from pure electronic terms, with the remainder from hybrid terms.

A two-state model including local field corrections has been previously described and applied to a series of organic crystalline materials [5,17]. In this model, the measured electrooptic coefficient is compared to the electronic contribution as calculated from the measured second-harmonic coefficient (assuming the second-harmonic coefficient is electronic in origin), using

$$r^{el}_{ij,k}(-\omega;\omega,0) = -\frac{4d_{kij}}{n_i^2(\omega)n_j^2(\omega)} \times \frac{f_{ii}{}^\omega f_{jj}{}^\omega f_{kk}{}^0}{f_{kk}{}^{2\omega'} f_{ii}{}^{\omega'} f_{jj}{}^{\omega'}}$$

$$\times \frac{(3\omega_0^2-\omega^2)(\omega_0^2-\omega'^2)(\omega_0^2-4\omega'^2)}{3\,\omega_0^2(\omega_0^2-\omega^2)^2} \tag{19}$$

where ω is the frequency of the electrooptic experiment, ω' the fundamental frequency in the second-harmonic generation experiment, and f_{uu}^ω are appropriate local field factors. When Eq. (19) was applied to electrooptic and second-harmonic measurements of organic crystals, it was found that the electrooptic coefficient of many crystals were due to pure electronic terms, but some crystals with smaller nonlinear optical coefficients had hybrid or pure vibrational contributions similar to that found in KDP. When Eq. (19) was applied to guest-host systems, the data suggest that electronic terms dominate the electrooptic coefficient in side-chain polymers, but some orientational contribution may have been present in doped polymer systems [18,19].

III. ORGANIC MATERIALS

Interest in organic materials for second-order nonlinear optics first arose in the 1960s [20], but the true promise of these materials was not appreciated until Davydov et al. [21] discovered the connection between charge transfer and large nonlinear susceptibilities.

This triggered activity which proceeds to the present mainly on molecular crystals for parametric processes. This work includes the elucidation and optimization of the molecular properties as well as the factors of crystal structure and growth. Much of the work along these lines up to 1987 is described in the volume by Chemla and Zyss [3]. Meredith et al. [22] and Williams [23] introduced the possibility of polymeric materials for second-order nonlinear optics through the concept of electric field poling. The potential of these polymeric materials for electrooptics was then realized [5,24]. Interest in polymeric materials has developed rapidly, and their use in devices was rapidly demonstrated. In this section we will briefly review second-order nonlinear optics with emphasis on developments since the Chemla and Zyss volume and on polymeric materials.

Organic materials possess many important attributes for applications involving high-speed processes. These attributes include large nonresonant susceptibilities, fast response times, and low dielectric constants. Low scattering and absorption losses and high optical damage thresholds are critical factors in those applications where high-power signals are being transmitted. The inherent tailorability of organic compounds—that is, the ability to alter material properties to fit specific needs—is another advantage of this material class.

Polymeric systems offer mechanical, chemical, and thermal robustness, adhesion to a variety of materials (other polymers, inorganics, metals), processability into films, fibers, waveguides, and large-area structures, integration with electronics and optical sources and detectors, low cost, and potential for long-range orientation.

Combining these classes of materials may yield new material systems possessing many of these desirable characteristics. A discussion of the microscopic origin of the second-order optical nonlinearity in organic molecules follows. Next, a brief description of the major measurement techniques used to explore this nonlinearity is given. Recent studies exploring the factors contributing to the microscopic susceptibility are described, followed by approaches to realize this nonlinearity in bulk systems.

A. Molecular Processes

In analogy with Eq. (5), a molecular second-order nonlinear optical polarization can be written as

$$p_i(-\omega;\omega_1,\omega_2) = \beta_{ijk}(-\omega;\omega_1,\omega_2)F_j(\omega_1)F_k(\omega_2) \tag{20}$$

where $\beta_{ijk}(-\omega;\omega_1,\omega_2)$ is the second-order microscopic susceptibility and the F_m are local fields. In second-order organic nonlinear optical materials, the molecules do not interact substantially so that a molecular identity is retained in the bulk, and β_{ijk} is related to the bulk susceptibility χ_{ijk} through an oriented gas model. This model has been applied to crystals by Chemla et al. [25]:

$$\chi_{\underline{IJK}}^{(2)}(-\omega;\omega_1,\omega_2)$$

$$= \frac{1}{V}f_I(\omega)f_J(\omega_1)f_K(\omega_2)\sum_s\sum_{ijk}\cos\theta_{\underline{Ii}}^{(s)}\cos\theta_{\underline{Jj}}^{(s)}$$

$$\cos\theta_{\underline{Kk}}^{(s)}\ \beta_{ijk}(-\omega;\omega_1,\omega_2) \qquad (21)$$

where V is the volume of the unit cell, the f's are appropriate local field factors (here diagonal in the crystal frame), and the cosines describe the transformation between the Lth crystal axis and the lth molecular axis. The sum over s takes into account multiple molecules within the unit cell. This oriented gas model has been extended to the general case of noncrystalline materials which may or may not possess innate axial order [18]. This includes the cases of electric-field poled nematic and smectic liquid crystals and polymers, and Langmuir-Blodgett films [26]. In these cases, the electronic contribution to the bulk second order susceptibility $\chi_{\underline{ijk}}^{(2)}$ is related to the microscopic susceptibility by

$$\chi_{\underline{ijk}}^{(2)}(-\omega;\omega_1,\omega_2) = N <\beta_{\underline{IJK}}^*(-\omega;\omega_1,\omega_2)> ijk \qquad (22)$$

where N is the number density of molecular species, $\beta_{\underline{IJK}}^*(-\omega;\omega_1,\omega_2)$ the local field corrected electronic molecular susceptibility, and the angle brackets denote a statistical average. In the development of new second-order nonlinear optical materials, the challenge then is to optimize both the magnitude of the molecular susceptibility as well as alignment of those species responsible for the nonlinear optical response inorder to obtain not only a noncentrosymmetric material but one whose molecular orientation is optimized for practical use [18,26]. These oriented gas models have generally been shown to be valid in a wide range of organic materials [5].

The molecular nonlinear optical susceptibility can be calculated using a molecular version of Eqs. (15) and (16) provided that an appropriate set of wave functions exist. By examining the features of the electronic states that contribute to the large terms in the perturbation expansion, the molecular features contributing to $\beta_{ijk}(-2\omega;\omega,\omega)$ can be assessed. Lalama and Garito first applied these methods to substituted benzene compounds using molecular Hartree-Fock theory to calculate the electronic states [27]. Their results showed good agreement with experiment and confirmed that the major contribution to β_{ijk} arises from the lowest-lying excited states in strong donor- or acceptor-substituted benzene compounds. These calculations confirmed the two-level model proposed by Oudar and Chemla [28].

In this model, the major contributions to the second-harmonic molecular susceptibility $\beta(-2\omega;\omega,\omega)$, where the second harmonic is below the first excited state ($2\omega < E_1$, with ω the fundamental frequency and E_1 the ground state to first excited state transition energy), are the transition moment μ_{1g} and the change in dipole moment between the ground state and first excited state of the molecule $\Delta\mu$, and are related by

$$\beta(-2\omega;\omega,\omega) \propto \frac{\mu_{1g}^2 E_1^2 \Delta\mu}{(E_1^2 - \hbar^2\omega^2)(E_1^2 - 4\hbar^2\omega^2)} \qquad (23)$$

Thus, the larger the transition moment and change in dipole moment, the larger the microscopic susceptibility. Approaches to increasing this susceptibility that have proved successful include the placement of strong electron-donor and electron-acceptor groups at opposite ends of a conjugated π-electron system and increasing the conjugation length [29-31]. Examples of such molecular systems are 2-methyl-4-nitroaniline (MNA) and the azo dye Disperse Red 1, shown in Fig. 4.

B. Measurement Techniques

Second-harmonic generation and electrooptic coefficient measurements are the most common techniques for determining $\chi^{(2)}$. By far, the most widely used technique to explore the nonlinear optical properties of organic materials is powder second-harmonic generation. Kurtz and Perry were the first to systematically investigate second-harmonic generation from powders [32]. Nicoud and Twieg compiled second-harmonic generation efficiency measurements on a substantial variety of organic compounds [33]. Nicoud and Twieg cite some enhancements made in this technique since the initial work of Kurtz and Perry [34,35]. This technique does not directly probe the microscopic susceptibility, β, but is an important method for examining bulk nonlinearities. Care must be taken in comparing second-harmonic generation from powders, since particle size and linear optical properties affect the interpretation of the results [34,35]. Though this method may be useful for screening compounds, orientation-dependent measurement of second-harmonic generation from single crystals or thin films is the preferred method for quantitative determination of bulk nonlinear optical properties.

Several techniques for measuring the linear electrooptic effect have been described [5,19]. Interferometric techniques are especially useful in that individual tensor components can be easily determined.

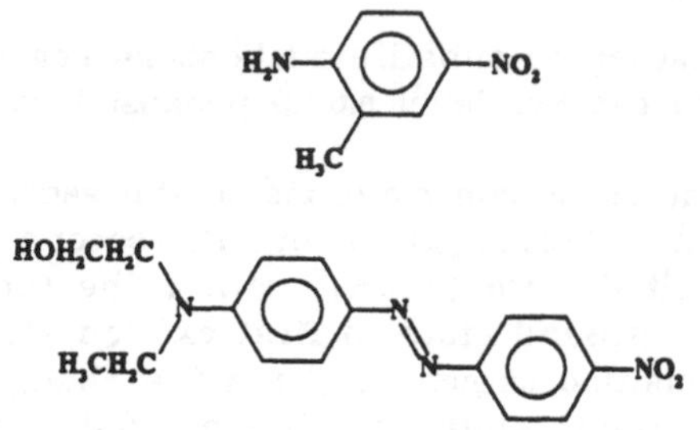

MNA

Disperse Red 1

Figure 4.   Examples of optically nonlinear organic molecules.

Additionally, material coefficients can be determined from measurements in waveguide geometries [36].

The technique most suited for investigating microscopic nonlinearities is electric-field-induced second-harmonic (EFISH) generation (dc-induced second-harmonic generation) in solution [29,37-39]. Details of this measurement technique, currently used in our group, have recently been described [29]. A strong electric field is used to temporarily align the molecules of interest in a dilute solution where intermolecular interactions are negligible. Thus the second-harmonic signal generated can be directly related to the molecular susceptibility.

C. Molecular Properties

Until recently, the compounds exhibiting the largest nonlinear optical susceptibilities were substituted nitroanilines and extended π-electron systems with amino electron-donor and nitro electron-acceptor groups. In addition to these aromatic systems, some quinoidal compounds systems possess large microscopic nonlinearities [40,41]. New systems with substantially larger nonlinearities have now been reported. The use of cyanovinyl electron-acceptor groups and dithiolylidene-methyl electron-donor groups to enhance the second-order microscopic susceptibility β was introduced by Katz et al. [42]. The relative strengths of these groups is evident from the data contained in Table 1. Larger values of $\beta\mu$ (greater than fivefold) are obtained in substituted aniline systems in going from nitro (DMNA) to dicyanovinyl (DCV-DMNA) to tricyanovinyl (TCV-DMNA) acceptor groups. Use of the 1,3-dithiole group as a donor (DTM-NB) more than doubles $\beta\mu$ as compared to the dimethylamino group (DMNA). Combining the 1,3-dithiole and tricyanovinyl groups yields a molecule (DTM-TCVB) with a value of $\beta\mu$ eight times that of the nitroaniline.

Singer et al. extended this work with a detailed determination of the nonresonant second-order microscopic susceptibilities of a series of 16 conjugated systems [29]. In addition to the effects of the electron-donating and electron-withdrawing substituents dis-

Table 1. Second-Order Microscopic Susceptibilities

Compound	Structure	$\beta\mu$ (10^{-30} cm^5 D/esu)
DMNA	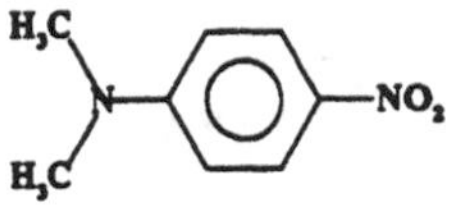	149
DCV-DMNA	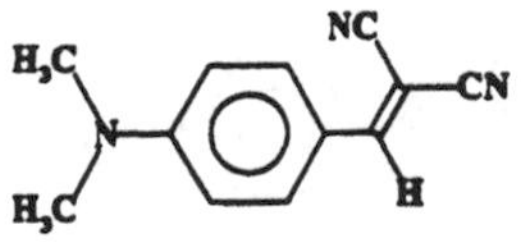	270
TCV-DMNA	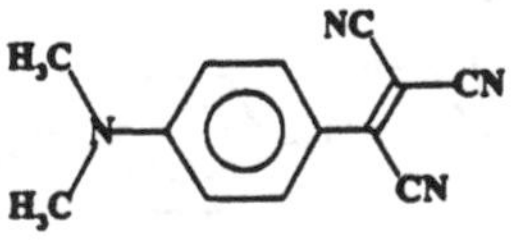	850
DTM-NB	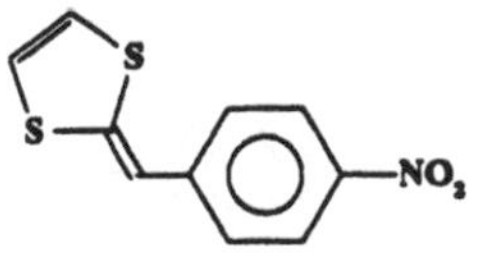	359
DTM-TCVB	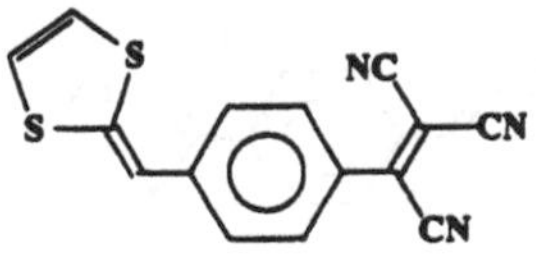	1200

Source: Ref. 42.

cussed above, effects of conjugation length, molecular configuration, π-electron system, and dispersion were investigated. The susceptibilities were determined using dc-induced second-harmonic generation in solution. The effect of conjugation length on the microscopic susceptibility can be seen from information presented in Table 2, where the dispersion-corrected susceptibility, β_0, introduced by Singer et al., is used. An increase of approximately fourfold is achieved in extending the conjugation between donor and acceptor groups: DMNA compared with DMA-NS or DMA-NAB.

Table 2. Dispersion-Corrected Second-Order Microscopic Susceptibilities

Compound	Structure	$\beta_0 (10^{-30}$ cm^5/esu)
DMNA	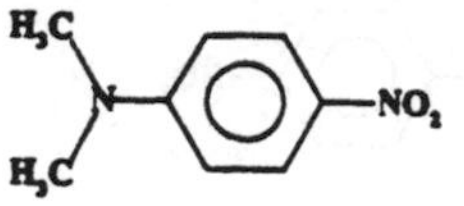	12
DMA-NS	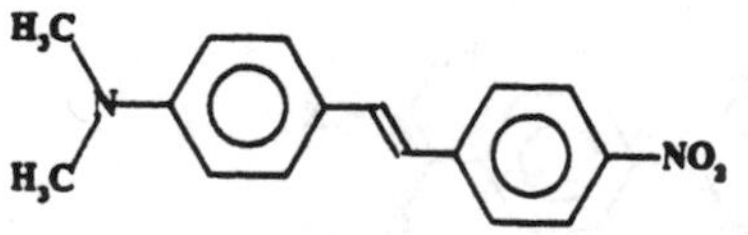	52
Disperse Red 1	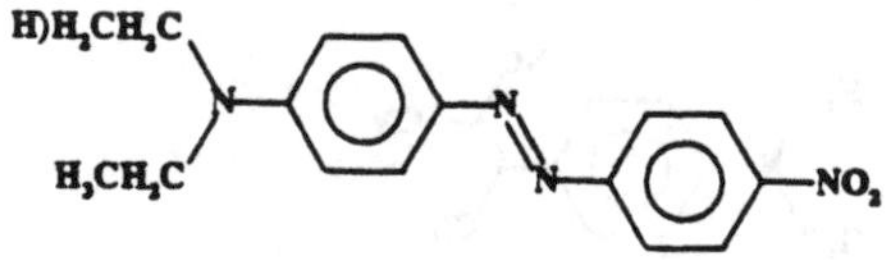	47
NB-DMAA	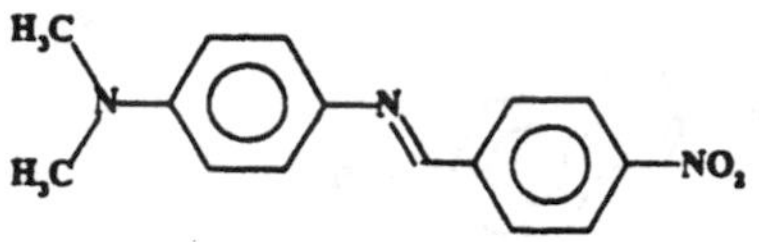	37
DMA-DCVS	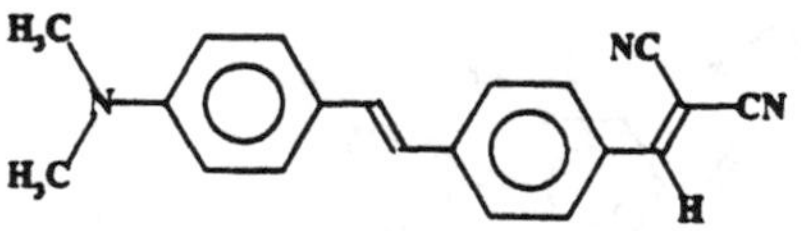	133
DEA-TCVAB	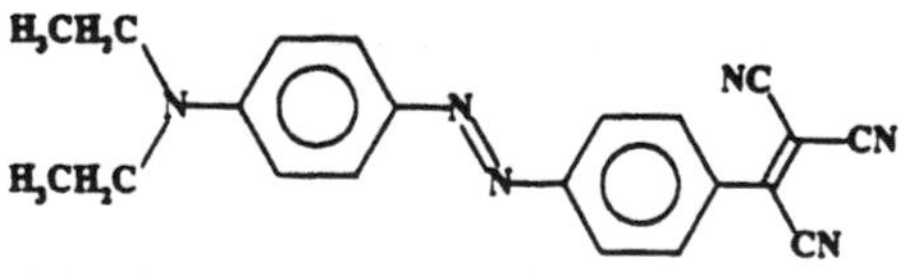	154

Source: Ref. 101.

It was found that azo (DMA-NAB) and ethylene (DMA-NS) π-electron systems are nearly equivalent in contributing to the microscopic susceptibility. However, azomethine systems (NB-DMAA) are less effective in contributing to β, owing to a loss of coplanarity of the two benzene rings.

Greatly enhanced values of β_0 are realized when cyanovinyl acceptors are used in azobenzene (DEA-TCVAB) or stilbene (DMA-DCVS) systems. The use of DMA-DCVS in a poled guest-host system has resulted in a bulk material with a practical electrooptic coefficient (*vide infra*).

The length dependence of the second-order hyperpolarizability was studied in conjugated systems bearing methoxy electron-donating groups and nitro electron-accepting groups [43]. EFISH from dioxane solutions of the chromophores yielded the results given in Table 3. After correcting for resonance enhancement using the two-level model, β is found to vary as the third power of the length of the conjugated system, in agreement with model predictions relating the hyperpolarizability to the mesomeric moment. The length dependence is in qualitative agreement with calculations reported by Morley et al. [44]. However, the predicted leveling off of β with increasing chain length [44] is not evident in the data.

D. Bulk Materials

Construction of a bulk material possessing a large bulk susceptibility $\chi^{(2)}$ requires not only molecular constituents with large microscopic susceptibilities but a noncentrosymmetric system where the orientation of the molecular species results in constructive additivity of the molecular susceptibilities. A variety of approaches to construct bulk materials with large nonlinearities are being pursued. These include crystal engineering, Langmuir-Blodgett film preparation, electric-field poling of polymeric systems, development of molecular composites and aggregates, and the preparation of fibers with crystalline organic materials as the core. We now describe recent efforts in these areas.

Crystals

Zyss and Oudar, using the oriented gas model as described in Eq. (21), discussed the relationship between the microscopic and bulk second-order nonlinearities in molecular crystals and provided guidelines for "crystal engineering"—that is, which crystal point groups are more favorable for realizing large bulk nonlinearities [45]. Zyss and Chemla presented the optimization of the second-order bulk nonlinearity in molecular crystals as exemplified by the crystals MAP, POM, and NPP (Fig. 5) [46]. Both MAP and NPP are examples

Table 3. Second-Order Microscopic Susceptibilities

Compound	Structure	$\beta\mu$ (10^{-46} esu)
MNB		0.30
MNS	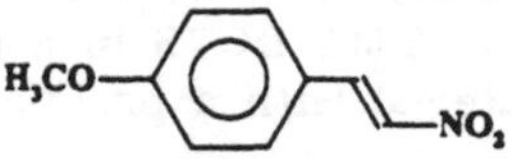	0.64
MPNP1	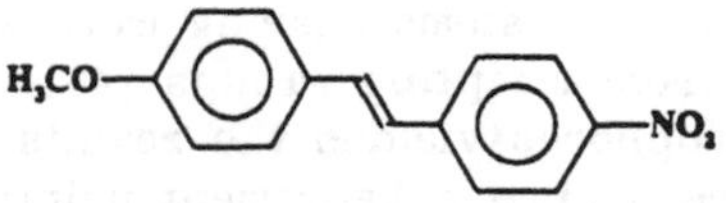	4.6
MPNP2	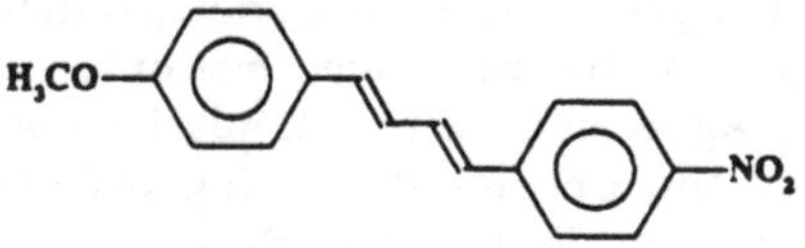	8.1
MPNP3	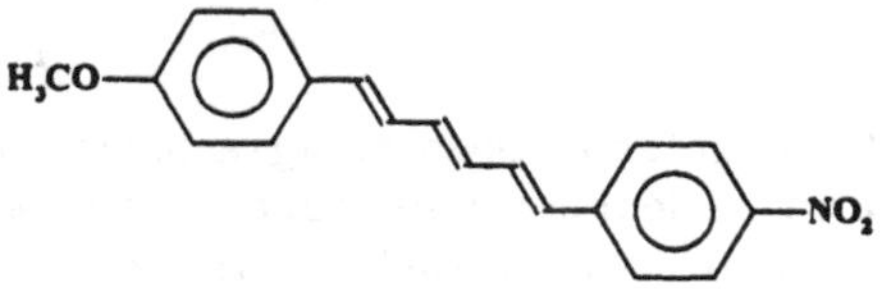	18.1
MPNP4	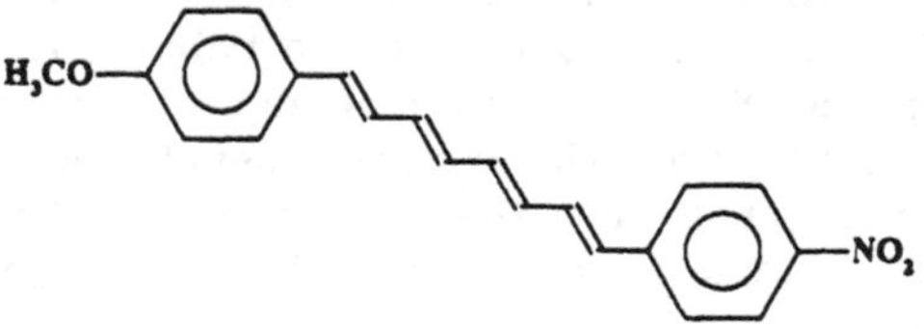	24.6
MPNP5	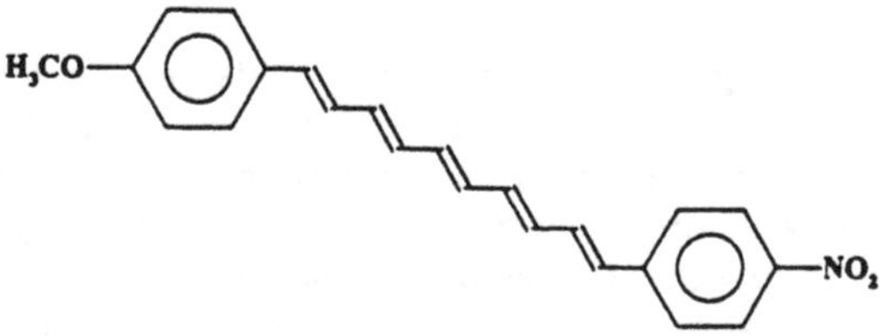	43.6

Source: Ref. 43.

MAP

POM

NPP

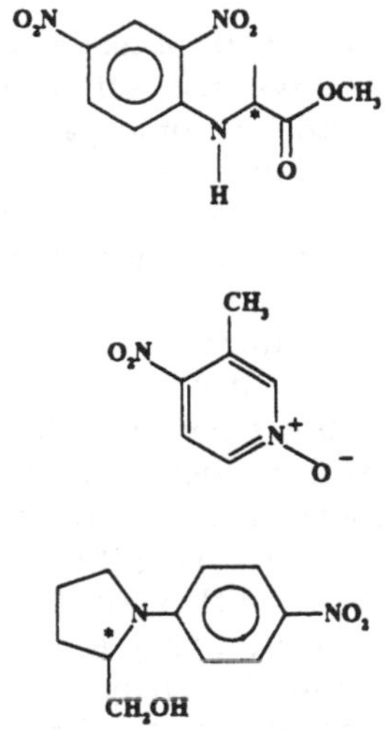

Figure 5. Molecularly engineered crystals.

of the use of chirality (* in Fig. 5 indicates the chiral carbon) to
obtain an optically nonlinear crystal. The use of chiral materials
ensures noncentrosymmetric crystals but does not ensure that those
regions giving rise to the nonlinear optical response are optimally
aligned. Minimization of the ground state dipole moment to allow
for more favorable molecular alignment in the crystal is epitomized
in POM. Dipole-dipole cancellation of the nitro and N-oxide groups,
and use of the methyl substituent, breaks the centrosymmetry of
4-nitropyridine-1-oxide, yielding the noncentrosymmetric crystal
POM.

Nicoud has recently described attempts at control of noncentrosymmetry in organic crystals [47]. Some of the techniques used
to control noncentrosymmetry in crystals include chirality, intermolecular hydrogen bonding, vanishing ground state dipole moments,
and the presence of other polar groups. Nicoud suggests that the
most effective strategy for crystal engineering is chirality, which
guarantees crystallization into a noncentrosymmetric point group.
However, this approach does not assume that the maximum bulk
nonlinearity will be realized. By this method, second-harmonic
efficiencies of 150 times that of urea are obtained (nitrophenylprolinol compounds).

Several groups have studied the nonlinear optical properties of
the crystalline material 4-(N,N-dimethylamino)-3-acetamidonitrobenzene

(DAN), a substituted nitroaniline. Norman et al. measured second-harmonic generation from powder [48]. Conversion efficiency of 20% was reported for phase-matched second-harmonic generation from a fundamental wavelength of 1.064 μm in a crystal 2 mm thick. This efficiency is a "lower limit" since no means were taken to optimize coupling of power into the crystal, and the natural growth facet was not of high optical quality. Crystal structure analysis shows that the important molecular axis (defined by the nitrogen of the amino group and the nitrogen of the nitro group) is inclined at 70.6° to the symmetry axis, considerably away from the optimal value of 54.74° [45]. Even so, this alignment yields 78% of the maximum macroscopic nonlinearity. Baumert et al. characterized the linear and nonlinear optical properties of DAN [49]. The molecular hyperpolarizability, determined using EFISH with a fundamental frequency of 1.064 μm, is $\beta = 30 \times 10^{-30}$ esu, using a dipole moment value of 8.1 D. The acetamido group is found to play an important role in promoting noncentrosymmetric crystal growth through hydrogen bonding, but does not contribute significantly to the microscopic nonlinearity. The effective nonlinear optical coefficient, measured at phase-matching conditions, is $d_{eff} = 27 \pm 3$ pm/V.

Recently, two groups reported the characterization of the linear and nonlinear optical properties of the crystal MBANP (2-(α-methylbenzylamino)-5-nitropyridine) [50,51]. The structure of MBANP is shown in Fig. 6. The hyperpolarizability is comparable to other nitroaniline compounds, while the charge transfer absorption band is significantly shifted to shorter wavelength, making MBANP a promising crystal for frequency-doubling applications, where the doubled frequency falls in the visible portion of the spectrum. The absorption cutoff for MBANP crystals is ~450 nm, with λ_{max} = 360 nm. A comparison of MBANP, p-NA, and MNA is given in Table 4 [51]. Kondo determined the d_{22} second-harmonic coefficient

MBANP

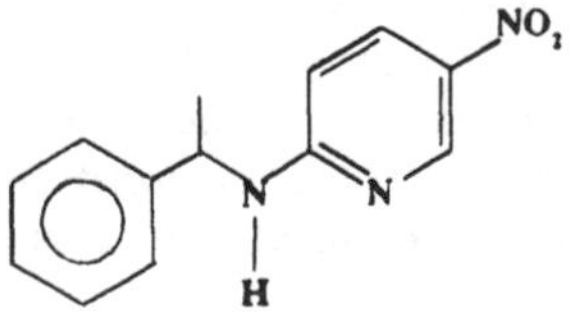

Figure 6.   Structure of the organic crystal MBANP.

Table 4.   Comparison of MBANP, *p*-NA, and MNA

Compound	λ_{max} (EtOH)	$\beta(10^{-30}$ esu)
p-NA	370	12.6
MNA	373	14.2
MBANP	360	15.0

Source: Ref. 51.

to be 126 times that of the d_{11} component of quartz and 1.46 times
that of the d_{33} component of LiNbO$_3$. Bailey found the d_{22} component
to be 83 times that of d_{11} of quartz. Type I phase-matched second-
harmonic generation was observed with an efficiency of ~50%.

Phase-matched second-harmonic generation from 1.064-μm fundamental has been reported by Guha et al. in 4-aminobenzophenone
crystals [52]. Powder efficiencies from the aminobenzophenone as
well as other benzophenones are given in Table 5 (efficiencies relative
to ADP). An efficiency of 14% for phase-matched second-harmonic
generation was observed in 4-aminobenzophenone; this crystal has
a cutoff frequency of 394 nm.

The aminonitropyridine, 2-cyclooctylamino-5-nitropyridine
(COANP), possesses sizable nonlinear optical susceptibilities [53].
The amphiphilic nature of the molecule allows for a layered crystalline
structure, with hydrogen bonding occurring between the amino
hydrogen and an oxygen atom of the nitro group. Both critical
and noncritical phase-matching properties were identified. The nonlinear optical susceptibilities determined are d_{32} = (32 ± 16) pm/V
and d_{33} = (13.7 ± 2) pm/V. These values are also listed in Table 9.
The structural dependence of the nonlinearities indicates that a
one-dimensional molecular model [45], where the one-dimensional
charge transfer axis is defined by the nitrogen atoms of the nitro
and amino groups, can be applied with qualitative success to COANP.

N-Monoalkylation of *p*-nitroaniline yields compounds exhibiting
second-harmonic generation activity [54]. The compound in this
series exhibiting the largest activity is *N*-butyl-*p*-nitroaniline, with
an efficiency 14 times that of urea and is phase-matchable. It was
also observed that the solvent used to recrystallize the materials
has a strong influence on the second-harmonic efficiency. The efficiency of the butyl compound drops to half that of urea when re-
crystallized from ethanol rather than a cyclohexane/ether mixture.
The only other compound in the series (from propyl through octyl)
exhibiting a nonzero efficiency is the octyl compound, as crystallized

Table 5.   SHG Efficiencies of Various Benzophenones

Compound	Structure	SHG efficiency (ADP = 1.0)
Benzophenone	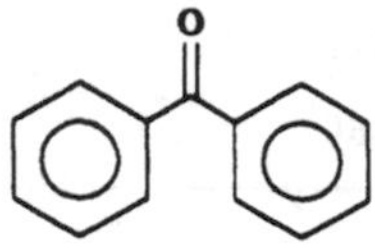	1.9
4-Aminobenzophenone	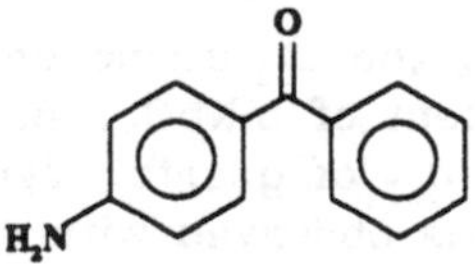	360
4-Nitrobenzophenone	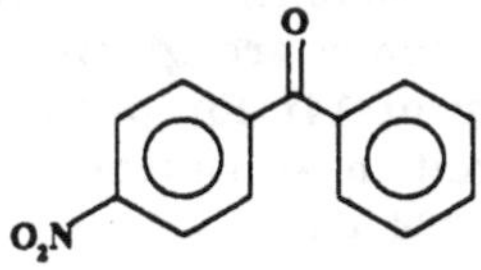	0.01
4-Amino-3-nitrobenzophenone	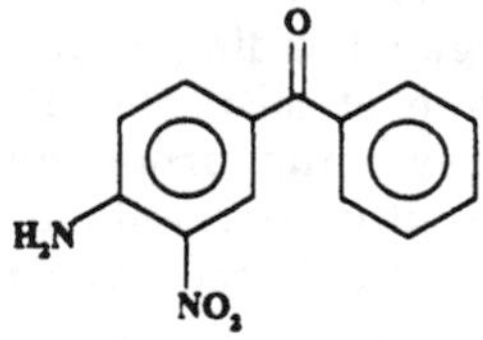	33

Source: Ref. 52.

from ethanol, with an efficiency 0.05 that of urea. These results
are given in Table 9. The interplay of recrystallization solvent
and alkyl substituent dramatically affect the nonlinear optical properties in this series of crystals. The authors also suggested that
composite films of *N*-alkylated nitroanilines in PMMA may prove
promising for nonlinear optical applications. This suggestion is
based on the results of Daigo et al. (*vide infra*) where a composite
film of *p*-nitroaniline (*p*-NA) in PMMA shows second-harmonic intensity 10 times that of urea, and the observation that introduction
of alkyl chains to spiropyrans results in smooth, transparent films
[55].

Water-soluble organic salts have been investigated by Velsko et al. for their second-harmonic generating capabilities [56]. Their systems comprised amino acids and their salts, and salts of tartaric, malic, and lactic acids. Approximately 10% of the crystals possessed second-harmonic efficiencies greater than that of KDP.

A series of nitrobenzofurazans has been examined for second-harmonic generation efficiencies [57]. In these systems, the direction of polarization due to photoexcitation is perpendicular to that of the electronic transition moment, since the electron-acceptor group, for nonlinear optical effects, is not the nitro group, but rather the oxadiazole group. Using the method of Kurtz and Perry [32], powder second-harmonic generation measurements on the 7-substituted-4-nitro compounds, where the 7-substituent is "para" to the nitro group; gave efficiencies on the order of that exhibited by urea. These results are given in Table 9. The π-electron system of 4-nitrobenzofurazan is thought to be comparable to dinitrobenzene, and strong electron donors, such as amino, are thought to result in too much of a shift of the absorption maximum to longer wavelength than that desired for frequency doubling applications. Thus a more moderate electron donating group (chlorine) is used to obtain efficient nonlinearities.

The effect of substituents, donor and acceptor chromophores as well as bulky alkyl groups, on the nonlinear optical properties of biphenyls was studied by Takagi et al. [58]. The crystal structure of 3,5-di-*t*-butyl-4-hydroxy-2',4'-dinitrobiphenyl indicated that the molecular arrangement is favorable for second-harmonic generation. A similar arrangement of the molecular long axis with polar 2_1-axis is found in PAN crystals (1-(1-pyrrolidino)-2-acetylamino-4-nitrobenzene). This compound exhibited second-harmonic efficiency 4.4 times that of urea at a fundamental wavelength of 1.064 μm. The analogous 3,5-dimethyl compound also exhibited a second-harmonic efficiency greater than that of urea. These results are given in Table 9. The absorption maxima in these systems range from 340 to 360 nm.

Green et al. determined the powder second-harmonic generation of ferrocenyl-substituted-(nitrophenyl)ethylenes [59]. The reported value for Z-ferrocenyl,*p*-nitrophenylethylene is about 62 times that of a urea reference sample. The rationale for investigating an organo-transition metal group as an electron donor is their facile redox ability, thought to lead to a large hyperpolarizability. The E-ferrocenyl compound exhibits no powder second-harmonic signal, suggesting that it crystallizes into a centrosymmetric structure. These results are listed in Table 9.

Investigators from the California Institute of Technology studied the powder second-harmonic generation (SHG) of asymmetric donor-

acceptor substituted acetylenes [60]. The general trend observed
is that the lower the ground state dipole moment, through either
reduction in the length of the conjugated π-electron system or
decreasing the strength of the donor and/or acceptor group, the
larger the SHG intensity, suggesting less tendency for cancellation
of microscopic nonlinearities in the bulk. The compound exhibiting
the largest value in this series, approximately 200 times that of
urea, is (p-thiomethyl)phenyl-p-nitrophenylacetylene. The efficiency
of second-harmonic generation of a Z,ferrocenyl,p-cyanophenylacetylene
was found to be 4 times that of urea, about 15 times lower than the
corresponding ethylenic compound. These results are given in
Table 9.

Nogami et al. determined powder SHG efficiencies of compounds
containing 1,3-dithiole rings as electron-donor sites [61]. The compounds studied exhibit absorption maxima around 370 nm with cutoffs
near 415 nm (methanol solutions). These materials are shown in
Fig. 7, with R = methyl, ethyl, n-propyl, n-butyl, and i-propyl.
The value of β for compound 1, R = ethyl, as determined by dc-
induced SHG is 13×10^{-30} esu, comparable to the microscopic susceptibilities of NMA and p-NA. The largest powder SHG value, obtained
from compound 3, R = ethyl, when crystallized from methanol, is
24 times that of urea. This value is listed in Table 9. Most of the
compounds studied, however, exhibited little or no SHG.

Frazier et al. determined the second-harmonic generation efficiencies of a series of transition-metal-organic compounds of the type
group VI metal carbonyl arene, pyridyl, or chiral phosphine complexes [62]. Four of the complexes possess second-harmonic efficiencies equal to or better than ammonium dihydrogen phosphate (ADP).
It was found that those features that contribute to second-order
optical nonlinearities in organic compounds are important in transition-
metal-organic compounds as well.

A new class of organic crystals, substituted diphenyl sulfides
and its chalcogen analogues, have been reported by workers at
Johns Hopkins University [63,64]. 4-Amino-4'-nitrodiphenylsulfide
exhibits a second-harmonic generation efficiency 20 times that of
urea, which is comparable to efficiencies shown by nitroaniline
compounds. The large nonlinear response was attributed to a molecular hyperpolarizability enhanced by sulfur to nitro group charge
transfer as well as a favorable alignment of molecules within the
unit cell. EFISH measurements of the oxygen, selenium, and tellurium
analogues show that the sulfur, selenium, and tellurium compounds
have the same microscopic susceptibility, while that of the oxygen
ether is about half that of the others.

Aratani et al. have used an ionized cluster beam technique
to deposit thin crystalline films of MNA onto glass substrates [65].

1

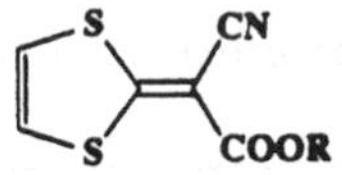

2

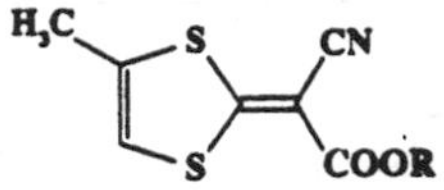

3

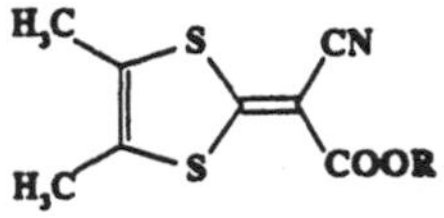

Figure 7. Molecules with 1,3-dithiole donor groups.

These films are found to be denser than those that are vacuum-deposited, and adjustment of the substrate temperature allows for control of orientation within the film. However, the films prepared to date do not possess the necessary orientation for exploiting the nonlinear optical properties of MNA.

Langmuir-Blodgett Films

The second-order nonlinear optical properties of mono- and multi-layers of substituted azobenzenes and polyenes were investigated by Ledoux et al. [66]. Second-order microscopic susceptibilities (β's) were determined from second-harmonic generation measurements on monolayers as shown in Table 6. The higher values obtained from the polyene systems were attributed to greater polarizability in these molecules as compared to the azobenzene derivatives. In all the systems studied dialkylamino groups were the electron-donating species; however, the electron-accepting group was not common to all compounds. The much weaker alkyl, ester, and carboxylic acid groups were present in the azobenzene compounds, while the polyenes possessed the much stronger cyanoacrylic acid functionality. Thus, direct comparison among the π-electron systems is difficult. Comparing the susceptibilities of the three polyenic compounds indicates

Table 6.   Second-Order Microscopic Susceptibilities as Determined
from Monolayers

Compound	Structure	$\beta(10^{38}$ SI)
108782	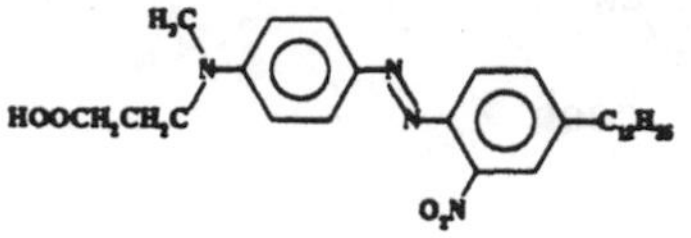	15 ± 1.5
109054	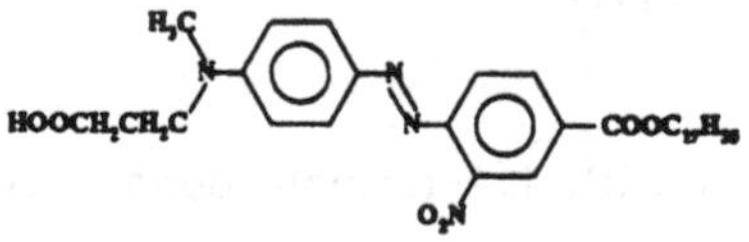	11 ± 1.5
107474	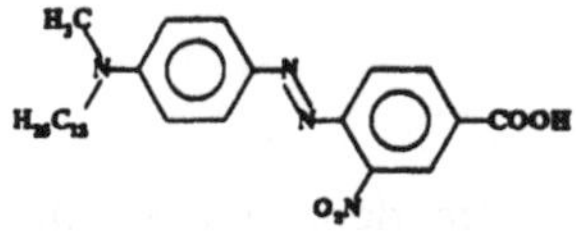	12 ± 1.6
109224	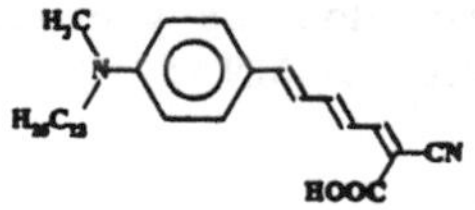	27 ± 4
111177	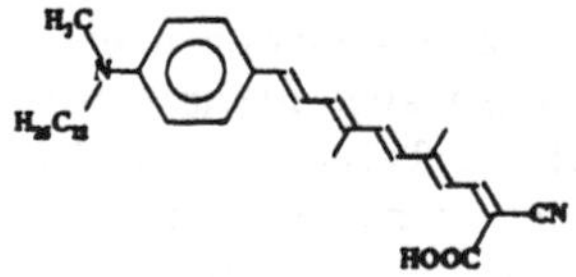	35 ± 5
114355	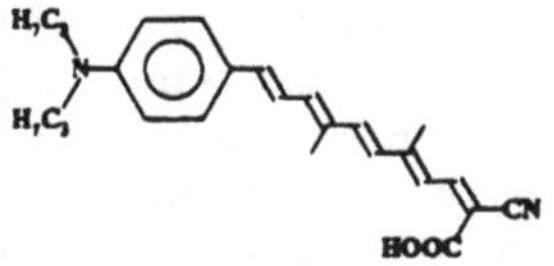	11 ± 1.5

Source: Ref. 66.

that factors other than those arising solely from the extended
chromophore are affecting the susceptibility. The material 114355
would be expected to have a larger microscopic susceptibility than
109224, owing to the presence of two more double bonds in the
π-electron system. The opposite is the case, which may be attributed
to different order present in the monolayer, perhaps arising from
the difference in substituents on the amine nitrogen. One would
also expect to see differences between compounds 108782 and either
109054 or 107474, since the electron-accepting ability of the carboxylic
acid or ester is greater than that of an alkyl group. Perhaps structural differences in the monolayer are responsible for the lack of
difference in the observed susceptibilities.

Using the technique of Y-deposited active-active alternate layers,
where each layer contains an optically nonlinear species and these
species are arranged so that their microscopic susceptibilities are
additive in the bulk, $\chi^{(2)}$ values of ~10^{-10} SI were obtained from
systems with 11 layers. The temporal stability in these active-active
alternating systems is found to be superior to Z-type systems,
with stability of the second-harmonic signal in excess of one year,
compared to the Z-type systems where the signal disappears after
a few months.

Second-harmonic generation from Langmuir-Blodgett films of
retinal, retinal Schiff bases, and bacteriorhodopsin have been studied
by Lewis et al. [67]. Bacteriorhodopsin is the only protein found
in the purple membrane of halophilic bacteria and contains retinal
linked to a lysine unit through a Schiff base. The second-order
molecular hyperpolarizabilities and the chemical structures of these
chromophores are given in Table 7. The values of the nonlinearities
shows the substantial amount of charge transfer, from the β-ionone
ring to the end groups, that occurs in these systems. The larger
value observed for the protonated material is explained by the
stabilization of the π-electron system provided by the proton bonded
to the nitrogen atom. Second-harmonic generation from films of
purple membrane in poly(vinyl alcohol) (PVA) was also reported.
The results indicate that the dipole moment change upon excitation
of the retinal chromophore in the purple membrane-PVA film is
nearly twice that of the free chromophore.

Using LB techniques, Allen et al. have prepared alternating
Y-type multilayer structures comprised of donor- and acceptor-
substituted polyenes [68]. Using the two compounds shown in Fig. 8,
where the hydrophobic group is on the electron-donor end in one
case and on the acceptor end in the other, allows for Y-type deposition where the microscopic nonlinearities will be additive. While
bulk nonlinearities up to 2×10^{-6} esu (χ_{113}) are observed in multilayer films immediately after preparation, virtually no second-harmonic

Table 7. Second-Order Microscopic Susceptibilities as Determined from Monolayers

Compound	Structure	$\beta(10^{-28}$ esu)
Retinal		1.4 ± 0.4
Retinylidene Schiff base		1.2
Protonated Schiff base		2.3

Source: Ref. 67.

generation is observed after several days, indicating substantial molecular rearrangement has occurred over time. The reported absorption maxima for these compounds—(I) λ_{max} = 538 nm (ethanol); (II) λ_{max} = 525 nm (chloroform)—suggest that these compounds do not exist in a completely extended structure. For comparison, the dicyanovinyl azo dye DMA-DCVS, which has one more benzene ring but four less double bonds and a stronger electron acceptor (cyano versus carboxyl), has an absorption maximum of 525 nm in dimethyl sulfoxide.

Poled Polymers

The advantageous properties of both organic compounds and polymeric systems can be combined if the nonlinear optical species can be aligned in the polymer matrix. Orientational order can be imparted to a polymeric system containing dipolar species by imposing an external electric field under conditions when the dipolar species are free to rotate in the polymer matrix [22,23,69]. The ordering energies are comparable to or less than the thermal disordering energies (kT, where k is the Boltzmann constant). A statistical model has been developed that relates the bulk nonlinear optical susceptibility to the microscopic susceptibility, the molecular dipole moment, the poling field, and microscopic order parameters. For

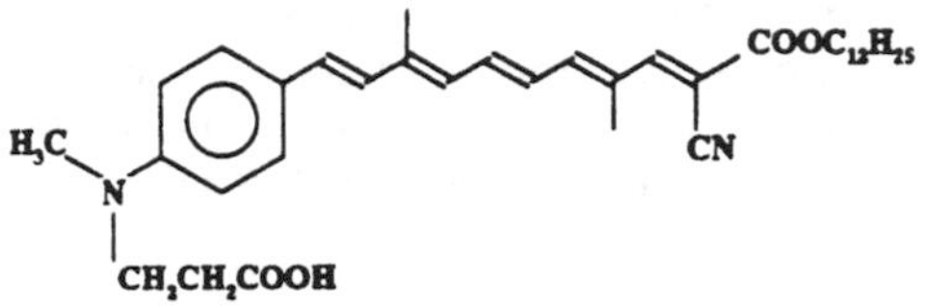

Figure 8. Molecules used in Y-type Langmuir-Blodgett deposition.

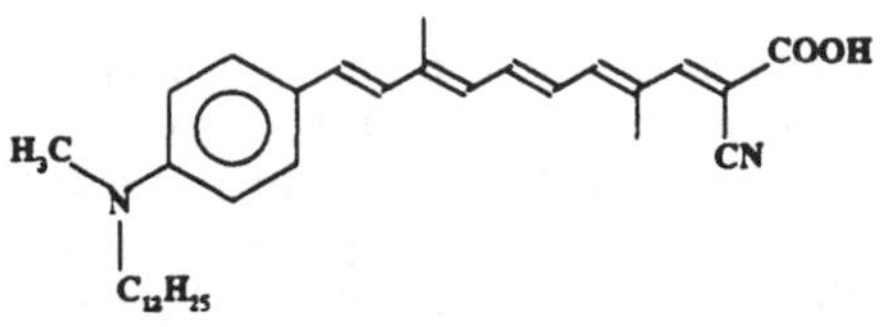

a one-dimensional molecule, that is m_z^* and B_{zzz}^* the only nonzero components, the susceptibility is [18]:

$$\chi_{333}^{(2)} \sim N\beta_{zzz}^* \; \frac{m_z^* E_p}{kT} \; \times \left\{ \frac{1}{5} + \frac{4}{7}<P_2> + \frac{8}{35}<P_4> \right\} \tag{24}$$

and

$$\chi_{311}^{(2)} = \chi_{113}^{(2)} = \chi_{131}^{(2)} \sim N\beta_{zzz}^* \; \frac{m_z^* E_p}{kT}$$

$$\times \left\{ \frac{1}{15} + \frac{1}{21}<P_2> - \frac{8}{70}<P_4> \right\} \tag{25}$$

where N is the number density of molecules, b_{zzz}^* the local-field-corrected molecular susceptibility, m_z^* the molecular dipole moment including corrections arising from the local poling field, E_p, and the $<P_i>$ are the microscopic order parameters used in systems with other ordering forces (such as liquid crystalline), where the mean field approximation is applied. In systems that are isotropic before poling, both $<P_2>$ and $<P_4>$ vanish and $\chi_{333}^{(2)}$ is three times larger than the other components. In liquid crystalline phases, the order parameters can be substantial, and as the order parameters approach unity, the $\chi_{333}^{(2)}$ component becomes a factor of 5 bigger than the isotropic case. The nonlinearities are thus determined by the order imposed by the poling field as well as order arising from intermolecular forces as quantified by the order parameters $<P_i>$.

Persistent second-harmonic generation from electric-field-poled films of Disperse Red 1 in PMMA was reported by Singer et al. [69]. These amorphous polymer glasses possess second-harmonic coefficients larger than that of KDP. Thin films of Disperse Red 1 in PMMA were spin-coated onto indium tin oxide-coated glass followed by evaporation of a semitransparent gold layer onto the polymer. Application of electric fields of 0.2-0.6 MV/cm at ~ 100°C (above the glass transition temperature of the guest-host system) partially aligned the optically nonlinear molecules. Cooling well below the transition temperature with the field still applied yields a system where orientational order is now present. A thermodynamic model, based on Eq. (23), relates the second-harmonic coefficient d to the number density N of the nonlinear optical groups, the microscopic susceptibility β, the molecular dipole moment μ, and the poling field E_p, using the third-order Langevin function $L_3(p)$ (Eqs. (26)-(28)) [69,70].

$$d_{33}(-2\omega;\omega,\omega) = Nf^{2\omega}f^{\omega}f^{\omega}\beta_{zzz}(-2\omega;\omega,\omega)L_3(p) \tag{26}$$

where

$$L(p) = \frac{p}{5} - \frac{p}{105} + \cdots \tag{27}$$

where

$$p = \left[\frac{\varepsilon(n^2 + 2)}{(n^2 + 2\varepsilon)} \right] \frac{\mu E_p}{kT} \tag{28}$$

where ε is the static dielectric constant and n is the index of refraction.

Electrooptic phase modulation and second-harmonic generation have been measured in corona-poled polymer films [19]. Both the magnitude and temporal stability of the nonlinear optical susceptibility are increased, as compared to the corresponding guest-host system, by covalent attachment of the nonlinear optical chromophore to the polymer backbone. The materials studied are guest-host systems of Disperse Red 1 in PMMA (DR1/PMMA), a dicyanovinyl-substituted stilbene (DMA-DCVS from Table 2) in PMMA (DCV/PMMA), and the single component system of a copolymer of methyl methacrylate and a 4-dicyanovinyl-4'-(dialkylamino)azobenzene-substituted methacrylate, DCV-MMA (Fig. 9). The effects of using a chromophore with a larger susceptibility and covalent attachment to the backbone can be seen from the results shown in Table 8. N is the number density of chromophores, d_0 the second-harmonic coefficient determined immediately after poling, and d_f the second-harmonic coefficient after initial decay (~ 1 month). Using a chromophore with a larger nonlinearity (DR1 compared with DCV) results in a proportionally larger second-harmonic coefficient in agreement with the model described above (Eqs. (25)-(27)). In the DCV/PMMA guest-host system, decay of the nonlinearity is substantial—decreasing to 25% of the initial value. Attachment of the chromophore to the polymer backbone (DCV-MMA) gives a system that, when poled, retains

DCV-MMA Copolymer

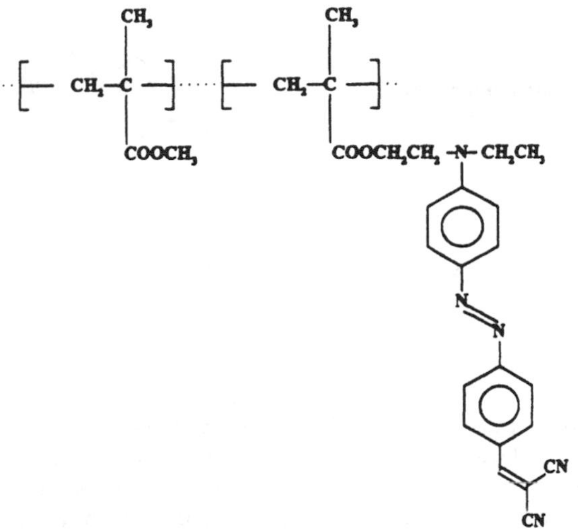

Figure 9. Copolymer of methyl methacrylate and a 4-dicyanovinyl-4'-(dialkylamino)azobenzene-substituted methacrylate.

Table 8. Nonlinear Optical Properties of Corona-Poled Films

Compound	N	$d_0(10^{-9}$ esu)	$d_f(10^{-9}$ esu)
DR1/PMMA	2.3	$d_{33} = 20$ $d_{31} = 7$	
DCV/PMMA	2.3	$d_{33} = 74$ $d_{31} = 25$	$d_{33} = 19$ $d_{31} = 6$
DCV-MMA	-8	$d_{33} = 51$ $d_{31} = 17$	$d_{33} = 46$ $d_{31} = 15$

Source: Ref. 19.

the majority of the imparted polarization: $d_{33} = 51$ at $t = 0$ and $d_{33} = 46$ at $t = 1$ month. The ratio $d_{33}/d_{31} = 3$ confirms the simple thermodynamic model, in which the only potential acting to orient the dipoles is the poling potential [71]. These results provide guidance for realizing polymeric systems with sufficient optical non-linearities for device applications.

Marks, Wong, et al. have prepared single-component systems where the optically nonlinear chromophores are covalently linked to a polystyrene-like backbone [72]. The polystyrene was chosen for both its optical transparency and its excellent film-forming

Pyridinium chromophore

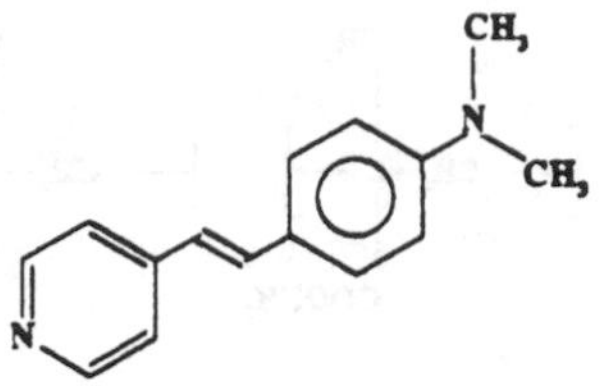

Figure 10.   Pyridinium chromophore appended to polystyrene back-bone.

abilities. Functionalization was achieved by first preparing 4-iodomethyl-derivatized polystyrene, and then coupling the chromophore. The two systems studied were 4.5% level of functionalization of the backbone with the pyridinium compound shown in Fig. 10, and 12.5% functionalized backbone with the azo dye, Disperse Red 1, as the chromophore. The nonlinear optical properties, imparted by electric field poling using transparent electrodes, were determined by second-harmonic generation in transmission. The value of d_{33} for the azo dye functionalized material, poled at 0.3 MV/cm, is 2.7×10^{-9} esu, exceeding the value of KDP and comparable to that observed in guest-host systems of Disperse Red 1 in PMMA [69]. The magnitude of d_{33} was found to vary linearly with the strength of the poling field, in accord with a thermodynamic model [69]. The ionic nature of the pyridinium system results in diminished values of d_{33}, compared to those expected from the microscopic susceptibility, and saturation at poling fields above 0.3 MV/cm.

Using the nitrophenylprolinol, NPP, as the chromophore, Marks, Wong et al. studied the temporal stability of poled, functionalized poly(p-hydroxystyrene) [73]. Annealing the films prior to poling was found to enhance the ultimate poling fields achievable, as high as 1.8 MV/cm, and the temporal stability of the resulting second-order nonlinear optical properties. A fairly linear dependence of the functionalization level on d_{33} was found as predicted by the Langevin function model [69]; however, a departure of the linear dependence of the poling field on d_{33} was observed at higher fields, where now higher-order terms in the model become important. The behavior of the second-harmonic signal with time shows at least two decay mechanisms: one a short-term loss with a half-life of 1.5 days, the other a long-term process with a half-life of 195 days for annealed samples. It was also noted that poling is effective at temperatures more than 20°C below the glass transition tempera-

ture signifying the important role secondary relaxations play in
both alignment and temporal stability.

   Physical aging and dopant size effects on second-harmonic generation in poled, doped amorphous polymers was studied by Torkelson
et al. [74]. The polymeric hosts used in this investigation were
bisphenol A polycarbonate and PMMA; the dopants were DMA-NS,
Disperse Red 1, and Disperse Orange 25 (a cyano group substituted
on the methyl group of Disperse Red 1). In general, it is found
that increasing dopant size, as well as physical aging (annealing),
improves the temporal stability of the second-order properties in
these guest-host systems.

   Eich et al., from IBM Almaden Research Center, recently reported
in situ measurements of corona-poled induced second-harmonic generation from the amorphous homopolymer PPNA (Fig. 11) [75]. This
system, basically p-nitroaniline bonded to a polyethylene backbone,
does not show the rapid loss of polarization found in solid solutions
of nitroanilines dissolved in amorphous hosts. d_{33} values ranged
from 19 to 31 pm/V, the larger value determined at t = 0 (immediately
after poling) and the smaller after five days. Analysis of the decay
in the nonlinearity indicates that two processes contribute to the
relaxation of the induced polarization-charge deposition and molecular
reorientation.

   Workers at Thomson-CSF have studied the poling process and
determined the nonlinear optical properties of azo dye-doped PMMA
and copolymers in which the dye is covalently bound to the methacrylate backbone [76]. The azo dye concentration ranged from 1 to
19 mol% in the copolymers. Second harmonic intensity is found to
increase with the square of the poling field as predicted by the
thermodynamic model [69]. In studying the effect of the poling
temperature on the second-harmonic coefficient using in situ second-
harmonic generation, researchers found a maximum in intensity
slightly below the glass transition temperature of the guest-host

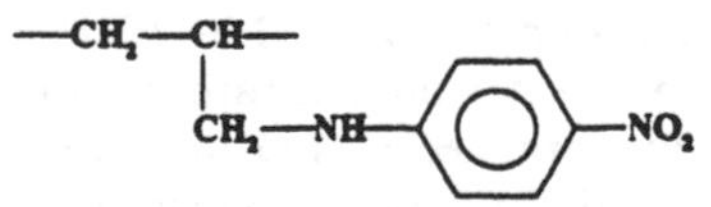

Figure 11.   Repeat unit of the amorphous homopolymer PPNA.

system. This maximum was explained as a balance between the viscous forces retarding orientation of the optically nonlinear moieties and the thermal randomization at higher temperature. When the dye molecules are covalently linked to the polymer backbone, different results are obtained. At lower temperatures the dye groups are more constrained as evidenced by a much sharper increase in second-harmonic signal with increasing temperature; this increase occurring at a higher temperature relative to the guest-host system.

The nonlinear optical properties of corona-poled thin films of the azo dye Disperse Red 1 dissolved in a PMMA matrix were also studied by Mortazavi et al. [77]. The optically nonlinear chromophores were oriented at temperatures above the glass transition temperature of the guest-host system (~100°C) using corona-poling techniques. It was noted that the color of the thin films changed from light orange to red-violet upon exposure to the corona, and this process is reversible by removing the corona. This result was attributed to corona-induced dipole moments and dipole orientation during poling. The bulk nonlinearity decayed rapidly after poling, but leveled to a value about 10 times that of quartz, and remains at that value for at least eight months.

Workers at Lockheed have developed a procedure for selectively poling regions in a polymer film and have demonstrated devices fabricated using this technique [78,79]. The procedure involves photolithographically defining an electrode pattern on substrate, overcoating the electrodes with a buffer layer to isolate the wave-guide layer from the electrodes, deposition of the waveguide material, deposition of the top buffer layer, and evaporation of a planar electrode. Application of an electric field results in poling only in those regions between the defined bottom electrodes and the planar top electrode. Patterning of the planar top electrode yields the device. With proprietary materials and this technique, a variety of devices have been demonstrated, including interferometers, directional couplers, and traveling wave modulators.

Molecular Composites

Efficient second-harmonic generation from dye aggregates in thin polymer films has been reported by Wang [80]. Thin films of aggregated thiapyrylium dye, with either tetrafluoroborate or hexafluorophosphate counterions, in polycarbonate from ~1-5 μm thick were shown to exhibit second-harmonic efficiencies comparable to that exhibited by a film of urea 250 μm thick. Preparation involves spin-coating onto a substrate, yielding a film where the dye is essentially unaggregated. Upon swelling with methylene chloride, spontaneous aggregation was observed. The absorption maximum shifted from 590 to 690 nm. No second-harmonic generation is observed from films of the unaggregated dye. Some reabsorption of the second-harmonic by the dye was noted by comparing values of the second-harmonid light generated from fundamental wavelengths of 1.06 and 1.9 μm. These results suggest that "self dipolar alignment" occurs on exposure to methylene chloride.

Complexes of p-NA/poly(oxyethylene), crystallized in the presence of an electric field, exhibit second-harmonic generation intensities up to 91 times that of urea [81]. The solubility of p-NA in poly(oxyethylene) (POE) allows for complexes containing up to 20 mol% p-NA. A system of 1:8 p-NA:POE yielded SHG intensity 20-30 times that of urea, comparable to the value of MNA. It was found that the initial SHG intensity is independent of the cooling rate in the polarizing-freezing process used to impart order to these materials. The intensity persists for extended periods, reaching half the initial intensity after ~100 h. The decrease in intensity is attributed to decomposition of the p-NA/POE complex, verified by x-ray powder diffraction which shows an increase in peaks from p-NA accompanied by a decrease in those peaks arising from the complex. The fact that these systems are not simple solutions of p-NA in POE is supported by analyzing the system using the thermodynamic model developed by Singer et al. [69]. The model yields values of the second-harmonic coefficient one-half that of urea, much too small to account for the SHG intensity. Thus some interaction between p-NA and POE results in a complex that possesses larger nonlinearities than would be expected from simple additive effects. It was found that the SHG intensity reaches a maximum for compositions of 1:6 p-NA:POE. For compositions richer in p-NA, the SHG signal is diminished owing to phase separation of p-NA, which no longer contributes to SHG. These systems also possess good mechanical strength and show some resistance to optical damage.

Alignment of MNA microcrystals in a PMMA matrix was reported by Daigo et al. [82]. Second-harmonic generation from thin films of the composite yielded efficiencies 19 times that of urea. The thin films were prepared by spin coating, followed by application of an electric field at 80°C to align the crystallites. Composite films of p-NA in PMMA yielded SHG efficiencies 10 times that of urea. The induced polarization has rather remarkable stability, with reported near-constant SHG activity for about six months.

Tam, Wang et al. reported results of second-harmonic generation in organometallic, stilbene, and diphenylacetylene systems [83,84]. The organometallic compounds consist of either platinum or palladium

CMONS

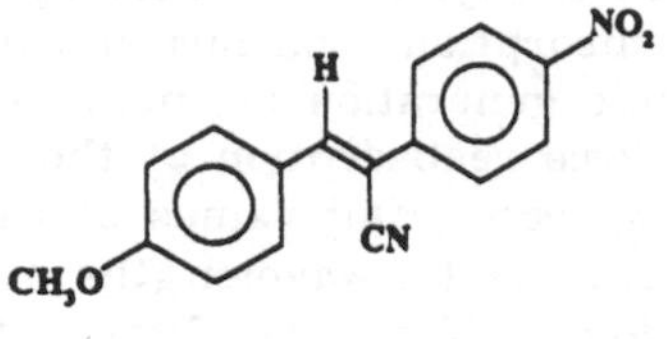

Figure 12.   Structure of the molecular crystal CMONS.

electron donors, and nitro, cyano, or formyl electron-acceptor groups
on a benzene ring. The second-harmonic efficiencies of powder
samples of these compounds range from 0.4 to 10 times that of
urea. The substituted stilbenes possess powder efficiencies up to
300 times that of urea. One, 2-cyano-*p*-methoxy-*p*'-nitrostilbene
(CMONS, see Fig. 12) has the largest powder SHG efficiency observed. The other compounds studied include MONS, where the
cyano group of CMONS is replaced by a hydrogen, BMONS, where
a bromine atom has replaced the cyano group of CMONS, and BONS,
where the donor group is bromine and the double bond has hydrogens
bonded to both carbons. Polymorphism is evident in CMONS: large
efficiencies are found in crystals grown from ethyl acetate or toluene
solutions (250-300 times urea), while crystals grown from the melt
show efficiencies 90 times urea, and those grown from dioxane show
only 0.15 times the efficiency of urea. MONS, BMONS, and BONS
also show polymorphism. SHG efficiencies from the diphenylacetylenes
range up to 136 times that of urea for bromo or iodo electron-
donating groups and a nitro acceptor group. It is important to
note that compounds possessing relatively weak electron donors
such as methoxy can exhibit efficiencies as large as the compounds
possessing stronger donors, such as amino, owing to favorable
orientation of the chromophores in the crystal.

    Eaton and Wang have used inclusion chemistry to prepare com-
plexes of a variety of optically nonlinear guests with hosts such
as β-cyclodextrin, thiourea, tris(*o*-thymotide), and deoxycholic
acid [85]. Powder SHG measurements showed efficiencies up to
four times that of urea (for a 1:1 complex of *p*-NA:β-cyclodextrin).
Other nitroaniline, aminonitropyridine, and aminobenzonitrile compounds showed efficiences less than that of urea. Several organometallic
complexes yielded values greater than urea: 3:1 benzenechromium
tricarbonyl:β-cyclodextrin (2.3 times urea), and 3:1 (fluorobenzene)-
chromium tricarbonyl (2.0 times urea). This technique is suggested
to be a general one for dipolar alignment of organic as well as

organometallic species. The rationale for this is the preferred head-to-tail alignment of dipoles in a lattice inclusion matrix so as to minimize electrostatic repulsion.

Stucky et al. have prepared inclusion compounds of p-NA and MNA in molecular sieve hosts [86]. This represents the first example of organic guest/inorganic host complexes possessing second-order nonlinear optical properties. With p-NA and centrosymmetric molecular sieve hosts, no second-harmonic generation was observed. Using an acentric host, ALPO-5 (a molecular sieve with a neutral framework consisting of alternating AlO_4 and PO_4 tetrahedra linked by oxygen bridges), SHG intensities up to 10 times that of p-NA in an organic host are realized. The SHG intensity reaches a maximum value for 13 wt% p-NA. The decrease of intensity at higher loadings is attributable to dilution with p-NA external to the zeolite host. An opposite situation occurs with MNA in ALPO-5. Only when MNA exists outside of the host does the SHG intensity increase. It was hypothesized that the methyl group in MNA restricts orientation in the channels of the zeolite and prevents bulk dipolar alignment.

E. Crystal-Cored Fibers

Organic crystal-cored fibers have been investigated for over a decade as a means for exploiting waveguide geometries [87]. The advantages of waveguide structures over bulk crystals, enumerated by Badan et al. are (1) natural material dispersion is replaced by modal dispersion, which can be tailored by the size and composition of the substrate and superstrate; (2) interactions can take place in the guide or the surrounding media; (3) low material dispersion and high bulk birefringence are not required for phase matching; and (4) large power densities can be confined in the waveguide. Compounds such as m-nitroaniline, m-dinitrobenzene, 2-bromo-4-nitroaniline, m-dihydroxybenzene, benzil, N-(4-nitrophenyl)-(L)-prolinol (NPP), and nitro-4-phenyl-N-(methylcyanomethyl)-amine (NPAN) as crystal-cored fibers were discussed by Badan et al. Recently, Tomaru and Zembutsu described attempts on growth of 4-(N, N-dimethylamino)-3-acetamidonitrobenzene (DAN) in glass capillaries [88]. Kerkoc et al. have characterized DAN crystal-cored fibers grown in glass capillaries [89]. They obtained a second-harmonic generation efficiency of $\sim 10^{-4}$ with a fundamental wavelength of 1.064 μm in a 20-mm-long fiber with a diameter of 7 μm.

F. Comparison of Nonlinear Optical Properties

Nicoud and Twieg have tabulated SHG powder test data and EFISH hyperpolarizability date for a large number of organic materials [33].

Table 9. Compilation of SHG Results 1986–1989

Compound	Powder efficiency	Reference
Z-ferrocenyl,p-nitrophenylethylene	62 × urea	59
E-ferrocenyl,p-nitrophenylethylene	0	59
(p-thiomethyl)phenyl-p-nitrophenylacetylene	200 × urea	60
Z-ferrocenyl,p-cyanophenylacetylene	4 × urea	60
styrenetricarbonylchromium(0)	1.8 × ADP	62
4-amino-4'-nitrodiphenyl sulfide	20 × urea	63
4-amino-4'-nitrodiphenyl selenide	10 × urea	64
4-amino-4'-nitrodiphenyl telluride	0	64
4-amino-4'-nitrodiphenyl ether	0	64
3,5-di-t-butyl-4-hydroxy-2',4'-dinitrobiphenyl	4.4 × urea	58
3,5-dimethyl-4-hydroxy-2',4'-dinitrobiphenyl	2.2 × urea	58
N-butyl-p-nitroaniline (from cyclohexane/ether)	0.5 × urea	54
N-butyl-p-nitroaniline (from ethanol)	1.4 × urea	54
N-octyl-p-nitroaniline (from cyclohexane/ether)	0	54
N-octyl-p-nitroaniline (from ethanol)	0.05 × urea	54
2-cyclooctylamino-5-nitropyridine (COANP)	$d_{23} = 32 \pm 16$pmN $d_{33} = 13.7 \pm 2$pmN	53
7-chloro-4-nitrobenzo-2-oxa-1,3-diazole	6 × urea (1.06 m) 3.2 × urea (1.9 m)	57
4-(N,N-dimethylamino)-3-acetamidonitrobenzene (DAN)	$d_{eff} = 27 \pm 3$pmN	49
[cyano(ethoxycarbonyl)methylene]-2-ylidene-4,5-dimethyl-1,3-dithiole	24 × urea	61
2-cyano-p-methoxy-p'-nitrostilbene (CMONS)	300 × urea	84

These compilations cover SHG data reported through 1985, and EFISH results through 1983. Table 9 lists some recent SHG results.

IV. DEVICE CONSIDERATIONS

In this section, we will discuss material aspects of device operation with special attention paid to the applicability of the various forms of organic and polymeric materials. The section begins with a description of the propagation of light in nonlinear optical materials, which is then followed by a discussion of the operation of bulk and guided-wave parametric devices. The section closes with an analysis of the applicability of organic and polymeric materials to electrooptic devices.

Any description of nonlinear optical devices takes its starting point from the formulation of the propagation of light in a nonlinear optical medium. This begins with Maxwell's equations connecting with the appropriate nonlinear constitutive relations [46]. Neglecting the magnetization, the dependence of the polarization on the magnetic field, and real charges and currents, the appropriate wave equation is

$$\nabla \times \nabla \times E_{\omega j} + \frac{\varepsilon_{\omega j}}{c^2} \frac{\partial^2}{\partial t^2} E_{\omega j} = - \frac{4\pi}{c^2} \frac{\partial^2}{\partial t^2} P_{\omega j}^{(2)} \tag{29}$$

We take monochromatic plane waves propagating in the positive z-direction of the form given in Eq. (6) where $\underline{E}(\underline{k}, \omega) = \underline{E}_{\omega} e^{ikz}$. In order to solve the inhomogeneous equation (Eq. (29)), the slowly varying amplitude approximation is invoked, namely

$$\frac{\partial^2 E_j(z)}{\partial z^2} \ll k_j \frac{\partial E_j(z)}{\partial z} \tag{30}$$

For small losses and the polarization given by Eq. (5), Eq. (29) leads to the coupled wave equations

$$\frac{d}{dz} E_{\omega} = -i\kappa E_{\omega_1} E_{\omega_2} \exp(i\, \Delta kz)$$

$$\frac{d}{dz} E_{\omega_1} = i\kappa_1 E_{\omega} E_{\omega_2}^* \exp(i\, \Delta kz)$$

$$\frac{d}{dz} E_{\omega_2} = i\kappa_2 E_{\omega} E_{\omega_1}^* \exp(i\, \Delta kz) \tag{31}$$

where the coupling coefficients are $\kappa_j = (2\pi\omega_j/n_j c)[e_j \cdot \chi^{(2)}(-\omega_j; \omega_i, \omega_k) : e_i e_k / (z \times e_j)^2]$. The factor $[e_j \cdot \chi^{(2)} : e_i e_k]$ is effective susceptibility, d_{eff}, describing the transformation from the crystal to the field coordinate system. The phase factor $\Delta k = k - k_1 - k_2$.

The coupled wave equation can be solved for any sum or difference interaction. However, in order to discuss general material requirement for devices, we find it useful to examine the case of second-harmonic generation. Zyss and Chemla have discussed issues concerning organic nonlinear optical crystalline devices in some detail, which we summarize here for purposes of a more general discussion on the potential application of various types of materials.

The intensity of second-harmonic light from Eq. (31) is given by

$$I_{2\omega} = \frac{128\pi^3\omega^3}{c^3} \frac{d_{eff}^2}{n_\omega^2 n_{2\omega}} z^2 I_{\omega_1}^2 \left[\frac{\sin(\kappa z/2)}{\kappa z/2}\right]^2 \tag{32}$$

The phase mismatch results in the intensity being a periodic function of distance with the characteristic distance called the coherence length, $l_c = \lambda/4(n_{2\omega}-n_\omega)$. Due to the natural energy dispersion in materials, the coherence lengths are typically on the order of tens of microns or less. Thus the amount of incident light converted to the second harmonic is severely limited by this phase mismatch. For typical nonlinearities and incident intensities the second-harmonic intensity is many orders of magnitude smaller than the incident intensity. This limitation is overcome by finding a phase-matching condition where the coherence length becomes very long. In this case the incident beam becomes depleted leading to a photon flux which varies as $N_{2\omega} = N_\omega(0)\tanh^2(gz)$, where g is the nonlinear gain proportional to the d_{eff}.

The efficiency of second-harmonic devices as well as the other second-order nonlinear optical devices (except electrooptic) depends not only on the material nonlinearity, but also on the attenuation, beam propagation characteristics, and the phase-matching conditions. As described in the previous section, organic and polymeric materials possess exceptionally large nonlinearities; and from that point of view promise to be the best candidates for nonlinear optical devices. Attaining the appropriate optical quality, beam propagation characteristics, and phase-matching conditions turns out to be the most challenging aspect of these devices.

Phase matching in crystals is attained by employing the crystal birefringence to cancel the normal dispersion of the crystal. This occurs if one of the waves is an ordinary wave and one an extraordinary wave as shown in Fig. 13 [16]. The figure depicts the orientational dependence of the extraordinary refractive index at the second-harmonic frequency, and the phase matching points in the xz-plane. In the negative uniaxial case depicted in the figure,

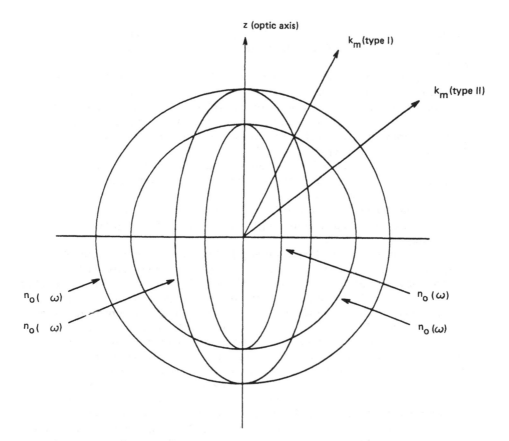

Figure 13. Section of index surfaces for an exaggerated negative uniaxial crystal showing dispersion and the type I and type II phase-matching conditions.

the propagation associated with the intersection of $n_e(2\omega) = n_0(\omega)$ corresponds to type I phase matching. The other case in the figure is type II phase matching, where $2n_e(2\omega) = n_0(\omega) + n_e(\omega)$. The efficiency of the device depends critically in many ways on the locus of points in a particular crystal where phase matching occurs, including the amount of coupling to a large nonlinear optical tensor component due to the relative orientation of the phase-matching locus to the susceptibility tensor, and the derivatives of the phase-matching locus with respect to the orientation, temperature, and wavelength. The sensitivity to orientation determines the acceptance angle, and, depending on the characteristics of the laser beam, can affect the fraction of the incident k-direction that gets converted. The sensitivity to wavelength determines the spectral bandwidth of the converted light.

The propagation of the beams also limits device performance. One consideration is the effect of beam walk-off, where the different propagation direction of the fundamental and harmonic wave limit the effective overlap or interaction region. Also, for short time pulse propagation the effective interaction length is also limited by the group velocity mismatch in the pulses [90]. In addition, the focal parameters of the propagating beam limit the useful interaction length in bulk crystalline devices. Many trade-offs may exist in these limiting factors. An advantage of organic materials is that the large susceptibilities may ease propagation limitations since a shorter propagation length is required for a given conversion efficiency.

In addition to the large susceptibility, a strong advantage of organic crystals, is the possibility to engineer materials with all of the device requirements in mind. To this end, Zyss and co-workers have enumerated the symmetry requirements implicit in Eq. (22), which simultaneously optimize both the nonlinear susceptibility and the phase-matching conditions in organic molecular crystals [45]. Once enumerated, they have completed an extensive body of work examining a series of candidate molecular crystals which has led to the first commercial introduction of organic nonlinear optical devices [91]. In the process they have found that the sharp dispersion near electronic resonances allows for noncritical phase-matching conditions, permitting new device applications not possible in other materials (PASS spectroscopy) [92]. Other practical problems which have been overcome in that work include the ability to grow high-optical-quality, stable, and robust crystals. In fact, this was certainly one of the more difficult challenges in the work, and remains a key factor in the applicability of other organic materials to a variety of applications.

Bulk crystalline materials are well suited for applications where the wavelength is not fixed, since the phase matching is easily tuned by orientation or temperature. In addition, the conversion efficiencies are certainly large enough for relatively high power (pulsed laser) applications, usually involving solid-state (not semiconductor) lasers. However, for low-power applications integrated optical approaches may be more appropriate since waveguide confinement greatly increases efficiency. These applications include frequency doubling a diode laser for optical data storage, fixed-frequency parametric oscillation, and amplification, for use in an all-optical amplifier or wavelength multiplexer.

In integrated optics, the light propagates within a waveguide. Since the light in a waveguide can be confined to a very small cross-sectional area over long lengths, more efficient interactions

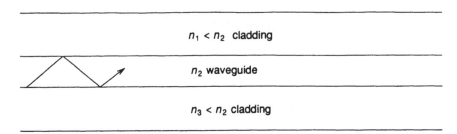

$n_1 < n_2$ cladding

n_2 waveguide

$n_3 < n_2$ cladding

Figure 14. Slab waveguide geometry including waveguide and cladding.

can take place. The advantage of waveguided devices over bulk devices can be as high as the ratio of the length of the waveguide to the wavelength of light, which can, in principle, be 10^4 [93]. The quantized properties of waveguide modes complicates the light propagation, which allows for more phase-matching schemes but may add practical difficulties in achieving phase matching in a real system. In addition, guided-wave applications require extremely high optical quality; even for the simplest applications 1 dB/cm is probably the upper limit on loss. Polymer glasses have been shown to possess such low losses and, from this point of view, would be desirable in guided-wave applications. The ability to deposit these glasses on semiconductor substrate and to use photo-lithographic techniques is also a promising trait.

Because guided waves must satisfy the mode conditions, the two polarizations propagate with different refractive indices even in an isotropic material, and the dispersion in the effective mode index of refraction is greater than that of the bulk index of refraction. These properties can be seen by considering the mode conditions for the slab waveguide geometry shown in Fig. 14. For modes with the electric-field vector in the plane of the guide (TE modes), the mode condition is

$$\tan(ht) = \frac{p + q}{h(1 - pq/h^2)} \tag{33}$$

where $h = (n_2^2 k_\omega^2 - n_{eff}^2)^{1/2}$, $q = (n_{eff}^2 - n_1^2 k_\omega^2)^{1/2}$, $p = (n_{eff}^2 - n_3^2 k_\omega^2)^{1/2}$, $k_\omega = \omega/c$, t the waveguide thickness, n_{eff} the propagation constant or effective index of refraction, and the indices of refraction are defined in Fig. 14. Similarly, for the modes propagating with their electric-field vector perpendicular to the plane of the waveguide (TM modes).

$$\tan(\bar{h}t) = \frac{h(\bar{p} + \bar{q})}{h^2 - \bar{p}\bar{q}} \tag{34}$$

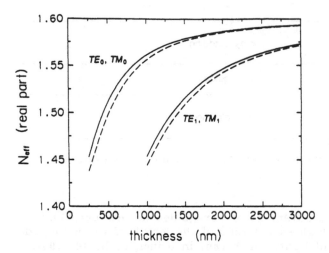

Figure 15. Mode index dispersion for an isotropic slab waveguide. Transverse electric (TE) and magnetic (TM) modes are shown for $n_1 = n_3 = 1.4$ and $n_2 = 1.6$.

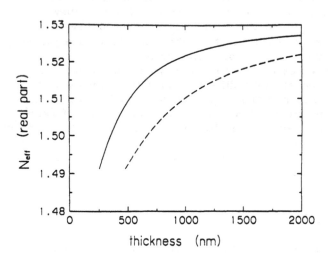

Figure 16. Mode dispersion when bulk birefringence and dispersion vanish ($n_1 = n_3 = 1.48$ and $n_2 = 1.53$).

where $\bar{q} = qn_2^2/n_1^2$ and $\bar{p} = pn_2^2/n_3^2$. Representative mode curves satisfying Eqs. (33) and (34) are shown in Fig. 15. The various mode indices (0, 1, etc.) correspond to integral multiples of π of the inverse tangent function or to constructive interference in the sense of geometrical optics [46]. Figure 16 depicts the mode dispersion for guided modes where the bulk birefringence and dispersion are zero. Thus, even when the bulk dispersion vanishes, addition dispersion due to the explicit dependence of n_{eff} on wavelength must be overcome to obtain phase-matched second-harmonic generation. Conversely, the residual birefringence in n_{eff} in isotropic films can be utilized to obtain phase-matched operation of modes of different order.

Several methods for obtaining phase matching in integrated optical structures have been proposed [94,95]. However, in addition to phase matching, efficient operation requires low-loss waveguides where the guided-wave energy is highly confined to the nonlinear optical waveguide so that a large fraction of energy is converted, and the fundamental and second-harmonic waves need to be confined in such a way that much of their energy interferes constructively.

The electric field profiles for the first few lowest-order modes for both polarizations shown in Fig. 17. All the modes exhibit confinement in the central guiding region and an evanescent field in the cladding region. The fraction of energy confined in each depends on the thickness of the guiding region, the relative bulk indices in the regions, the wavelength, and the order of the mode. For nonlinear waveguides, the efficiency is optimized with the lowest-order mode since it yields the most energy in the waveguide. For nonlinear cladding, a higher-order mode will yield more energy in the cladding.

In most schemes, both the fundamental and second-harmonic waves are confined to the waveguide, and the efficiency is related to the overlap integral

$$I_R \propto \int_{-t}^{0} [E_\omega^m(x)]^2 \, E_{2\omega}^n(x) \, dx \tag{35}$$

Phase matching can be achieved in some cases in a waveguide fabricated from isotropic materials by matching, for instance, a $TM_1^{2\omega}$ mode to a TE_0^ω mode. However, when the integration in Eq. (35) is carried out by using the field profiles in Figure 17, the efficiency is rather low. Therefore, it is still desirable to employ birefringence phase matching of the lowest-order modes of both polarizations to achieve phase matching. In this case off-diagonal tensor elements can be phase-matched in a similar manner as the bulk case, but with thickness instead of orientation being the tuning

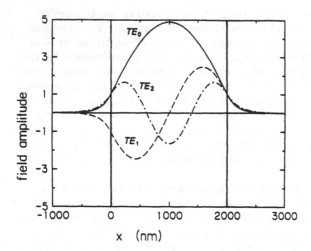

Figure 17.   Electric field profiles for lowest-order guided modes.

element. Work along these lines has been reported in slab waveguides. In the case of the slab waveguides of MNA, the waveguide was tapered in thickness to ensure a phase-matched condition [96]. This approach would not be applicable to channel waveguides which are needed for truly efficient devices.

In trying to phase-match using birefringence, the critical thickness condition may make specific frequency operation difficult due to device processing tolerances. Several approaches to attaining less critical phase matching have been described. First, thickness noncritical phase matching can be attained using zero-order modes near cutoff in a thin, nearly symmetric waveguide [97]. In this case the mode curves shown in Fig. 18 are tangent due to favorable first and second derivatives of the mode curves. However, to attain this noncriticality in thickness, a critical constraint has been placed on the bulk refractive indices of the materials comprising the layers. A liquid top cladding was used to achieve the critical cladding refractive index. Langmuir-Blodgett films and spin-deposited polymers are particularly suited to making multilayer structures, so may be useful for such structures if the refractive index can be controlled to the desired precision. Another multilayer method involves placing an additional linear optical guiding layer over the nonlinear optical guiding layer so that the interference is confined to the linear optical layer [98]. This increases the coherent field overlap substantially, and has resulted in the most efficient conversion reported.

This method does not require a trade-off with another critical factor, and so may be promising, especially for the flexible fabrication one can attain in organics. A drawback for optical data storage frequency doubling would be that a higher-order second-harmonic mode would be focused at the device output, so that the gain in diffraction limitation by going to shorter wavelength may be lost. Similar system penalties may be paid in other devices due to the use of a higher-order mode. It is also possible to employ a waveguide where the second harmonic is generated in a radiated rather than a guided mode and is confined to a cladding [99]. The phase-matching condition is attained in a noncritical way for certain wave vectors in the radiated mode. The efficiency is reduced, but by using the large nonlinearity in organic materials it may be possible to attain useful devices. In this method, a similar penalty in focusing to the diffraction limit as in the multilayer method above is present. This type of interaction has been observed in crystal-cored wave-guides containing organic materials [100]. Another method for achiev-ing coherent interaction is the use of counterpropagating beams in a waveguide [94].

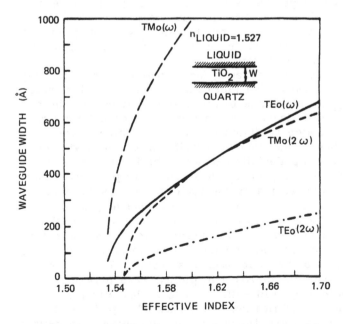

Figure 18. Noncritical waveguide phase matching near cutoff. (From Ref. 97, used with permission.)

An approach which would take advantage of the fabrication flexibility of poled polymers involves a periodic modulation of the guided-wave interaction by using a grating overlying the waveguide [101-103]. This geometry can facilitate phase matching in two different ways. In one way, the effective index is periodically modulated by the grating, which adds a component of wave vector in the guide which can be used to attain phase matching. In addition, a grating in the nonlinearity can be induced which periodically changes the phase of the nonlinear interaction. If the grating is chosen properly, a quasi-phase-matched condition is obtained. For a channel waveguide, the grating fabrication is critical to efficiency.

Another method of phase matching which is particularly suited to guest-host polymer materials and may yield efficient devices is to use anomalous dispersion to achieve phase matching [104]. In this case, the pump and generated wavelengths lie on either side of the energy of an excited state of the nonlinear optical guest. The linear polarizability of the guest exhibits anomalous dispersion, which, when added to the normal dispersion of the host can lead to zero dispersion at the appropriate concentration. This approach is applicable both to crystals and waveguides, but in waveguides additional anomalous dispersion is necessary in order to cancel the modal dispersion. This method has the advantage that any tensor component can be phase-matched, but has the disadvantage of critical thickness and/or concentration. In addition, finding an appropriately transparent guest will be challenging.

Although many potential schemes for waveguide phase matching exist, all involve critical device fabrication steps mostly involving thickness (or n_{eff}), which, due to processing control, may limit the efficiency of the devices. The schemes which do not require a critical fabrication step (radiated modes and multiple layers) produce energy profiles, which may be a limitation of the device in systems. The processing flexibility of guest-host polymers and Langmuir-Blodgett films may be useful in realizing phase-matching schemes. To date, optical-quality waveguides have not been demonstrated in Langmuir-Blodgett films, but have been in poled polymer films. However, the loss in susceptibility one suffers in guest-host systems due to number density, transparency requirements in the visible for parametric devices, and the Boltzmann energy in poling make the advantages of this approach less certain. There seems to be no clearly favorable approach to guided-wave parametric devices, and the necessary trade-offs need to be investigated. The inherent synthetic and fabrication flexibilities of organics and polymers, however, may eventually lead to interesting guided-wave parametric devices.

Electrooptic devices differ from parametric devices mainly in that phase-matching considerations are not important. The dielectric properties of the material at the modulating frequency is an important property. In addition, electrooptic devices involve only one optical frequency, so transparency requirements are generally less stringent than in all-optical devices. Like parametric devices, electrooptic devices may be made both in bulk and guided-wave, thin-film form. Bulk electrooptic devices are used generally for shutters; usually formed from inorganic crystals such as potassium dihydrogen phosphate or lithium niobate ($LiNbO_3$). It is not apparent that organic crystals will significantly improve on the performance of such devices, and so it is not expected that organic materials will find application in bulk electrooptic devices.

In guided-wave applications, organic materials may have a significant impact due to many of their favorable properties. A guided-wave technology already exists ($LiNbO_3$), but the potential advantages of organic materials are definitely compelling enough to justify a strong effort to apply organic materials in this area. These advantages include higher electrooptic coefficients and lower dielectric constants [105]. Organic crystals, while possessing these attractive features, do not have the additional attractive features of easy waveguide definition and integrability with semiconductor substrates that thin-film approaches such as poled polymer films or Langmuir-Blodgett films do. However, high-optical-quality waveguides have not been demonstrated in Langmuir-Blodgett films, so that poled polymer films appear to be the most promising approach to integrated electrooptic device technology. In fact, the first devices made from guest-host polymers have been fabricated by using selective poling techniques [78]. The function of these devices with respect to their optical and electrooptic behavior has been as expected from basic material properties. Based on the progress in the development of these materials, it is expected that truly high-performance devices are not far off.

Guided-wave channel electrooptic devices include modulators and switches which can perform a variety of system functions, such as modulation, switching, deflection, multiplexing, and analog-to-digital conversion. These devices can be made by using various structures, of which two are depicted in Figs. 19 and 20. Figure 19 shows a Mach-Zehnder modulator, which acts as an amplitude modulator, and Fig. 20 a directional coupler switch, which can be used in a variety of functions including modulation and 2 × 2 switching. The Mach-Zehnder operates on the principle of interference between the split amplitudes traversing the waveguide. The amplitude at the output depends on the phase difference between the light in each

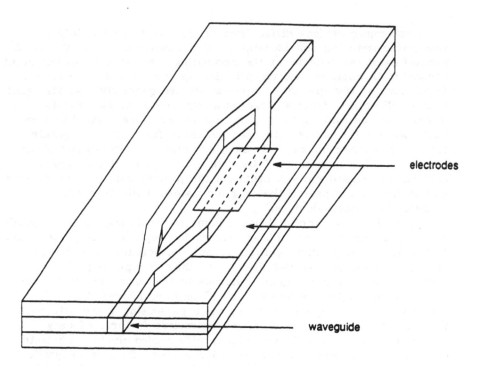

electrodes

waveguide

Figure 19.   Schematic of guided-wave electrooptic Mach-Zehnder modulator.

channel, which itself depends on the length and refractive index in each arm of the device. By applying voltage to one arm, the optical phase is adjusted by the index of refraction change from the electrooptic effect. By proper design and voltage application, the amplitude is modulated by selecting voltages corresponding to constructive or destructive interference between the two paths. In the directional coupler, two waveguides in proximity can exchange energy over their length through evanescent coupling of the guided modes. The length for energy transfer between the waveguides depends on the index of refraction, so the switching or modulating action is obtained by adjusting the optical phase using the electrooptic effect. This device is well developed in $LiNbO_3$ [106].

Another important device, used, for example, in optical gyroscopes is the transverse phase modulator [16], shown for two electrode configurations in Fig. 21. The behavior of this device is particularly simple and is suitable for discussing the material properties and device considerations which impact on all electrooptic devices. The phase change induced by the electric field is given by

$$\Delta\phi = \frac{\pi}{\lambda} n^3 r El \qquad\qquad (36)$$

where E is the applied electric field and l is the length of the electrode. For the electrode geometry in Fig. 21b, the electric field is V/t, where V is the applied voltage and t is the thickness. The tensor nature of the electrooptic coefficient has been ignored for simplicity. The amount of modulation for a given voltage is determined by the geometry, the index of refraction, and the electrooptic coefficient. In general, the electrooptic coefficients of organic and polymeric materials compare favorably with inorganic materials. The index of refraction is generally somewhat smaller in organics, but a larger electrooptic coefficient will produce comparable phase shifts and index-match more closely to optical fibers. Further, the trends in poled polymer films indicate that the electrooptic coefficients will increase substantially in the future. An important aspect of an electrooptic film technology which cannot be attained in a bulk crystal approach is the ability to form layered structures, which allow modulators in the configuration of Fig. 21b to be formed

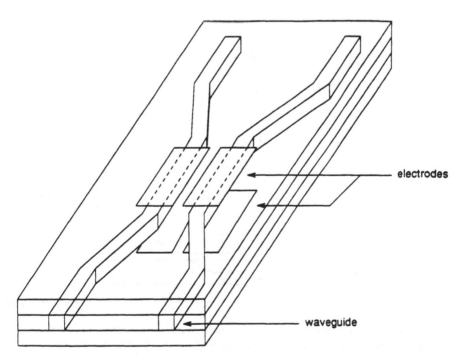

Figure 20.   Schematic of electrooptic directional coupler switch.

(a)

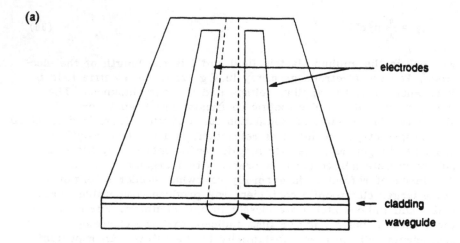

(b)

Figure 21. Schematic of transverse phase modulator: (a) bulk crystal electrode geometry; (b) thin-film electrode geometry.

as opposed to Fig. 21a. It has been shown that a several-fold increase in the phase shift can be obtained in the Fig. 21b geometry over the Fig. 21a geometry [107]. Thus, the geometric fabrication capability of polymers will improve device performance.

The true advantage of polymeric electrooptic materials is expected to come in high-speed performance. It has been shown previously that the electrooptic effect in organic crystals and polymers is mainly due to the electronic structure [5]. Thus, the nonlinear optical response is expected to be in the subpicosecond regime. The speed of electrooptic devices will be limited, then, by the modulating field. In the simplest case, the electrooptic device appears as a lumped RC element to the modulating circuit [16]. In Fig. 21b, one can show that the modulation power to attain a peak retardation Γ_m is given by

$$P = \frac{\Gamma_m^2 \, \lambda^2 \, A t \varepsilon \, \Delta \nu}{4 \pi l^2 \, n_0^6 \, r^2} \tag{37}$$

where A is the cross-sectional area of the electrodes, l is the length of the device, and $\Delta \nu$ is the modulating bandwidth. Thus, the modulating power per unit bandwidth is proportional to ε and inversely proportional to $n_0^3 \, r^2$. Since one wishes to minimize the modulating power, a figure of merit for high-speed modulation of a lumped modulator is $n_0^3 \, r/\sqrt{\varepsilon}$. Since the dielectric constant of many polymers is a factor of 10 smaller than ferroelectrics, such as $LiNbO_3$, polymers would be appropriately faster when in the lumped element regime. The reduction in bandwidth due to the higher capacitance of the geometry of Fig. 21b does increase the modulating power, but this is more than offset by the increased field of this geometry; thus the power is linear in the t. One must take care to minimize the electrode area A, though, in designing such a device.

True high-speed devices do not occur in the lumped element limit, and one must consider the transit time of both the optical and modulating waves [16]. If the modulating field is assumed to be uniform across the device during the transit of the optical wave, the modulating field will change during the transit of the light according to the time dependence of the modulating field. If the transit time is τ_d and the modulating field is sinusoidal with angular frequency ω_m, then the peak modulation is reduced by a factor $r = (1 - e^{-i\omega_m \tau d})/i\omega_m \tau_d$. This can limit useful retardation to the gigahertz regime. This can be overcome by designing a modulator where both the modulating and optical fields are traveling waves, so that the optical wave can, in principle, keep up with the modulating

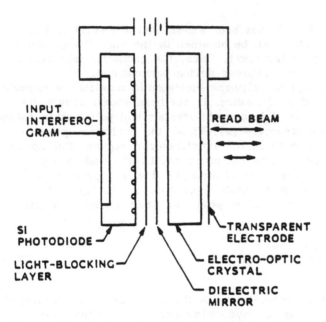

Figure 22.   Schematic of electrooptic spatial light modulator. (From Ref. 79, used with permission.)

wave. In this case, both waves are integrated over their propagation across the crystal. The retardation is then given by

$$\Gamma(t) = \Gamma_0 r = \Gamma_0 e^{i\omega_m t} \frac{e^{i\omega_m \tau_d(1-c/nc_m)} - 1}{i\omega_m \tau_d(1-c/nc_m)} \quad (38)$$

where c/n is phase velocity of the optical wave and $c_m = c/\sqrt{\varepsilon}$ is the phase velocity of the modulating wave (c is the speed of light in vacuum). If the maximum modulating frequency is taken as that for which $\omega_m \tau_d(1-c/nc_m) = \pi/2$, then

$$(\nu_m)_{max} = \frac{c}{4nl(1-c/nc_m)} \quad (39)$$

Based on Eq. (39), a figure of merit for the bandwidth or useful device length is $1/(\sqrt{\varepsilon}-n)$. This figure can be an order of magnitude larger for polymers than for $LiNbO_3$.

Another potential application of organic or polymeric materials is an electrooptic spatial light modulator as depicted in Fig. 22 [79]. This employs a photoreceptor whose accumulated charge modulates light through the adjacent electrooptic material. The low dielectric constant of polymers and organics will allow maximum voltage for a given charge, thus leading to a more efficient device.

V. CONCLUSION

Much of the early investigation of the underlying mechanisms responsible for the nonlinear optical susceptibilities of organic molecules and crystals has yielded insight which has allowed optimization of both molecular and crystal structures for second-order nonlinear optical materials. No doubt new insights await discovery which will lead to new avenues of investigation. In parallel with the rich variety of molecular and material structures which wait for further study, research into the behavior of organic and polymeric device structures is well underway. The first useful crystal parametric devices have been developed, as well as the initial poled polymer waveguide electrooptic devices. These two promising lines of investigation will no doubt continue. In addition, the use of both crystals and poled polymers in guided-wave optical parametric devices needs to be investigated, as no efficient devices have yet been demonstrated. The inclusion of device researchers in the interdisciplinary groups now working in the field will produce fruitful interactions benefiting the science and technology of organic nonlinear optical materials.

ACKNOWLEDGMENTS

The authors gratefully acknowledge useful comments from J. Zyss, M. G. Kuzyk, and H. C. Ling.

REFERENCES

1. "Nonlinear Optical Properties of Materials," C. M. Bowden, J. Haus, eds. *J. Opt. Soc. Am. B, 6* (1989).
2. A. M. Glass, *Mat. Res. Bull., 13*:16 (1988).
3. D. S. Chemla and J. Zyss, eds., *Nonlinear Optical Properties of Organic Molecules and Crystals*, Academic Press, Orlando, 1987.
4. A. J. Heeger, J. Orenstein, and D. R. Ulrich, ed., *Nonlinear Optical Properties of Polymers*, Vol. 109, Materials Research Society, Pittsburgh, 1988.
5. K. D. Singer, S. J. Lalama, J. E. Sohn, and R. D. Small, *Nonlinear Optical Properties of Organic Molecules and Crystals*, D. S. Chemla and J. Zyss, eds., Academic Press, Orlando, 1987, p. 437.

6. C. Flytzanis. In *Quantum Electronics: A Treatise*, H. Rabin and C. L. Tang, eds., Academic Press, New York, 1975, p. 9.
7. Y. R. Shen, *The Principles of Nonlinear Optics*, Wiley, New York, 1984.
8. M. Schubert and B. Wilhelmi, *Nonlinear Optics and Quantum Electronics*, New York, 1986.
9. J. F. Nye, *Physical Properties of Crystals*, Oxford University Press, London, 1957.
10. D. A. Kleinman, *Phys. Rev.*, *126*:1977 (1962).
11. I. P. Kaminow, *An Introduction to Electrooptic Devices*, Academic Press, New York, 1974.
12. T. K. Ye and T. K. Gustafson, *Phys. Rev.*, *A 18*:1597 (1978).
13. M. G. Kuzyk, Ph.D. dissertation, University of Pennsylvania, 1985.
14. B. J. Orr and J. F. Ward, *Mol. Phys. 20*:513 (1971).
15. J. F. Ward and P. A. Franken, *Phys. Rev.*, *133*:A183 (1964).
16. A. Yariv, *Quantum Electronics*, Wiley, New York, 1989.
17. M. Sigelle and R. Hierle, *J. Appl. Phys. 52*:4199 (1981).
18. K. D. Singer, M. G. Kuzyk, and J. E. Sohn, *J. Opt. Soc. Am.*, *B 4*:968 (1987).
19. K. D. Singer, M. G. Kuzyk, W. R. Holland, J. E. Sohn, S. J. Lalama, R. B. Comizzoli, H. E. Katz, and M. L. Schilling, *Appl. Phys. Lett.*, *53*:1800 (1988).
20a. R. M. Rentzepis and Y. H. Pao, *Appl. Phys. Lett.*, *5*:156 (1964).
20b. G. H. Heilmeir, N. Ockman, R. Braunstein, and D. A. Kramer, *Appl. Phys. Lett. 5*:229 (1964).
21. B. L. Davydov, L. D. Derkacheva, V. V. Dunina, M. E. Zhabotinskii, V. F. Zolin, L. G. Koreneva, and M. A. Samokhina, *JETP Lett. (Eng.)*, *12*:16 (1970).
22. G. R. Meredith, J. G. Van Dusen, and D. J. Williams, *Macromolecules, 15*:1385 (1982).
23. D. J. Williams, *Angew. Chem. Intl. Ed. Engl.*, *23*:690 (1984).
24. K. D. Singer, J. E. Sohn, S. J. Lalama, and R. D. Small, *SPIE Proc.*, *704*:240 (1986).
25. D. S. Chemla, J. L. Oudar, and J. Jerphagnon, *Phys. Rev. B, 112*:4534 (1975).
26. J. D. Legrange, M. G. Kuzyk, and K. D. Singer, *Mol. Cryst. Liq. Cryst.*, *150b*:567 (1987).
27. S. J. Lalama and A. F. Garito, *Phys. Rev. A, 20*:1179 (1979).
28. J. L. Oudar and D. S. Chemla, *J. Chem. Phys.*, *66*:2664 (1977).
29. K. D. Singer, J. E. Sohn, L. A. King, H. M. Gordon, H. E. Katz, and C. W. Dirk, *J. Opt. Soc. Am. B, 6*:1339 (1989).
30. C. C. Teng and A. F. Garito, *Phys. Rev. Lett.*, *50*:350 (1983).
31. H. E. Katz, K. D. Singer, J. E. Sohn, C. W. Dirk, L. A. King, and H. M. Gordon, *J. Am. Chem. Soc.*, *109*:6561 (1987).

32. S. K. Kurtz and T. T. Perry, *J. Appl. Phys.*, *39*:3798 (1968).
33. J. F. Nicoud and R. J. Twieg, *Nonlinear Optical Properties of Organic Molecules and Crystals*, D. S. Chemla and J. Zyss, eds., Academic Press, Orlando, 1987.
34. J. P. Dougherty and S. K. Kurtz, *J. Appl. Cryst.*, *9*:145 (1976).
35. A. Coda and F. Pandarese, *J. Appl. Cryst.*, *9*:193 (1976).
36. W. R. Holland, to be published.
37. J. L. Oudar, *J. Chem. Phys.*, *67*:446 (1977).
38. K. D. Singer and A. F. Garito, *J. Chem. Phys.*, *75*:3572 (1981).
39. B. F. Levine and C. G. Bethea, *J. Chem. Phys.*, *63*:2666 (1975).
40. S. J. Lalama, K. D. Singer, A. F. Garito, and K. N. Desai, *Appl. Phys. Lett.*, *39*:940 (1981).
41. J. E. Sohn, K. D. Singer, and M. G. Kuzyk, *Polymers for High Technology—Electronics and Photonics*, M. J. Bowden and S. R. Turner, eds., American Chemical Society, Washington, D.C., 1987, p. 401.
42. H. E. Katz, K. D. Singer, J. E. Sohn, C. W. Dirk, L. A. King, and H. M. Gordon, *J. Am. Chem. Soc.*, *109*:6561 (1987).
43. R. A. Huijts and G. L. J. Hesselink, *Chem. Phys. Lett.*, *156*:209 (1989).
44. J. O. Morley, V. J. Docherty, and D. Pugh, *J. Chem. Soc. Perkin Trans. II*, *6*:1351 (1987).
45. J. Zyss and J. L. Oudar, *Phys. Rev. A*, *26*:2028 (1982).
46. J. Zyss and D. S. Chemla, *Nonlinear Optical Properties of Organic Molecules and Crystals*, D. S. Chemla and J. Zyss, eds., Academic Press, Orlando, 1987.
47. J. F. Nicoud, *SPIE Proc.*, *971*:68 (1988).
48. P. A. Norman, D. Bloor, J. S. Obhi, S. A. Karaulov, M. B. Hursthouse, P. V. Kolinsky, R. J. Jones, and S. R. Hall, *J. Opt. Soc. Am. B*, *4*:1013 (1987).
49. J.-C. Baumert, R. J. Twieg, G. C. Bjorklund, J. A. Logan, and C. W. Dirk, *Appl. Phys. Lett.*, *51*:1484 (1987).
50. R. T. Bailey, F. R. Cruickshank, S. M. G. Guthrie, B. J. McArdle, G. W. McGillivray, H. Morrison, D. Pugh, E. A. Shepherd, J. N. Sherwood, C. S. Yoon, R. Kashyap, B. K. Nayar, and K. I. White, *SPIE Proc.*, *971*:76 (1988).
51. T. Kondo, N. Ogasawara, S. Umegaki, and R. Ito, *SPIE Proc.*, *971*:83 (1988).
52. S. Guha, C. C. Frazier, and W. Chen, *SPIE Proc.*, *971*:89 (1988).
53. Ch. Bosshard, K. Sutter, P. Gunter, and G. Chapuis, *J. Opt. Soc. Am. B*, *6*:721 (1989).
54. D. Chen, N. Okamoto, and R. Matsushima, *Opt. Commun.*, *69*:425 (1989).

55. S. Hayashida, H. Sato, and S. Sugawara, *Jpn. J. Appl. Phys.*, *24*:1436 (1985).
56. S. P. Velsko, L. E. Davis, F. Wang, and D. Eimerl, *SPIE Proc.*, *971*:113 (1988).
57. T. Kurihara, T. Kaino, S. Matsumoto, and S. Tomaru, *Jpn. J. Appl. Phys.*, *27*:2082 (1988).
58. K. Takagi, M. Ozaki, K. Nakatsu, M. Matsuoka, and T. Kitao, *Chem. Lett. 173* (1989).
59. M. L. H. Green, S. R. Marder, M. E. Thompson, J. A. Bandy, D. Bloor, P. V. Kolinsky, and R. J. Jones, *Nature, 330*:360 (1987).
60. J. W. Perry, A. E. Stiegman, S. R. Marder, D. R. Coulter, D. N. Beratan, D. E. Brinza, F. L. Klavetter, and R. H. Grubbs, *SPIE Proc.*, *971*:17 (1988).
61. T. Nogami, H. Nakano, Y. Shirota, S. Umegaki, Y. Shimizu, T. Uemiya, and N. Yasuda, *Chem. Phys. Lett.*, *155*:338 (1989).
62. C. C. Frazier, M. A. Harvey, M. P. Cockerham, H. M. Hand, E. A. Chauchard, and C. H. Lee, *J. Phys. Chem.*, *90*:5703 (1986).
63. H. Abdel-Halim, D. O. Cowan, D. W. Robinson, F. M. Wiygul, and M. Kimura, *J. Phys. Chem.*, *90*:5654 (1986).
64. D. W. Robinson, H. Abdel-Halim, S. Inoue, M. Kimura, and D. O. Cowan, *J. Chem. Phys.*, *90*:3427 (1989).
65. S. Aratani, M. Isogai, and A. Kakuta, *Chem. Lett.*, 741 (1988).
66. I. Ledoux, D. Josse, J. Zyss, T. McLean, P. F. Gordon, R. A. Hann, and S. Allen, *J. Chim. Phys.*, *85*:1085 (1988).
67a. J. Huang, A. Lewis, and Th. Rasing, *J. Phys. Chem.*, *92*: 1756 (1988).
67b. J. Y. Huang, Z. Chen, and A. Lewis, *J. Phys. Chem.*, *93*:1756 (1989).
68. S. Allen, T. D. McLean, P. F. Gordon, B. D. Bothwell, P. Robin, and I. Ledoux, *SPIE Proc.*, *971*:206 (1988).
69. K. D. Singer, J. E. Sohn, and S. J. Lalama, *Appl. Phys. Lett.*, *49*:248 (1986).
70. S. Kielich, *IEEE J. Quant. Electron. QE-4*:562 (1969).
71. K. D. Singer, M. G. Kuzyk, and J. E. Sohn, *Nonlinear Optical and Electroactive Polymers*, P. N. Prasad and D. R. Ulrich, eds., Plenum, New York, 1988, p. 189.
72. C. Ye, T. J. Marks, J. Yang, and G. K. Wong, *Macromolecules*, *20*:2324 (1987).
73. C. Ye, N. Minami, T. J. Marks, J. Yang, and G. K. Wong, *Macromolecules, 21*:2901 (1988).
74. H. L. Hampsch, J. Yang, G. K. Wong, and J. M. Torkelson, *Polym. Commun.*, *30*:40 (1989).
75. M. Eich, A. Sen, H. Looser, D. Y. Yoon, G. C. Bjorklund, R. Twieg, and J. D. Swalen, *SPIE Proc. 971*:128 (1988).

76. S. Esselin, P. LeBarny, P. Robin, D. Broussoux, J. C. Dubois, J. Raffy, and J. P. Pocholle, *SPIE Proc.*, *917*:120 (1988).

77. M. A. Mortazavi, A. Knoesen, S. T. Kowel, B. G. Higgins, and A. Dienes, *J. Opt. Soc. Am.*, B *6*:733 (1989).

78. J. Thackara, M. Stiller, G. F. Lipscomb, A. J. Ticknor, and R. Lytel, *Appl. Phys. Lett.*, *52*:1031 (1988).

79. R. Lytel, G. F. Lipscomb, M. Stiller, J. Thackara, A. J. Ticknor, *SPIE Proc.*, *917*:218 (1988).

80. Y. Wang, *Chem. Phys. Lett.*, *126*:209 (1988).

81a. T. Watanabe, K. Yoshinaga, D. Fichou, and S. Miyata, *J. Chem. Soc., Chem. Commun.*, *250* (1988).

81b. T. Watanabe, K. Yoshinaga, D. Fichou, Y. Chatani, and S. Miyata, *Mat. Res. Soc. Symp. Proc.*, *109*:339 (1988).

82. H. Daigo, N. Okamoto, and H. Fujimura, *Optics Commun.*, *69*:177 (1988).

83. Y. Wang, W. Tam, S. H. Stevenson, R. A. Clement, and J. C. Calabrese, *Chem. Phys. Lett.*, *148*:136 (1988).

84. W. Tam, Y. Wang, J. C. Calabrese, and R. A. Clement, *SPIE Proc.*, *971*:107 (1988).

85a. Y. Wang and D. F. Eaton, *Chem. Phys. Lett.*, *120*:441 (1985).

85b. D. F. Eaton, A. G. Anderson, W. Tam, and Y. Wang, *J. Am. Chem. Soc.*, *109*:1886 (1987).

86. S. D. Cox, T. E. Gier, G. D. Stucky, and J. Bierlein, *J. Am. Chem. Soc.*, *110*:2986 (1988).

87. J. Badan, R. Hierle, A. Perigaud, and P. Vidakovic, *Nonlinear Optical Properties of Organic Molecules and Crystals*, D. S. Chemla and J. Zyss, eds., Academic Press, Orlando, 1987.

88. S. Tomaru and S. Zembutsu, *Prepr. 2nd Soc. Polym. Sci. Jpn. Intl. Polym. Conf. 2*, 145 (1986).

89. P. Kerkoc, Ch. Bosshard, H. Arend, and P. Gunter, *Appl. Phys. Lett.*, *54*:487 (1989).

90. I. Ledoux, J. Badan, J. Zyss, A. Migus, D. Hulin, J. Etchepare, G. Grillon, and A. Antonetti, *J. Opt. Soc. Am.*, B *4*:987 (1987).

91. I. R. Peterson, *Angew. Chem. Int. Ed. Engl.*, *27*:1215 (1988).

92. I. Ledoux, J. Zyss, A. Migus, D. Hulin, and A. Antonetti, *J. Appl. Phys.*, *64*:3309 (1988).

93. J. Zyss, *J. Molec. Electron*, *1*:25 (1985).

94. G. I. Stegeman and C. T. Seaton, *J. Appl. Phys.*, *58*:R57 (1985).

95. G. I. Stegeman, C. T. Seaton, and R. Zanoni, *Thin Solid Films*, *152*:23 (1987).

96. H. Itoh, K. Hotta, H. Takara, and K. Sasaki, *Appl. Optics*, *25*:1491 (1986).

97. W. A. Burns and R. A. Andrews, *Appl. Phys. Lett.*, *22*:143, 554 (1973).

98. H. Ito and H. Inaba, *Opt. Lett.*, *2*:139 (1978).
99. K. I. White and B. K. Nayar, *J. Opt. Soc. Am. B*, *5*:317 (1988).
100. B. K. Nayar, *ACS Symp. Ser.*, *233*:153 (1983).
101. S. Somekh and A. Yariv, *Opt. Commun.*, *6*:301 (1972).
102. B. F. Levine, C. G. Bethea and R. A. Logan, *Appl. Phys. Lett.*, *26*:375 (1975).
103. J. P. van der Ziel, M. Ilegems, P. W. Foy, and R. M. Mikulyak, *Appl. Phys. Lett.*, *29*:775 (1976).
104. P. Cahill, K. D. Singer, and L. A. King, *Optics Lett.*, *14*: 1137 (1989).
105. R. D. Small, J. E. Sohn, K. D. Singer, and M. G. Kuzyk, *Photonic Switching*, T. K. Gustafson and P. W. Smith, eds., Springer-Verlag, Berlin, 1988.
106. R. C. Alferness, *IEEE J. Quant. Electron.*, *QE-17*:191 (1981).
107. R. V. Mustacich, *Appl. Opt.*, *27*:3732 (1988).

3

Polymeric Materials for Third-Order Nonlinear Optical Susceptibility

Luping Yu, David W. Polis, Malcolm R. McLean, and Larry R. Dalton
University of Southern California, Los Angeles, California

I. OVERVIEW

In this chapter we focus upon third-order nonlinear optical materials.
This is an intriguing topic and certainly one that it is difficult to
do justice to in the space available here. Fortunately, entire books
devoted to this topic are now available or in preparation and will
provide a more in-depth coverage [1-4]. However, writing this
chapter is a challenge not only because of the number of concepts
(theory of optical nonlinearity, material synthesis, material processing, material characterization, measurement theory, application theory,
device fabrication) but also because of the state of the field. Unlike

second-order materials, the prognosis for commercial application of third-order materials is not clear [1]. It is generally conceded that substantial improvement in material properties such as optical nonlinearity, optical transparency, laser damage characteristics, and processibility must be achieved before serious device development can occur. Definition of material quality factors for various applications such as optical computing or sensor protection has proven controversial, with various factors being ignored or at least downplayed. If the device designers are uncertain as to their minimum requirements, the situation involving characterization of optical nonlinearity is even more confusing. Very little quantitative theoretical background exists for the various measurement techniques (degenerate four-wave mixing (DFWM), third-harmonic generation; (THG), etc.); thus, comparison of data from different experiments is often difficult. Moreover, few extensive data sets with measurements made as a function of frequency, pulse conditions, or material morphology exist. Thus, when optical nonlinearity is discussed, it is not clear what mechanisms are operative or what relative contributions are due to resonant and nonresonant processes. With only limited data available, it is impossible to make a meaningful survey of the field at this point in time or to predict if and when materials with appropriate device characteristics can be realized. With this in mind, a focus of this chapter will be discussion of the difficulties encountered in preparation and characterization new nonlinear optical materials. However, significant advances in material properties have been realized in the past several years and these will be reviewed with emphasis into the insight that the results provide for the future development of improved materials. In the following, we provide a brief overview of the difficulties encountered in the design, development, and application of third order materials. After setting this perspective we turn our attention to a more detailed discussion of each of the concepts introduced, namely, theory, synthesis, characterization, processing, and device fabrication.

II. GENERAL CONSIDERATIONS

Ideally, quantum mechanics would guide the design of new nonlinear optical materials. However, for polymeric materials containing delocalized π-electrons the computational problem is too large to be handled by ab initio methods. Moreover, it can be argued that both electron-phonon and electron Coulomb correlation terms will give rise to optical nonlienarity, e.g., multiphoton effects in the case of Coulomb correlation. Thus, the theorist is forced to adopt an approximate treatment of the problem with realization that the nature of the approximation will preclude possible mechanisms of optical nonlinearity. The two most general approaches have been (1) to consider a static nuclear lattice (Born-Oppenheimer adiabatic approximation) and attempt to treat Coulomb correlations at various levels of approximation defined by the size of the system, and (2) to consider structural relaxation within the framework of complete neglect of Coulomb effects. It is interesting to note that these two methods can give rise to opposite signs for the third-order susceptibility. Thus it is important to consider sign as well as magnitude when comparing theoretical with experimental data. Moreover, it can be noted that both of these approaches typically neglect intermolecular electronic interactions and thus fail to predict some excitonic contributions. Since a given theoretical calculation cannot be considered satisfactory without detailed correlation with experimental data, care must be taken to make experimental measurements in a way to permit meaningful correlation with theory. For example, resonant and nonresonant contributions to third-order susceptibility must be defined by measurements performed as a function of frequency. Since third-order susceptibility is an additive property, an effort must be made to separate intra- and intermolecular contributions to optical nonlinearity and to define structural disorder effects. This is particularly true for measurements made near the band edge where excitonic, $\pi-\pi^*$, $n-\pi^*$, and charge transfer transitions may overlap. It is also important to define real and imaginary contributions to optical nonlinearity. Various approximate theoretical approaches have been useful in defining certain structure/function relationships. For example, third-order susceptibility has been predicted to increase with the length of the π system [1-13]. Moreover, it is generally believed from theoretical work that this increase does not occur without limit due to self-localization in the π-electron system (arising from the competition of electron-phonon and electron Coulomb interactions). Within a given theoretical framework it is possible to investigate substituent effects, e.g., to consider the effect of adding electron donating or withdrawing groups such as is shown in Fig. 1. Such calculations may be useful in alerting the synthetic chemist to potentially interesting derivatizations, but quantitative accuracy should be suspect because of the theoretical approximations involved.

The calculations of de Melo and Silbey [14] in turn indicate the potential changes in optical nonlinearity anticipated with changes in the redox state of the polymer. While preliminary qualitative agreement has been found between theory and experiment in predicting the above dependences, e.g., third-order susceptibility is observed to increase from 4.0×10^{-10} to 2.4×10^{-9} esu in going from poly-p-phenylenevinylene (PPV) to dimethoxy-PPV [15], there are

Acceptor Position	β	γ*
1	190	64.3
2	59.9	30.4
3	115	24.8

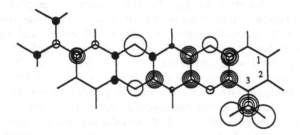

Figure 1.   Variation of hyperpolarizabilities β and γ for aminonitro-substituted ladder oligomers as a function of the position of the nitro substituent.

some disturbing discrepancies; e.g., Prasad has observed third-order susceptibility to decrease (rather than increase as predicted by theory) as the polaron states of polythiophene are populated [16]. Thus, while theory may suggest trends to be investigated, quantitative validity will be only as good as the particular approximate theoretical approach.

It has long been recognized that π-electron containing materials exhibit interesting optical nonlinearity. It is intuitively reasonable that the π-orbitals will also affect polymer intermolecular interactions (van der Waal forces) and thus polymer physical properties such as solubility. This speculation was quickly confirmed by early synthetic efforts producing linear polyene and heteroaromatic polymers. Poor solubility has created a number of problems, including making characterization and processing of these polymers difficult to impossible. When such electroactive polymers are prepared by polycondensation syntheses, pool solubility results in low-molecular-weight polymers which in turn yield films characterized by poor mechanical properties. It is interesting that one of the claimed advantages of polymers relative to inorganic materials is that polymers can be easily fabricated into a variety of forms. This concept derives from the consideration of traditional polymers, but good processibility is not necessarily a property of electroactive polymers. Indeed, for good solubility and processibility to be realized for delocalized π-electron polymers it is necessary to destabilize polymer-polymer interactions and/or to enhance polymer-solvent interactions. In developing materials with improved solubility it is also important to consider entropy

factors, i.e., the entropy of mixing. In the past several years
a number of synthetic approaches have been developed to effect
improved solubility. Derivatization of polymer backbones has been
utilized to improve solubility for polymers ranging from polythiophene
[17-19] to heteroatomic ladder polymers [20-24]. Another successful
approach has been the synthesis of electroactive polymers via soluble
precursor polymer routes followed by thermal conversion of the
precursor polymer to the final electroactive form [20-28]. An example
of such a synthetic scheme is shown in Figs. 2 and 3a.

In addition to improving processing options, solubility permits
improved structural characterization of polymers which is important
in defining structure/function relationships. Obviously, to develop
useful structure/function relationships, the polymer structure must
be known. New synthetic approaches have also proven useful in
providing control of properties ranging from optical transparency
to laser damage threshold. For example, the optical band gap can
be controlled by synthesizing copolymer systems containing delocalized
polymer segments of defined length [28] as shown in Fig. 3a. Such
an approach is useful in avoiding a distribution of band gaps
associated with different polymer lengths; this prevents tailing
of the optical spectra toward the infrared which is in turn useful
in avoiding the contribution of resonant optical nonlinearity for
measurements made within the band gap (Fig. 3b). The synthetic
versatility realized in the past several years is impressive and has
permitted an extensive variety of processing options to be pursued
and some insight into structure/function relationships to be explored.
Discussion of synthetic concepts shall be a major focus of this review.
In addition to discussing the synthesis of materials currently being
examined for optical nonlinearity, we shall review the synthesis of
prototypical polymers of interest for future investigation.

One of the major problems in discussing optical nonlinearity
is correlating results deriving from various experiments. Among
the most common means of measuring third-order susceptibility are
DFWM and THG. DFWM measures $\chi^{(3)}(\omega;\omega,-\omega,\omega)$ and is sensitive
to both electronic and thermal contributions. In this experiment
only a single wavelength is involved and one need consider optical
loss at only this one frequency. The time evolution of the DFWM
signal and its wavelength dependence provide insight into fundamental
nonlinear processes. It is also important to investigate the dependence
of signals upon pulse conditions, such as pulse widths, that must
be explicitly taken into account (for example, by density matrix
methods [29,30]) in quantitatively analyzing the DFWM signals.

Third-harmonic generation measures $\chi^{(3)}(3\omega;\omega,\omega,\omega)$ which is
directly related to the electronic component of $\chi^{(3)}(\omega;\omega,-\omega,\omega)$.
Obviously, two frequencies must be considered in the THG experi-

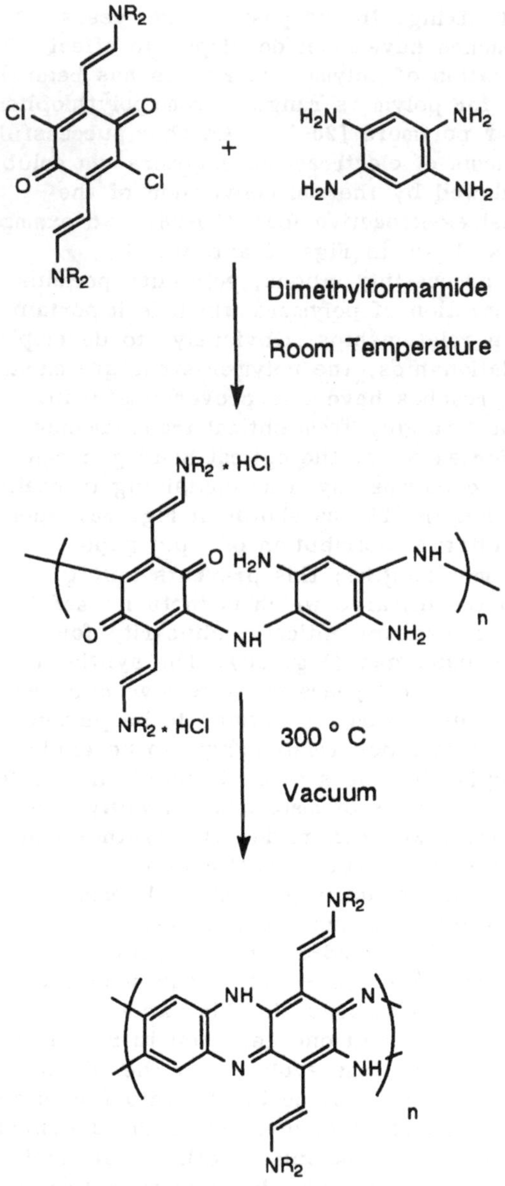

Figure 2. Condensation between a substituted quinone and tetra-aminobenzene to form a soluble precursor followed by thermal annealing to form the desired ladder polymer. R = ethyl, propyl, cyclohexyl, and tetrahydropyryl.

ment, and optical losses at both frequencies must be taken into consideration. If polymer films on substrates are being evaluated, it is often necessary to use different substrates for DFWM and THG experiments. For example, a glass slide may suffice for the DFWM experiment but may result in too much optical loss for the THG experiment. It is important to realize that DFWM signals depend critically upon instrumental as well as molecular conditions. Both details must be explicitly considered in theoretical analysis, and in general, instrumental conditions such as beam polarizations, pulse delays, and pulse widths (and shapes) must be controlled and varied to separate various contributions to the DFWM signal. In addition to DFWM [31,32] and THG [33-35] measurements of third-order susceptibility, a variety of experiments such as time-resolved pump-probe transmission [36], time-resolved Kerr [37], electric-field-induced second-harmonic generation (EFISH) [38,39], waveguide coupling [40], and four-wave mixing (FWM, nondegenerate) [41] experiments provide useful insight into optical nonlinearity.

Third-order susceptibility is a tensorial property and advantage can be taken of this orientation dependence providing solid-state order can be achieved [42]. Two of the most obvious approaches to achieving molecular order are to employ Langmuir-Blodgett methods to produce aligned thin films of polymers and to fabricate ordered thin films of polymers by processing from liquid crystalline solutions of polymers. Each of these approaches is characterized by a number of practical difficulties and most of the systems investigated to date have been examined as amorphous films.

No commercial applications of organic third-order materials have been realized yet and considerable controversy exists as to the types of devices which are likely to be the most amenable to the utilization of these materials. Controversy also exists as to the definition of material quality factors which can be used to guide the development of new materials. Most quality factors involve the ratio of optical nonlinearity (third-order susceptibility or index modulation) to optical loss. On the basis of this type of factor it has been argued that the best currently available nonlinear optical material is ordinary glass, as this yields a better quality factor than either multiple-quantum-well semiconductors or polymers such as polydiacetylene. Indeed, this might be the case for long waveguide structures; however, for applications such as optical computing or sensor protection where size limitations are serious device considerations, this argument is fallacious. Moreover, it is not clear what type of optical nonlinearity should be exploited for specific devices such as limiters. For example, should fast response nonresonant optical nonlinearity be utilized or should a phenomenon such as saturable absorption of materials (such as silicon naphthalocyanine as proposed by Garito and co-workers [13]) be used?

Where R = -(CH₂)₅- or -(CH₂-O-CH₂)₅-

(a)

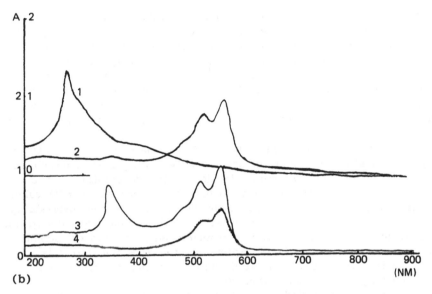

(b)

Figure 3. (a) Synthesis of prepolymer and polymer a and b.
(b) UV/Vis spectra of (1) the prepolymer, (2) polymer a,
(3) polymer b and (4) the model compound, 3,10-dimethoxy-7,9-
dichlorophenodioxazine, to show the control of optical band gap
by copolymer synthesis.

III. THEORY OF THIRD-ORDER OPTICAL NONLINEARITY

Reduced dimensionality and quantum confinement of delocalized
electrons have been subjects of increasing interest in condensed
matter theory. These concepts have been important for understanding optical nonlinearity for both semiconducting quantum-well hetero-
structures and delocalized electron polymer structures. In the case
of quantum-well structures, confined resonant excitons contribute
to a large intensity dependent refractive index. In the case of
π-electron systems of reduced dimensionality, the motion of electrons
is confined and the motion becomes highly correlated as a result
of the existence of electron Coulomb and electron-phonon interactions.
Coulomb interactions give rise to characteristic features such as
multiple quantum contributions to the third-order susceptibility.

Initial calculations of third-order optical nonlinearity employed
independent particle theories, such as Huckel theory [2,5-9]. Although such theories succeeded in predicting a dependence of
optical nonlinearity upon the extent of the interacting π-electron

system, it is very unlikely that such calculations would be quantitatively correct. For example, for linear polyenes such as polyacetylene, independent particle theories neglect the contribution made to optical nonlinearity by excited A_g states. Clearly this is not appropriate as the two-photon $2\,^1A_g$ state is observed for many polyenes to lie below the one-photon $1\,^1B_u$ state and it has been demonstrated that a proper account of electron correlation arising from electron Coulomb interactions is necessary to obtain the correct ordering of these states.

Ab initio self-consistent field (SCF) calculations of molecular hyperpolarizabilities can be classified in two categories, corresponding to different levels of approximation in the description of the perturbed systems. In the finite-field (FF) approach, an SCF calculation is made on the perturbed molecule and the hyperpolarizability is obtained by numerical differentiation of the dipole moment expectation value with respect to the appropriate electric-field components. In the sum-over-states (SOS) approach, the SCF calculation is made on the unperturbed molecule and the hyperpolarizability is calculated by perturbation theory.

Recently, Garito and co-workers [12,13] employing an SOS approach, have incorporated Coulomb interactions by considering all singly and doubly excited π-electron configurations within the framework of an all-valence electron SCF molecular orbital method in the standard, rigid-lattice complete neglect of differential overlap/spectroscopic (CNDO/S) approximation. In calculations performed on polyene (cis and trans molecular conformations) segments of various lengths, Garito has observed excited A_g states to make important contributions to optical nonlinearity. The hyperpolarizability tensor was found to exhibit the following dependence on polymer chain length (for both cis and trans conformations):

$$\gamma_{xxxx} \propto L^{4.6} \quad (= 1.55 \times 10^{-35} \text{ esu for } \textit{trans}\text{-octatetraene, OT})$$

Reasonable agreement is obtained with experimental values for short-chain polyenes. It is interesting to note that experimental values of $\chi^{(3)} = 10^{-10}$ esu observed for Shirakawa polyacetylene samples would suggest an effective π-electron delocalization length of 60 Å. Since the polymer chains are known to be longer than this, it is reasonable that self-localization occurs in longer-chain systems. This result is also suggested by electron nuclear double-resonance (ENDOR) magnetic resonance spectroscopic experiments [20,21], where finite hyperfine interactions define a self-localized valence electron wave function. The concept of a finite electron correlation length is relevant to the synthetic design of new electroactive polymers as it is not necessary to have long, uninterrupted conjugated segments to realize optimum optical nonlinearity. Indeed the π-electron portion of a copolymer need only exceed the self-localization length to yield optimum nonlinearity. Another relevant consideration when reviewing theoretical results for the purpose of designing new polymers is to keep in mind that optical band gap and third-order susceptibility are not independent. To work at a given wavelength it may be necessary to trade off some magnitude of optical nonlinearity to ensure adequate transparency [28,43,44].

Garito and co-workers [45] have also conducted a very important investigation of the effect on polymer symmetry on optical nonlinearity. Lowering of symmetry, for example by heteroatomic substitution, from centrosymmetric to noncentrosymmetric structures is expected to have an effect on optical properties by influencing both the electronic state structure and selections rules. To illustrate important concepts, Garito and co-workers have reported theoretical investigation of the nonlinear optical properties of 1,1-dicyano-8-N,N-dimethylamino-1,3,5,7-octatetraene (NOT). The lowered symmetry of NOT, as compared to OT, produces a new type of virtual excitation process which dominates the third-order hyperpolarizability; for example, at 0.65 eV (nonresonant limit) the third-order hyperpolarizability is found to be 17.3×10^{-35} esu, which is an order of magnitude larger than observed for centrosymmetric OT (1.55×10^{-35} esu). Analogous symmetry effects are demonstrated in Fig. 1. Clearly, such theoretical results provide important guidance to synthetic chemists in suggesting the importance of derivatization. Indeed, such substituent effects have already been observed for poly-p-phenylenevinylene, polythiophenes, and heteroaromatic ladder polymers.

Garito and co-workers [45] have also applied their theoretical approach to heteroaromatic (rigid-rod) polymers. For example, they calculated for $trans$-benzobisoxazole (t-BOZ), the fundamental unit of the poly-p-phenylene benzobisoxazole polymer, a value for the third-order hyperpolarizability, γ_{xxxx}, of 6.87×10^{-35} esu (at 0.65 eV). The virtual excitation series, $g \rightarrow 1\,^1B_u \rightarrow x\,^1A_g \rightarrow 1\,^1B_u \rightarrow g$, is the most important one for t-BOZ just as it was for the polyenes. Calculations such as those conducted by Garito and co-workers are useful in gaining insight into which electronic states and hence structural features are important for optical nonlinearity. It is interesting to note that such calculations may provide a useful guide for structure selection for the synthetic chemist. These results, together with the semiempirical theoretical results presented in Fig. 1, suggest that heteroaromatic polymers may be useful structures to consider.

Comparison of theoretical and experimental data is a difficult undertaking, as in many cases the relative contribution of resonant and nonresonant contributions are unknown. For example, Alfano and co-workers [46] report a third-order susceptibility of 10^{-9} esu for soluble polyacetylene (at 532 nm); however, this value clearly has a significant resonant component. As the amount of frequency-dependent susceptibility data increases, such correlation of theory and experiment becomes more realistic, and it is likely that currently available data support the contention that nonresonant third-order susceptibilities in the range 10^{-10} esu are quite realizable for organic polymers. Because of the increased size of the computational problem for organometallic polymers, very little useful quantitative theoretical guidance exists at this time, and one can at best extrapolate the concepts realized from the consideration of organic polymers.

One of the few theoretical investigations of organometallic systems reported to date is the second hyperpolarizability value of 0.61×10^{-35} esu calculated for ferrocene by Waite and Papadopoulous [47]. This value is less than the experimental value of 1.6×10^{-35} esu measured by DFWM by Prasad and co-workers [48] and considerably lower than the value determined by Winter et al. [49], using a power-limiter technique. The reasons for the discrepancies between theory and experiment are likely those alluded to above, namely, resonance contributions, contributions from thermal mechanisms, and the approximate nature of theory. One extremely important observation drawn by Prasad and co-workers in their study of a series of ferrocenes is that the metal is not effectively coupled to the π-conjugation; thus, ideas developed from organic π-electron systems appear applicable to these organometallic systems.

From the preceding discussion it should be clear to the reader that the work of Garito and others employing semiempirical and ab initio (for small molecules) methods is quite important. One further consideration, however, needs to be mentioned, namely, that the calculations are carried out in the rigid-lattice approximation. In low-dimensional materials, structural relaxation can potentially have profound effects upon electronic properties. Consider the electronic absorption spectra of the neutral and polaron/bipolaron states of a phenylpolyene shown in Fig. 4a. The former spectrum is charac-terized by a strong absorption in the visible and transparency in the infrared while the converse is true for the latter spectrum. These two spectra represent a fundamental mechanism of optical nonlinearity when it is realized that excitation across the band gap will produce electron-hole pairs which by structural relaxation can rapidly decay into polaronic species. Some qualitative support for this concept already exists from experimental observation of the dependence of optical nonlinearity on the redox state of polymers capable of supporting polaronic species. However, experimental data cannot be said to be in agreement with theory at this point

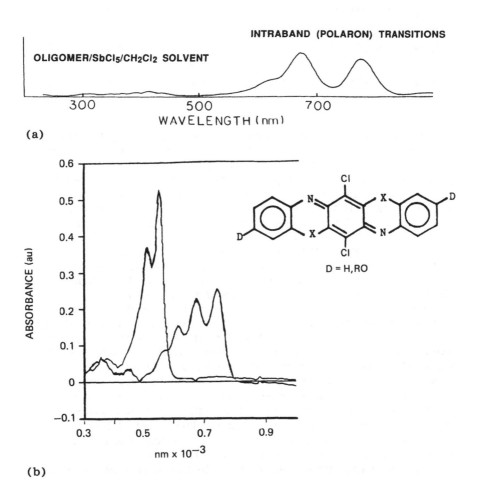

Figure 4. (a) Effect of chemical doping on the linear optical spectrum of a delocalized electron phenylpolyene oligomer: introduction
of visual transparency and generation of a charged(polaron/bipolaron)
lattice. (b) Electron absorption spectral of neutral and polaron/
bipolaron states of a ladder oligomer. (From Ref. 50.)

TABLE 1. Calculated Ionization Potential (IP), Bandwidth (BW), and Band Gap (Eg) for Selected Polymers

Polymer	IP (ev)	BW (eV)	Eg (eV)	Ref.
Polyacene	6.0	5.3	0.2	51
	5.8	5.9	0.0	52
Polypyridinopyridine	8.7	3.9	2.9	51
	8.7	4.3	3.0	52
Polypyrazinopyrazine	8.1	4.4	0.0	51
	8.1	5.1	0.0	53
Polyperinaphthalene	4.0	4.4	0.3	54

in time. An extreme case of this situation is given by the theoretical calculations of de Melo and Silbey [14], which suggest that third-order susceptibility increases on going from the neutral to the polaronic lattice and the experimental data of Prasad and co-workers [11], which suggests the opposite. Because of the approximate nature of current theoretical approaches and the limited availability of experimental data, little can be said other than the investigation of redox states of delocalized electron polymers is important for the experimental scientist and new techniques for considering both electron Coulomb interactions and structural relaxation must be developed by theorists before one can be confident of identifying all intramolecular mechanisms of optical nonlinearity.

We are still left with the effects of intermolecular interactions and structural disorder upon optical nonlinearity. Clearly, the potential for excitonic and charge transfer interactions to be important exists for π-electron polymers, and particularly for planar systems where close packing is sterically permitted. Structural disorder is a statistical reality in polymer systems by the very nature of the different synthetic processes. For polymers such as polyaniline where protonation can occur (and influence electronic properties) the situation is even more complicated. The effect of intermolecular interactions and structural disorder upon optical nonlinearity simply have not been adequately addressed at this time.

We would emphasize that we have in the preceding discussion provided an overview of theoretical methods and difficulties. The reader is simply referred to the growing literature on this subject for more detail and an introduction to the diversity of approaches being pursued. Also, the reader's attention is called to calculations of energy level structures of prototypical unsaturated polymers systems of interest for future theoretical and experimental NLO investigations. Some representative systems together with band gaps (related directly to electron delocalization) are given in Table 1.

IV. SYNTHESIS OF NEW ELECTROACTIVE MATERIALS

As already mentioned, one of the major problems which must be overcome in the development of new electroactive polymers is that of insolubility due to strong-interchain van der Waals interactions. Solubility is necessary both to permit processing of polymers into appropriate forms such as optical-quality thin films and to facilitate characterization of new polymers so that appropriate structure/function relationships can be defined. Three fundamental concepts have been utilized to overcome solubility problems in polymer systems ranging from polyacetylene [55] to heteroaromatic ladder polymers [20-24]. These are (1) the synthesis of soluble precursors followed by chemical or thermal treatment to attain an electroactive structure; (2) the use of steric interactions from substituent groups to destabilize polymer-polymer interactions relative to polymer-solvent interactions; and (3) the introduction of flexible-chain segments to improve the entropy of mixing.

One might ask if modifications of the polymer, for example, by side-chain derivatization, or alternative synthetic methods would dramatically affect the electronic properties. ENDOR measurements [20] on polyacetylene performed as a function of temperature establish that different polymer lattices are realized by various synthetic procedures; however, ENDOR spectra obtained at low temperatures (1-4 K) frequently correspond to comparable electron delocalization lengths, suggesting that these intrinsic lengths are most influenced by intrachain electron Coulomb and electron-phonon interactions. It should be kept in mind that the time scale of the electronic optical nonlinearity (determined by electronic rearrangement) is much shorter than the time scale of interactions modulating hyperfine interactions and determining high-temperature ENDOR spectra; thus, it is the intrinsic (or short time scale) electron localization length that is likely to be important in defining third-order susceptibility. From these arguments optical nonlinearity is not expected to vary with polymer lattice. Indeed, it has been experimentally observed that changes in morphology and interchain order do not significantly alter the nonlinear response of polyacetylene [46]. Moreover, we have observed that the attachment of substituents to ladder polymers does not perturb optical nonlinearity unless the electron-donating or -withdrawing nature of the substituent is varied. Based on these observations, the search for ways to modify materials that show

promising electronic properties to improve processibility can be
categorized by methods which perturb optical nonlinearity and those
which do not depend upon the electronic properties of the substituents.

Let us consider the three basic methods that have been used
to overcome solubility problems by examining some representative
examples of classical and modified synthetic routes for polyacetylene
and other electroactive polymers. One important class of electroactive
polymers, ladder polymers, will be considered separately.

A. Soluble Precursors

Shirakawa and co-workers [56] introduced one of the earliest and
most direct routes to polyacetylene (shown in Fig. 5) in which
acetylene is converted into polyacetylene employing a Ziegler-Natta
catalyst. The material so obtained was found to be insoluble and
intractable. Moreover, the material was contaminated by residual
catalyst and a variety of structural imperfections including varying
amounts of cis/trans isomeric segments. One of the earliest alternative synthetic approaches is that shown in Fig. 6, developed by
Feast and his co-workers [57] at Durham. In this scheme polyacetylene is synthesized by preparing a soluble precursor polymer
which can then be converted to an intractable polyacetylene by
thermal processing. Baker [55b], Grubbs [58,59], and Schrock
[60] also give alternative routes to polyacetylene. A study has
been made [61] comparing processibility and conductivity of poly-

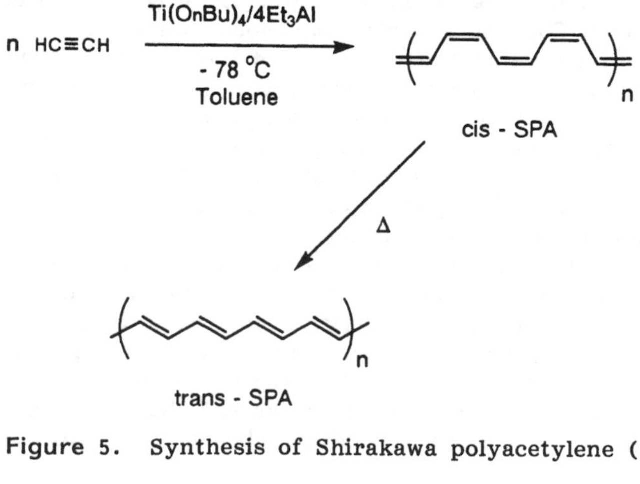

Figure 5.   Synthesis of Shirakawa polyacetylene (SPA).

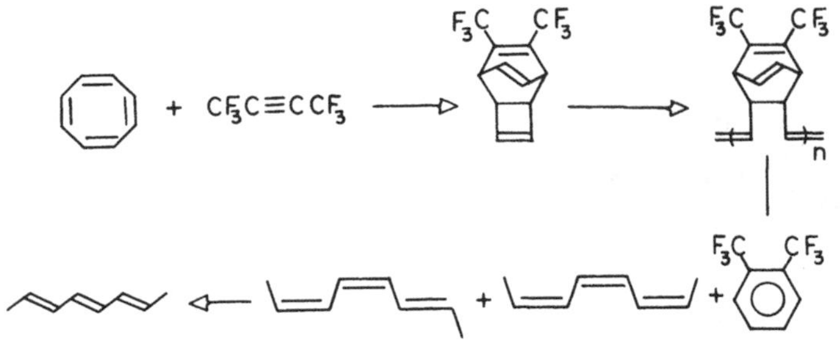

Figure 6.   Synthesis of Durham polyacetylene.

acetylene copolymers. The Naarman-Theophilou synthesis [62] leads
to polyacetylene samples that, when doped with iodine, have conductivities as high as 75,000 S/cm [63]. High-density and high-
mechanical-strength polyacetylene films (density > 1.0 g/cm^3; Young's
modulus and tensile strength of trans films stretched by seven or
eight times, 100 and 0.9 GPa, respectively) have been reported [64]
with a maximum conductivity of approximately 45,000 S/cm [65].

Another interesting class of polymers is the poly(phenylene
vinylenes) [25-27], which are synthesized via thermal elimination
reactions of soluble poly(p-xylene sulfonium chlorides) as shown
in Figs. 7a and b. Although the final NLO active product, poly(phenylene vinylene) is the same for each of the monomeric starting
materials, the cyclic sulfonium salts produce polymers with less
sulfur, greater electrical conductivity of the doped polymer, and
an assumed longer conjugation length. It was suggested that improved
conductivity was the result of a higher degree of orientation and
order permitted in the final product by the cyclic sulfonium groups.
Furthermore, it was felt that the dialkyl sulfonium groups participated
more readily than the cyclic sulfonium groups in side reactions that
did not lead to the conjugated polymer.

B.   Side-Chain Derivatization

Grubbs [66] has used ring-opening metathesis of trimethylsilycyclo-
octatetraene to synthesize poly(trimethylsilylcyclooctatetraene), a
polyacetylene with trimethylsilyl substituents. Other polymer systems
for which derivatization has been used to improve solubility include
polythiophenes [17-19,67], polyaniline [68], and polypyrrole [69].

There are many examples of derivatized polydiacetylenes; a
discussion of their synthesis and physical, electronic, and optical
properties has been published [70].

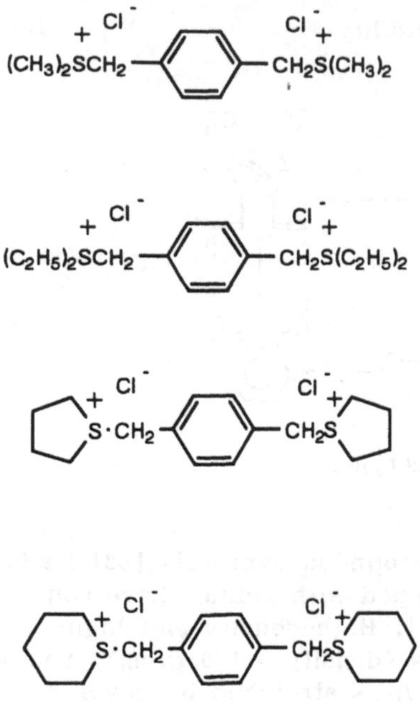

(a)

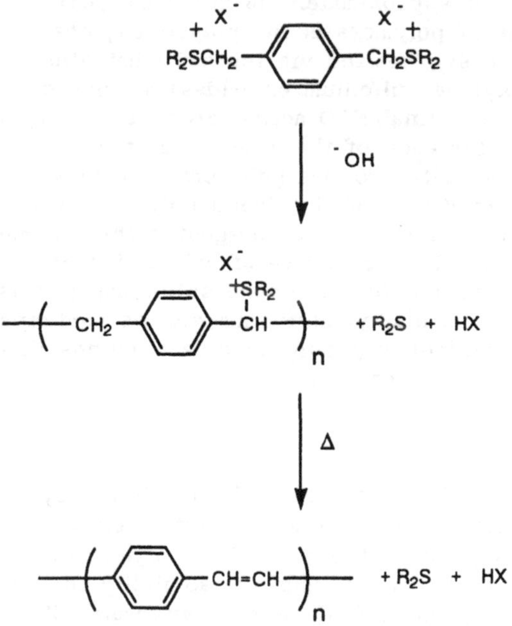

(b)

Figure 7. (a) Monomers used in the synthesis of poly(phenylene-vinylene). (b) Synthesis of poly(phenylenevinylene).

C. Block Copolymers

The synthesis of block copolymers has been used to improve solubility of ladder polymers (vide infra), polyacetylene, and polypyrrole [71]. For example, metathesis copolymerization of 1,3,5,7-cyclooctatetraene (COT) and 1,5-cyclooctadiene (COD) produces a block copolymer whose concentration of unsaturated units depends on the COT/COD ratio [72a]. For the COT/COD copolymers [73], the maximum reported $\chi^{(3)}$ is approximately 2×10^{-12} esu. Because of the much faster rate of ring opening for the strained norbornene, the copolymerization of norbornene and COT produces a long chain of polymerized norbornene with a long final block of the conjugated polyacetylene unit polymerized from COT [72b]. Flexible-chain phenylene vinylene and phenylpolyene copolymers are discussed elsewhere in this book.

A novel diacetylene copolymeric system [74] with high third-order susceptibility has been synthesized as shown in Fig. 8. The copolymer itself has virtually no detectable third-order NLO behavior; however, on thermal treatment the diacetylene units cross-link in order to form the conjugated, NLO active material.

Homopolymers of aniline and of thiophene have been extensively studied as conductive and NLO active materials, but fail to be useful due to high absorption characteristics [75]. Recently, copolymeric systems of aniline and thiophene have been synthesized which show a third-order NLO activity ($\chi^{(3)}$ of approximately 10^{-10} esu; Fig. 9), as well as interesting magnetic properties [76,77].

These materials can be cast into composite films with poly(vinylalcohol) which significantly reduces absorption and retains the optical nonlinearity of the pristine film [78].

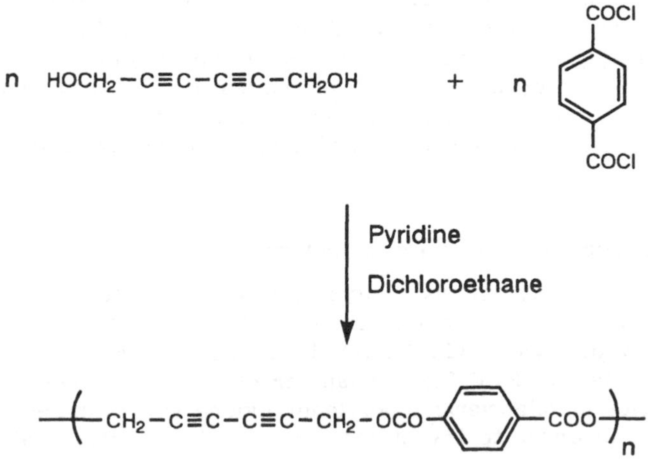

Figure 8. Diacetylene copolymer.

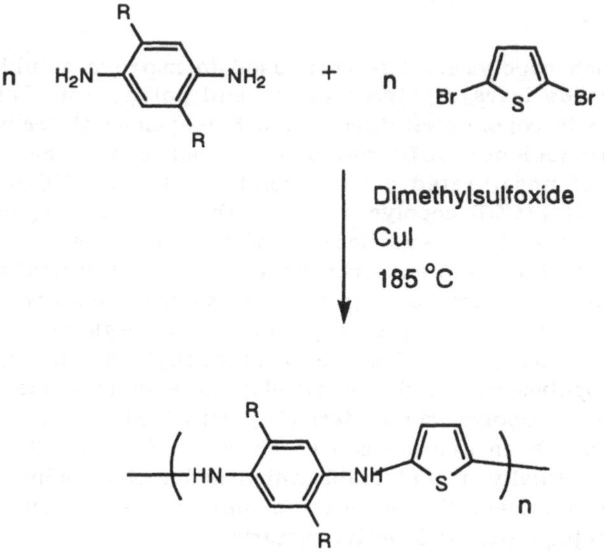

Figure 9.   Synthesis of aniline-thiophene copolymers.

Acridine and merocyanine dyes have been shown to possess NLO activity in polymeric composites [79a]. Subsequently, acridine and merocyanine units were incorporated into polymeric backbones by novel methods (Figs. 10a and b); their optical properties are currently under investigation [79b,79c].

Copolymers incorporating phenyl-capped polyene and phenylene-vinylene units have been synthesized [80,81] as shown in Figs. 11a and b. These systems are important due to the ability to control the electronic absorption by doping (as discussed above). In addition to incorporating polyene segments as main-chain repeat units in copolymeric systems, poly(vinylalcohol) composites of phenyl-capped polyene oligomers with various donor or acceptor end groups have been prepared and their conductivity and optical properties studied [81].

D.   Structure/Function Relationships of Polymers

In order to begin to quantitatively analyze the structure/function relationship of polymers, it is useful to study oligomeric structures. Related to the poly(phenylene vinylenes) discussed earlier is poly(thiophene vinylene), PTV [82]. Spangler et al. [83], have synthesized a series of oligomeric phenylenevinylene and thiophene-vinylene oligomers in order to investigate the effects of delocalization

[84] and polar substituent groups [85] on optical and electrical properties (Fig. 12). These oligomeric compounds allow the comparison of optical properties on three structural features: (1) the presence or absence of a heteroatom, (2) the effect of the position of the heteroatom, and (3) the effects of electron-donating substituents. Also, the utility of the synthetic method in preparing those types of oligomers allows future research to explore quantitatively the effect of chain length on optical nonlinearity. Table 2 gives the UV/Vis data for oxidatively doped (excess $SbCl_5$ in CH_2Cl_2) samples of the oligomers shown in Fig. 12.

Polar substituents (compare I and II; III and IV) in this series of oligomers enhance delocalization stability, in agreement with the work of de Melo and Silbey [14] that suggests that polar substituents would yield materials with high nonlinear polarizability. Spangler's work demonstrates that the appropriate substituent on mesomerically active oligomers can favorably influence the extent and/or the concentration of solitonic, polaronic, or bipolaronic domains. Current research aims at transferring these behavioral characteristics to polymeric systems.

All of the polyenes share one characteristic: because third-order susceptibility varies with q (where q is the dihedral angle), the presence of a nonzero dihedral angle in the polymer backbone will limit third-order susceptibility. For this reason, the development of polymer structures containing conjugated segments with a zero dihedral angle becomes important. Fused-ring polymers (ladder polymers) are therefore being intensely investigated at this time, although there are significant problems associated with their synthesis and processing into films [23,24,28].

E. Ladder Polymers

The synthesis of ladder polymers has been carried out by two different routes. The first route employs a two-step procedure in which a soluble precursor is synthesized and followed by either thermal or chemical treatment to obtain a ladder polymer. This route leads to materials in spectroscopically desired forms such as films, but the ladder structures are usually imperfect because of statistical effects and side reactions in the solid state. A second route is a direct one-step reaction in a solution or melt state, in which the ladder polymer is obtained in the form of powders which are difficult to process. In the following, we review some of the traditional approaches to ladder polymers which have not yielded materials appropriate for NLO characterizations and applications but provide insight into the synthesis of interesting prototype structures. We also review the modification of ladder structures to yield materials appropriate for NLO work.

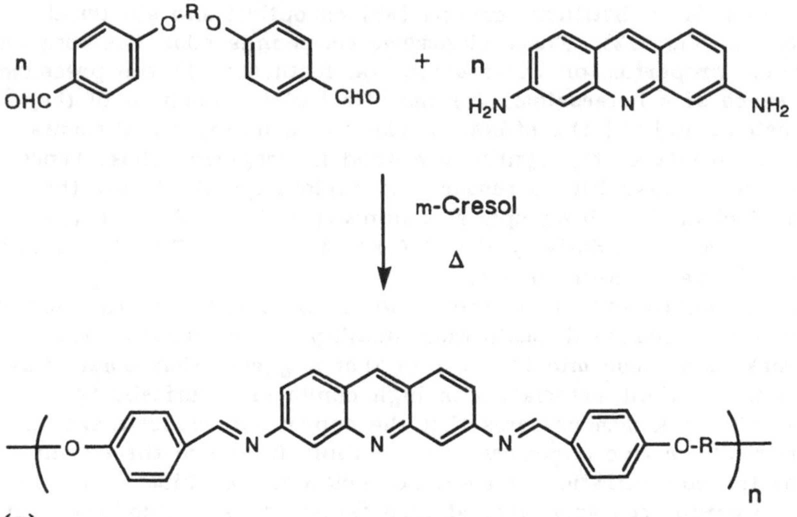

(a)

Figure 10.   (a) The synthesis of an acridine/flexible chain copolymer.
(b) The synthesis of a copolymer containing a cyanine dye unit.

Polyacene-type polymers can be prepared in multifunctional
polycondensation reactions by different methods [86–88]. Butadiyne
undergoes spontaneous polymerization when its vapor is left in
contact with an inert plastic film, such as polytetrafluoroethylene.
After five weeks in contact with butadiyne vapor, a colored coating
was observed which darkened further on heating. The final product
was paramagnetic with a free-electron g value and a linewidth of
10 gauss. The line shape was Lorentzian, and a spin density of
8×10^{19} spin/g was observed. An IR absorption at 885 cm^{-1} was
tentatively assigned to the acene C–H vibration. Pendant acetylene
group absorptions were also detected in the IR spectrum. The detailed chemistry or structure of the material is uncertain, but the
reaction shown in Fig. 13 has been proposed.

Another route to polyacene is via polymer precursors of substituted diacetylenes [89]. The diacetylene monomer undergoes polymerization by standard Ziegler–Natta catalysis. The conjugated linear
polymer obtained contains pendant ethylene units and can be thermally annealed, resulting in cyclization to the substituted polyacenes
(Fig. 14). The cyclized polymer remains as a black insoluble powder.
Both cyclized and uncyclized precursor polymers possess paramagnetic centers; the spin concentration increasing approximately one
order of magnitude upon thermal treatment of the precursor polymer.

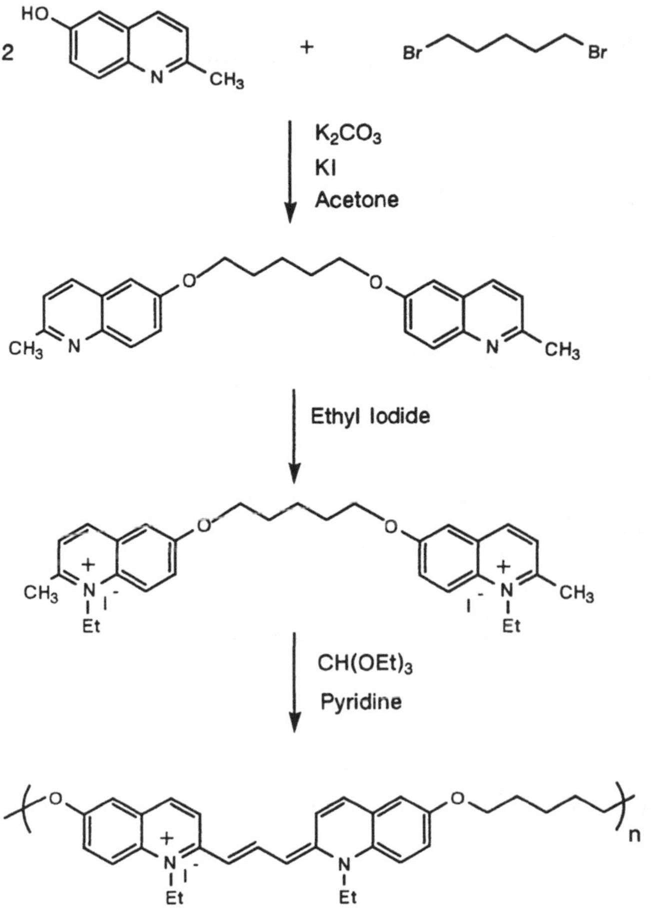
HO
2
CH_3
N
+
Br
Br
K_2CO_3
KI
Acetone
O
O
CH_3
N
N
CH_3
Ethyl Iodide
+
N
I-
Et
Et
$CH(OEt)_3$
Pyridine
n

(b)

Further evidence for the ladder structure is a red shift in the electronic absorption spectrum on thermal annealing. Initial conductivity measurements of the cyclized polymer were low (approximately 10^{-14} S/cm), indicating a rather amorphous material or perhaps cross-linking; indeed, the mechanism of thermolysis is complex, and structural changes are by no means definitive or straightforward.

Recently, a slightly different procedure has been used to obtain polyacene [90]. Triethylsilylbutadiyne has been polymerized using a Ziegler-Natta catalyst and the polymer obtained desilylated with Bu_4NF, resulting in a conjugated polymer with pendant acetylene. This can be thermally annealed to yield a highly conductive material

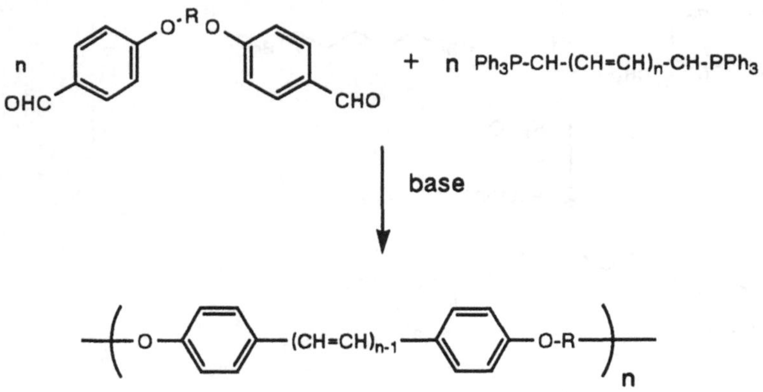

(a)

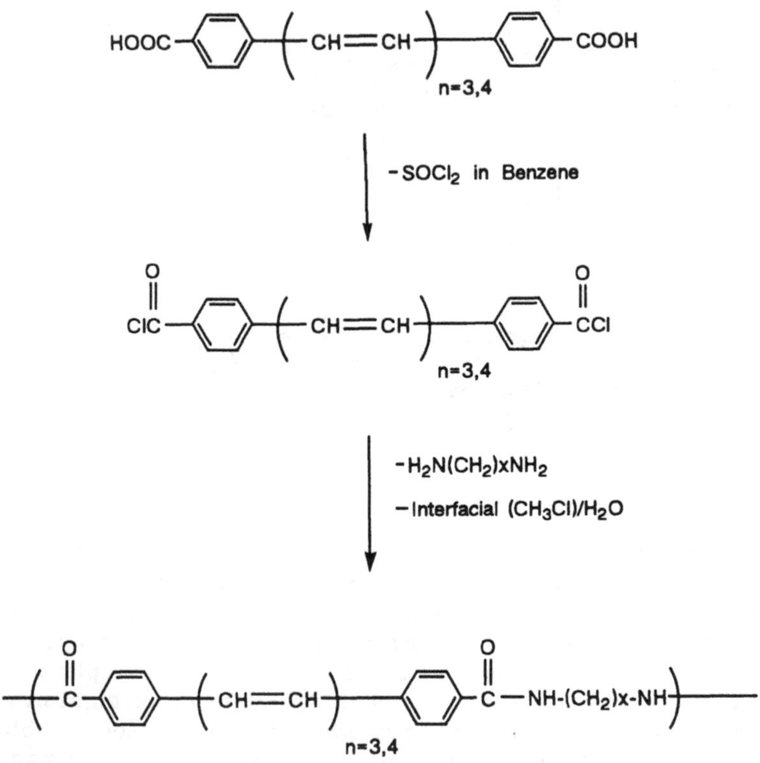

(b)

Figure 11. (a) The synthesis of a polyene/flexible chain copolymer.
(b) The synthesis of phenylpolyene copolymers.

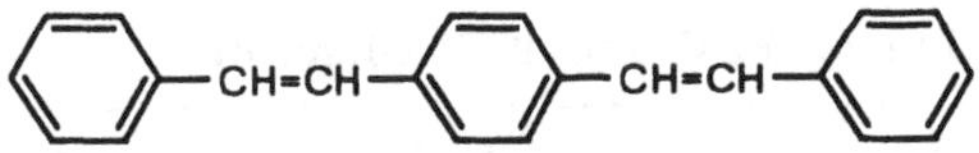

I

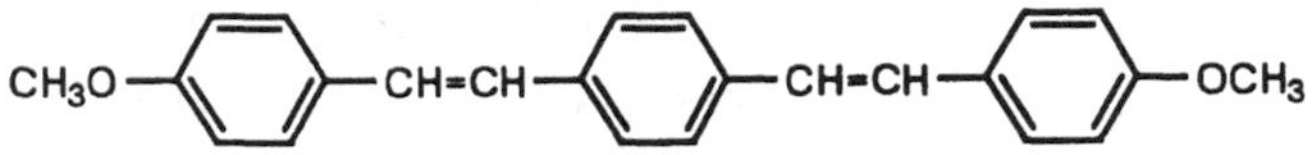

II

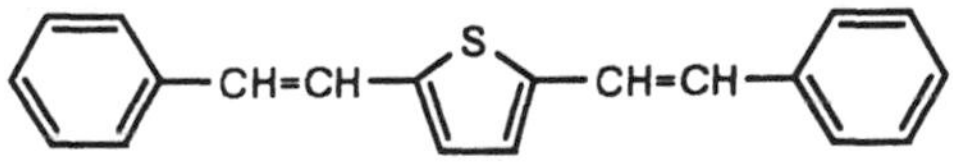

III

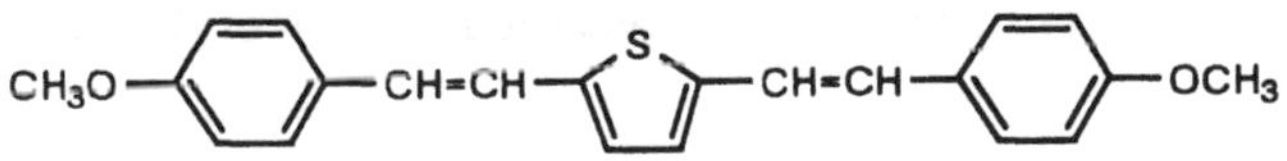

IV

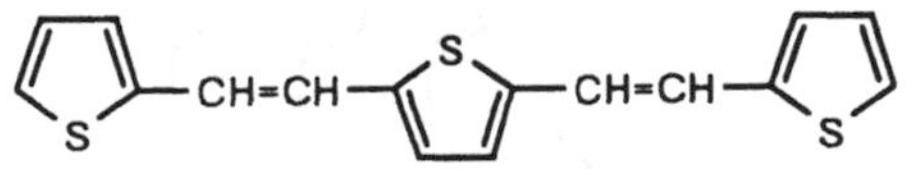

V

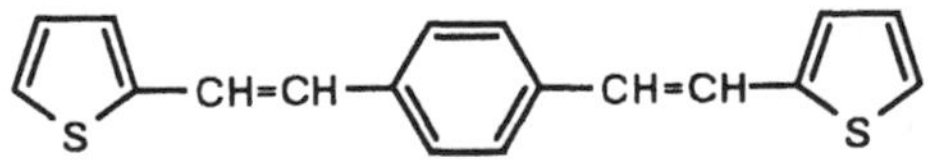

VI

Figure 12. Phenylenevinylene oligomers.

Table 2.   Oxidative Doping of PPV and PTV Oligomers

Compound	$\pi - \pi^*$ Transition (nm)	P Transition	BP Transition
I	350	604,542,496	Not observed
II	362	664,580,540	478
III	383	618,558	498
IV	396	820,(692)	744,676
V	410	688	568
VI	372,366	Not observed	480

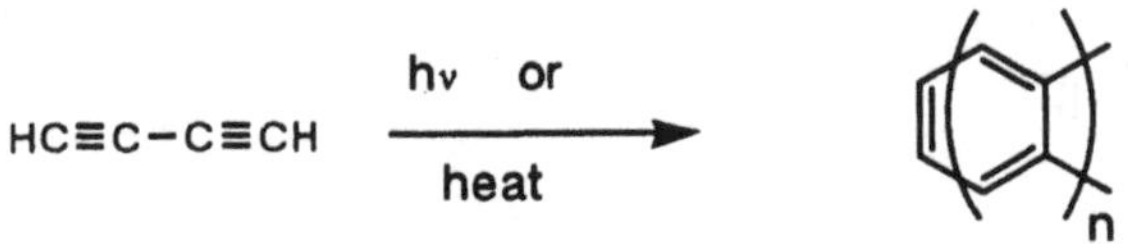

Figure 13.   Photo/thermal-initiated polymerization of diacetylene.

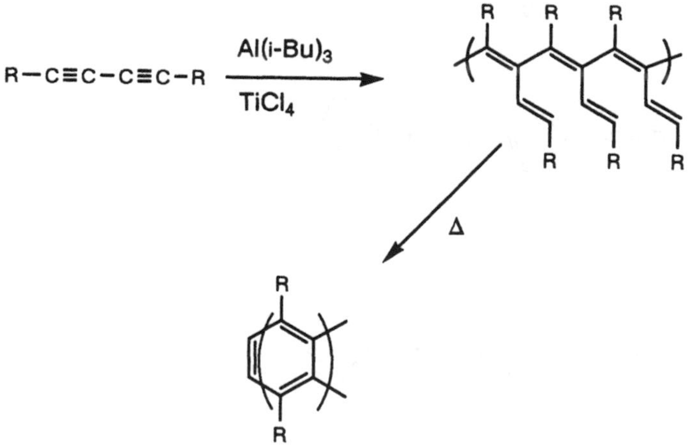

Figure 14.   Two-step procedure to polyacene from substituted diacetylenes.

(0.78 S/cm). The structure of the final material has been assigned
to polyacene (Fig. 15).

A route to polyacene from methyl-chlorovinylketone has been
described (Fig. 16) [91]. Again, the structural information is limited
as the material is insoluble and therefore difficult to characterize.
However, the polymerization is believed to proceed via a conjugated
polyketone intermediate which is subsequently pyrolysed to yield
polyacene.

In 1938, Marvel and Levesque [92] studied the cyclization of
polymethylvinylketone. When the polymer was heated to 300°C, water
was eliminated, resulting in a red fusible glass. Chemical analysis
showed that segments with a ladder-type structure were formed,
and the condensation was governed by a statistical law with only
86% of the ketone units participating in the condensation. The ladder
polymer studied by Marvel and co-workers, poly(dihydroacene),
can also be prepared by intramolecular aldol condensation of
poly(methylvinylketone) catalyzed by n-butyllithium, followed by
dehydration using trifluoroacetic acid (Fig. 17) [93].

When 3,4,9,10-perylenetetracarboxylic dianhydride is pyrolyzed
in an evacuated system, a highly conducting material can be obtained
[94]. Elemental analysis, along with infrared and Raman spectroscopy,
indicate a ladder-type polymer, poly(perinaphthalene) (PPN) [94,95]
(Fig. 18). Conducting films or ribbons can be produced by different
synthetic procedures [94-100]. These materials are chemically and
environmentally stable and have a density of 1.758 g/mL. The films
showed metallic ($\sigma_{RT}/\sigma_{4K} = 2$) and semiconductor behavior depending
on the reaction temperature. Pellets of PPN gave room temperature

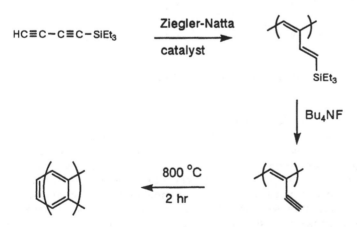

Figure 15.   An alternative route to prepare polyacene from diacetylene.

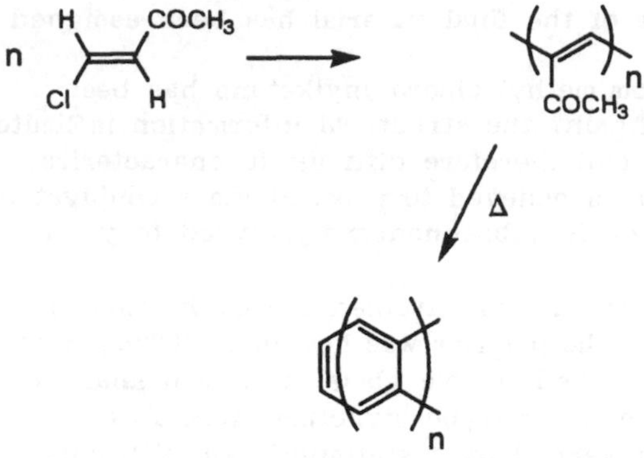

Figure 16.   Preparation of polyacene from polyketone.

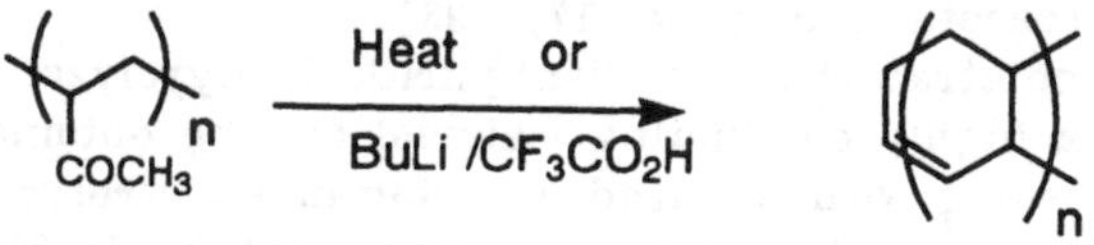

Figure 17.   Preparation of polydihydroacene from poly(methylvinyl-ketone).

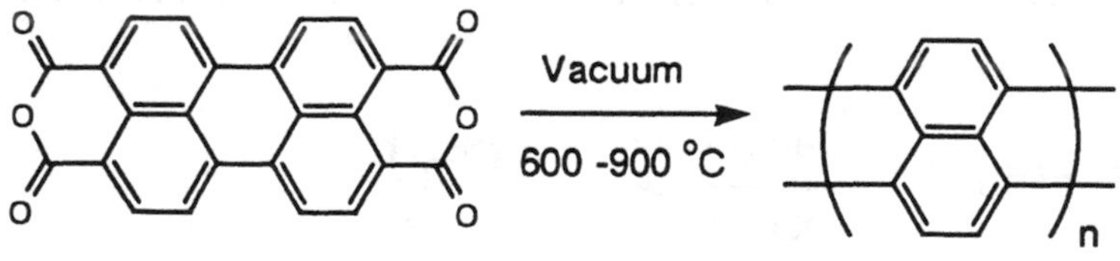

Figure 18.   Preparation of poly(perinaphthalene).

four-point probe conductivities between 10^{-2} S/cm at T_p (pyrolysis temperature) of 530°C to 15 S/cm at T_p of 800 and 900°C. The conductivities of PPN samples prepared at various temperatures were unchanged upon exposure to compensating agents such as NH_3 or benzoquinone, suggesting that the observed conductivities are intrinsic. Electron paramagnetic resonance (EPR) spectra showed a moderately strong line with g = 2.0025 and with a peak-to-peak linewidth of 3-7 gauss at room temperature. When the starting dian-hydride is mixed with niobium pentachloride and subjected to pyrolysis, mirrorlike films can be obtained, containing niobium but no chlorine. Conductivity measurements showed these films to be superconducting with an onset T_C of 12 K. When boric acid was mixed with the starting monomer, a p-type semiconductor film was obtained. When doped with zinc metal, an n-type semiconductor resulted.

Polyacrylonitrile (PAN), a commercially available polymer, can be readily polymerized in the solid state by thermolysis (Fig. 19). As heating progresses, the polymer develops a distinctive red color which has been ascribed to the formation of thermally stable ladder polymer segments with C=N and C=C conjugation [100-107].

There are two ways of thermolysis of PAN to poly(pyridinopyridine) (PPP), namely, a nonoxidative treatment under inert atmosphere, and a treatment in air. Both lead to formation of ladder polymers, but the structures thus obtained are quite different. The former possesses saturated heterocycles containing C=N conjunction, while the latter have fully aromatic heterocycles carrying hydroxyl groups. In addition, materials formed by oxidative treatment are more thermally stable. The presence of oxygen during the oxidative treatment increases the rate of cyclization. Infrared and elemental analysis showed that oxygen was incorporated into PPP as hydroxyl and carbonyl groups [104,108,109].

Cyanoacetylene can be polymerized into linear polymers at low temperature (Fig. 20) under anionic, Ziegler-Natta, and metathesis polymerization conditions [110,111]. The polymers are brown to black amorphous powders, soluble in polar organic solvents, with low molecular weights (on the order of several thousand). Polymers formed through Ziegler-Natta catalysis have a cis-transoid structure, while polymers formed by anionic polymerization are rich in trans-transoid structure. The latter structure is easily cyclized due to

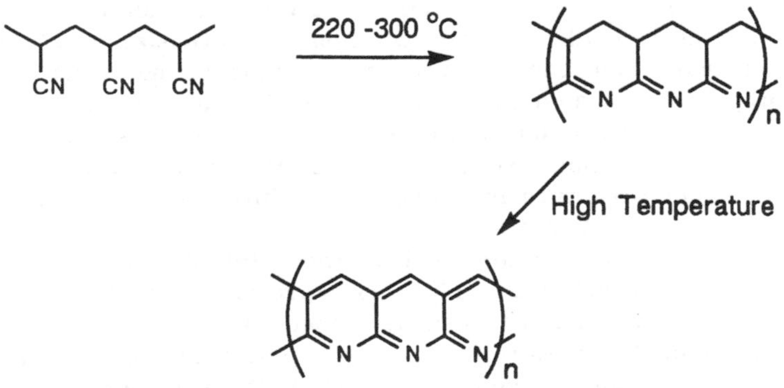

Figure 19. Thermal cyclization of polyacrylonitrile.

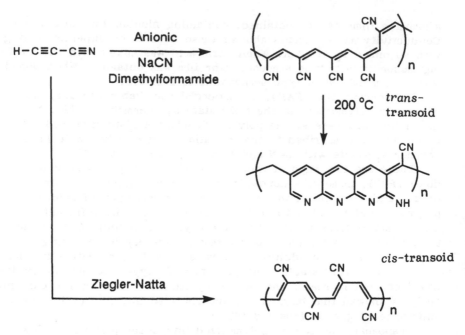

Figure 20. Polymerization of cyanoacetylene and thermal cyclization of polycyanoacetylene.

the steric availability of adjacent cyano groups. The final polymer is insoluble, suggesting a ladder structure containing PPP segments. Both polymers (cyclized) and prepolymer (open chain) are EPR active; the cyclized polymer can be doped by iodine to a room temperature conductivity of 10^{-3} S/cm.

Litt et al. [112–115], have utilized the reaction of *o*-diaminobenzenes with formaldehyde in acidic aqueous solution to synthesize poly(phenylenemethylenes) which can be cyclized in strong acidic media such as polyphosphoric acid (PPA) (Fig. 21). The poly(phenylenemethylenes) are low molecular weight (intrinsic viscosity = 0.15 dL/g) while the cyclized materials are completely intractable; however, the cyclized ladder structure is supported by infrared, UV/Vis, and elemental analysis characterization.

PXL polymers (with X determined by the heteroatom in the starting material; see below and Fig. 22) represent another class of ladder polymers which have been developed by Stille et al. [116–119]. Tetrafunctional monomers, such as aromatic tetraamines and 2,5-disubstituted quinones, have been used in the synthesis of these ladder polymers in acidic media such as polyphosphoric acid.

When tetraaminobenzene (TAB) reacts with 2,5-dihydroxy-*p*-benzoquinone in PPA or hexamethylphosphoramide (HMPA), a quinoxaline-type ladder polymer is obtained, which is soluble in 1,2-dichloro-1,1,2,2-tetrafluorethane. This polymer is thermally stable up to 520°C. When 1,4-dihydroxy-2,5-diaminobenzene polymerizes with 2,5-dichloro-*p*-benzoquinone in a two-step reaction, POL-type ladder polymers can be prepared. These ladder polymers have poor solubility in common organic solvents. The thermal stability of these ladder polymers is very poor: a slow decomposition starts at ca 275°C under nitrogen, which could be an indication of incomplete cyclization and low molecular weight. POL can also be prepared from 2,5-dichloro-*p*-phenylenediamine and 2,5-dihydroxy-*p*-benzoquinone in PPA at 250°C [120].

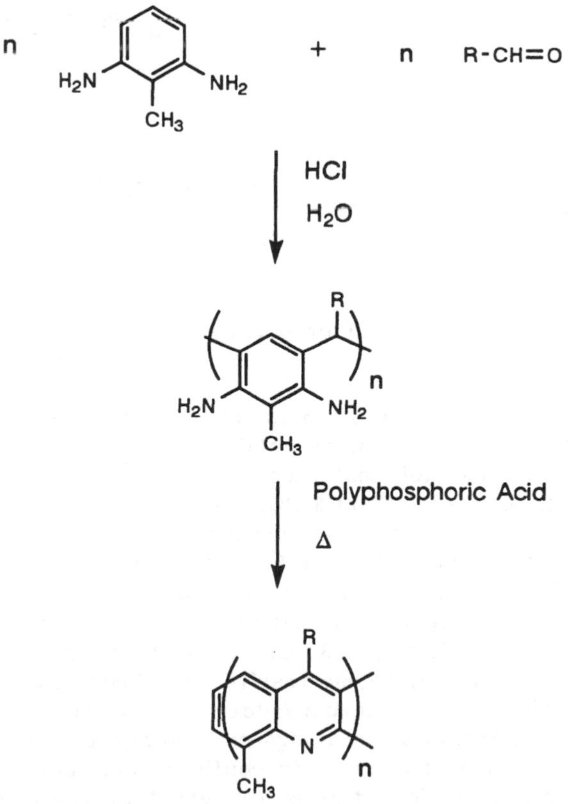

Figure 21. Preparation of ladder polymers from poly(phenylenemethylenes).

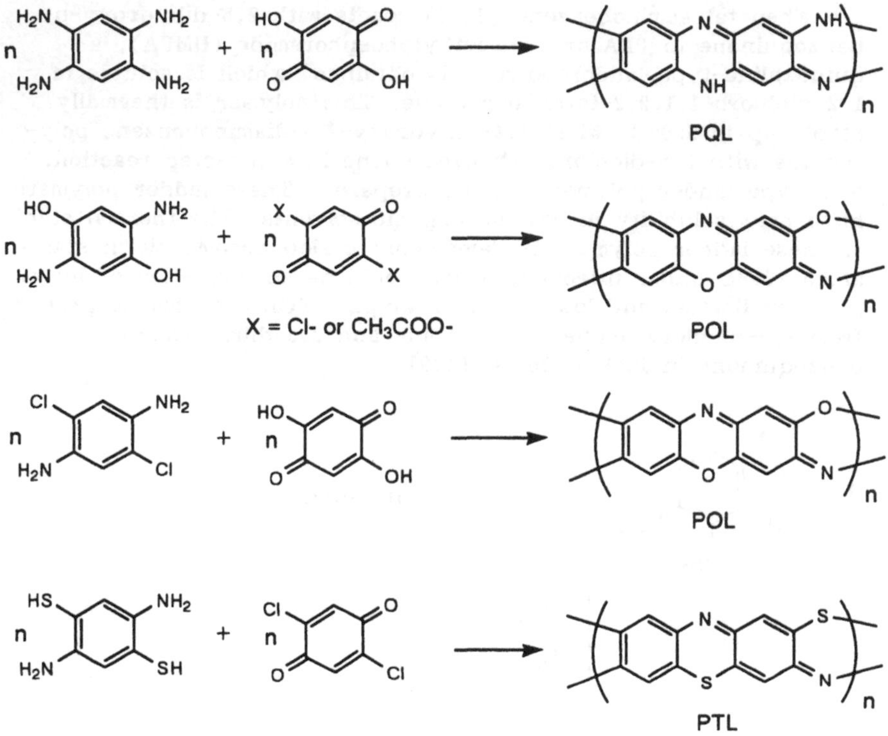

Figure 22.   Preparation of PXL-type ladder polymers.

When 2,5-diamino-1,4-benzenedithiol reacts with 2,5-dichloro-*p*-benzoquinone in PPA at 250°C, PTL can be obtained [120]. All of these polymers are black insoluble materials.

Brownish orange thin films of POL and PQL can be prepared by electropolymerization (Fig. 23) of *o*-phenylenediamine and *o*-aminophenol in acidic solutions [121,122]. A POL thin film had a conductivity of 4.0×10^{-7} S/cm at 298 K. The IR spectrum showed an absorption at 1645 cm^{-1}, which is assigned to C–N bond stretching, and absorption peaks at 1050 and 1235 cm^{-1}, which could be due to C–O–C stretching vibration. These results have been cited as evidence of the formation of phenoxazine ladder units in the polymer, but, a strong absorption around 3420 cm^{-1} suggests that the polymer has a large portion of open-chain configuration, and the molecular weight is relatively low; this is supported by thermogravimetric analysis ($T_{decomp.}$ = 300°C). PQL can be prepared by electropolymerization of phenylenediamine under similar conditions.

An incompletely cyclized polymer with low thermal stability has
been obtained in the form of thin films.

The structures of PXL type ladder polymers are of interest
because these materials are capable of supporting bond alternation
defects (solitons/polarons/bipolarons) upon chemical doping. These
materials are paramagnetic; EPR and ENDOR measurements indicate
highest occupied molecular orbital (HOMO) delocalization/localization
phenomena similar to that observed for polyacetylene [84a].

Electrical conductivity measurements indicate conductivities
of 4×10^{-6} S/cm and 1×10^{-8} S/cm for PTL and POL, respectively
[123]. These values, which are unusually high among conjugated
organic polymers, could arise from the contribution of contamination
by impurity residues such as PPA, etc. PTL can be doped either
by p-type dopants like AsF_5 or n-type dopants like potassium or
sodium naphthalide to increase conductivity by 2-5 orders of magnitude.

Because the structure of the ladder polymers promotes the
extensive electron delocalization essential for third-order susceptibility
[84a,23b], the development of alternative synthetic routes to improve
processibility has received significant attention.

Dichloroquinone monomers derivatized with vinylamine substituents
were prepared as shown in Fig. 24 [23,28]. These quinone monomers
A-D were reacted with tetraaminobenzene to prepare prepolymers
and polymers A-D respectively. Polar organic solvents such as
dimethylsulfoxide, dimethylformamide, dimethylacetamide, *m*-cresol,
pyridine, etc., have been used as polymerization media. The prepolymer solution was then used to cast films which can be thermally
treated at 300°C to obtain ladder polymer films. Fourier transform
infrared (FTIR), elemental analysis and nuclear magnetic resonance
(NMR) results support the synthetic scheme outlined earlier in
this chapter (Fig. 2).

Viscosity measurements of prepolymers dissolved in dimethylformamide (DMF) at 25°C gave intrinsic viscosities of 1.4 dL/g,
0.30 dL/g, 0.25 dL/g, and 0.15 dL/g for prepolymers A, B, C,
and D respectively (Fig. 24). This indicates that as the size of

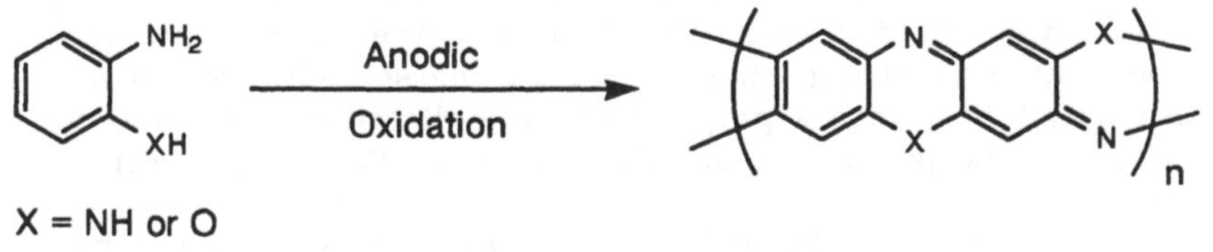

X = NH or O

Figure 23.   Electrochemical preparation of PXL-type ladder polymers.

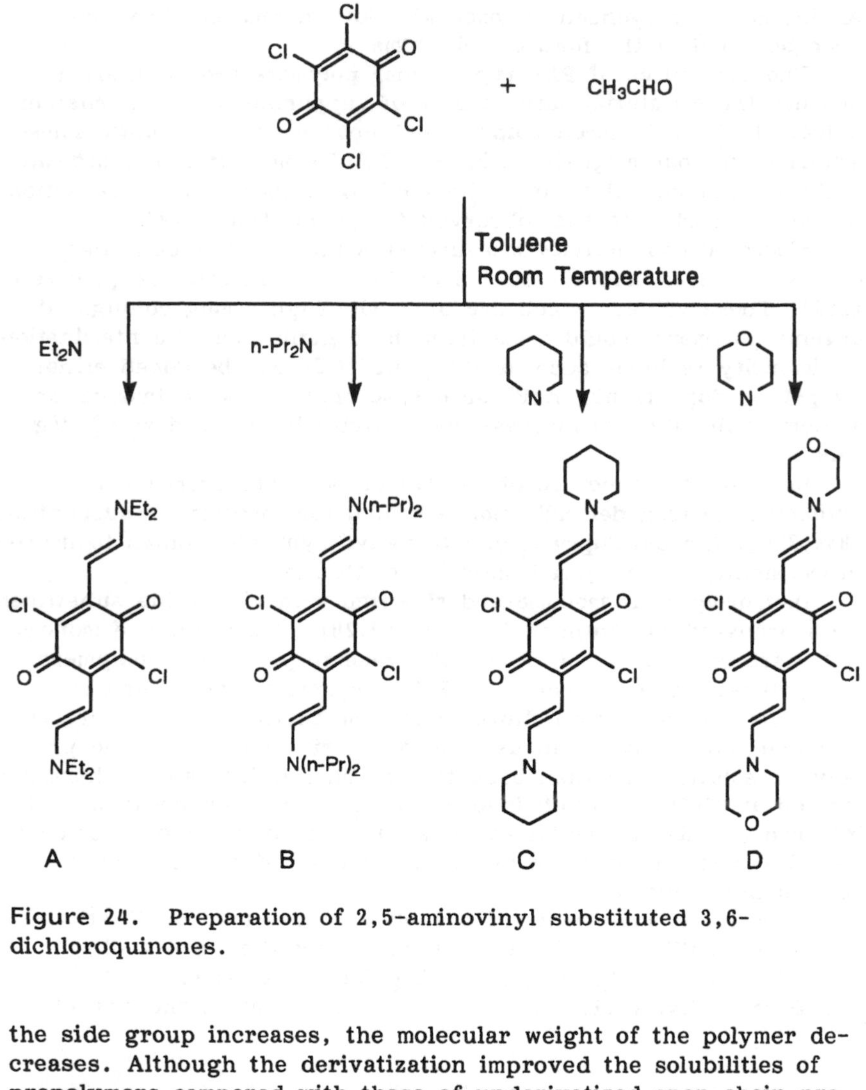

Figure 24.   Preparation of 2,5-aminovinyl substituted 3,6-dichloroquinones.

the side group increases, the molecular weight of the polymer decreases. Although the derivatization improved the solubilities of prepolymers compared with those of underivatized open-chain prepolymers, the van der Waals interactions in the prepolymers and polymers are still strong enough to limit the solubility of both. It was found that once a prepolymer precipitated out of solution, it could only be partially redissolved (40% in DMF for polymer A) in organic solvents.

UV/Vis spectra of the ladder polymers showed a lower band energy edge, at 1.56 eV compared with that of the prepolymer

Table 3. TGA Decomposition Temperature of Ladder Polymers

Polymer samples	A	B	C	D
T decomp (°C)	690	564	474	498

at 1.79 eV. This implies that the polymer system has a more extended π-electron conjugation. This is reasonable when one considers that a finite dihedral angle (between adjacent phenyl rings) can be expected for the prepolymer compared to a zero degree dihedral angle for the polymer; a zero dihedral angle, of course, implies maximum π-orbital overlap. Thermogravimetric analysis (TGA) of these polymers showed very high thermal stabilities. Table 3 summarizes the decomposition temperature of different polymers under nitrogen atmosphere. The high thermal stability of these polymers is consistent with high laser damage thresholds (greater than 1 GW/cm^2) observed in NLO characterizations.

Both prepolymers and polymers are EPR active. In the prepolymers, g values are shifted significantly from the free-electron value (2.0023), as might be expected for quinone radicals. The g values of the polymers are identical with those of underivatized ladder polymers reported in Ref. 84a and are near the free-electron value, implying significant electron delocalization. Linewidth (ΔH) changes from prepolymer to polymer also imply a greater electron delocalization in the latter (see Table 4). These polymers exhibit significant third-order nonlinear susceptibilities (see Table 5).

It can be seen that these polymers also exhibit high $\chi^{(3)}/\alpha$ values (this parameter is more relevant for device considerations). Polymers A through D showed very similar values for this ratio, which indicates that different alkyl substituents on the amine have only a minor influence on optical nonlinearity as long as we maintain the ladder structure with aminovinyl derivatization. This is not surprising, as it is the aminovinyl group that should characterize

Table 4. ESR Measurements of Ladder Polymers

Sample	Prepolymer		Polymer			
	A	B	A	B	C	D
ΔH	9.00	10.05	6.30	8.75	7.50	7.30
g	2.00515	2.00510	2.0034	2.00298	2.00312	2.00302

Table 5.   Optical Properties of Ladder Polymers

Properties	Polymers			
	A	B	C	D
n^a	1.96±0.02	1.92±0.02	1.92±0.02	1.92±0.02
α (10^4 cm^{-1})[b]	2.32±0.16	2.54±0.18	1.88±0.31	4.08±0.24
$\chi_{1111}^{(3)}$ (10^{-10} esu)[c]	9.78±3.0	8.90±2.5	7.40±2.1	15.5±5.1
$\chi_{1221}^{(3)}$ (10^{-10} esu)	1.8±0.5	1.9±0.5	1.6±0.5	3.0±0.9
$\chi_T^{(3)}$ (10^{-10} esu)	3.8±1.2	4.4±1.2	3.8±1.2	7.7±2.0
$\chi_{1111}^{(3)}/\alpha$ (10^{-14})	3.23	3.50	3.93	3.80

[a] n is the index of refraction.
[b] α is the absorption coefficient.
[c] $\chi^{(3)}$ is the third-order susceptibility; $\chi^{(3)}_{ijlk}$'s are the tensorial components of the third-order susceptibilities, and $\chi^{(3)}_T$'s are the thermal contributions to the third-order NLO processes.

Table 6.   Effects of Oxidation and Cyclization

Polymer	A	A (oxidized)	Prepolymer A
n	1.96	1.75	1.70
α (μm^{-1})	2.32	1.10	1.70
$\chi_{1111}^{(3)}$ 10^{-10} esu	9.78	0.60	1.00
$\chi_{1111}^{(3)}/\alpha$ (10^{-10} esu mm^{-1})	3.23	0.54	0.59

the inductive effect upon backbone electron delocalization and which should influence the extent of the π-system.

As shown in Table 6, oxidation of a sample reduced the third-order susceptibility by reducing electron delocalization. The same can be said for incomplete condensation as is seen by comparing values for prepolymers and polymers. Theoretical studies show that, in polyconjugated systems, π-electron delocalization is limited by electron-phonon and electron-electron interaction [12,14,124]. Consequently, the third-order NLO response will increase as the conjugation length increases for short conjugation length and then begin to saturate to some finite length. For example, in polyene systems, theory predicts that after 15 double bonds, the third-order susceptibility begins to saturate [124]. These results suggest that in order to have optimum optical nonlinearity, it is not necessary to have an infinitely extended conjugated system within the polymer backbone. Based on this observation, copolymers containing ladder units and flexible-chain segments have been synthesized [28]. Incorporation of finite-chain-length NLO active units into a processible polymer helps to overcome another problem, the smearing of the band gap that arises from the presence of conjugated polymers of varying lengths. Yu and Dalton [28] have attempted to circumvent problems associated with the production of a distribution of π-electron delocalization lengths by synthesizing copolymer systems containing defined lengths capable of electron delocalization separated by saturated segments. In addition to confining electron delocalization, saturated linkages have the advantage of improving solubility. The reaction scheme reported is shown in Fig. 3a.

The assignment of intermediates is supported by IR, NMR, and optical data. The realization of electron confinement is illustrated in Fig. 3b, which shows that the optical spectra for the molecule and polymer are virtually indistinguishable. Three peaks in the visible region (478, 515, 566 nm) can be observed. Both polymers a and b show sharply defined absorption band edges. This is an advantage for NLO studies as resonant and nonresonant contributions to third-order susceptibility can be more easily determined. DFWM measurements have been performed on polymers a and b. Films from polymer b exhibit substantial light scattering because of polymer aggregation. This feature results in a noisy DFWM signal. This problem does not exist for polymer a. According to our current analysis, these two polymers exhibit similar optical nonlinearity with $\chi^{(3)}$ and $\chi^{(3)}/\alpha$ values of 4×10^{-9} esu and 10^{-12} esu $\cdot$ cm at $\lambda = 585$ nm. The frequency dependence of the latter values demonstrates a resonant contribution to these values (see Fig. 25).

Polybenzimidazobenzophenanthroline (BBL) ladder polymer (first developed by Deusen et al.) [125] is synthesized from naphalenetetra carboxylic acid and tetraamino benzene in strongly acidic, dehydrating media such as PPA (Fig. 26). The synthesis of this type of ladder polymer is actually an extension of dye synthesis [125-131]. BBL is completely soluble in methane sulfonic acid (MSA) and can be cast into flexible films which possess an intense golden luster. A 0.3-mm film cast from MSA solution near 60°C and under pressure of 0.95 mm was found to have an initial modulus of 1.1×10^6 psi. Tensile strength averaged 16,600 psi with rupture elongation of 2.9%. This ladder polymer is found to be thermally stable up to 550 and 700°C in air and nitrogen, respectively. It is a semicrystalline polymer with a layered structure in two-dimensional arrays.

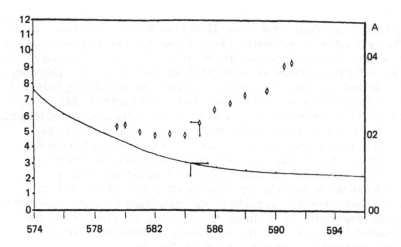

Figure 25. Spectral dispersion for the ratio of third-order susceptibility χ^3/α ($\times 10^{-13}$ esu-cm) to linear absorption (wavelength in nm)
for polymer a (Fig. 3a).

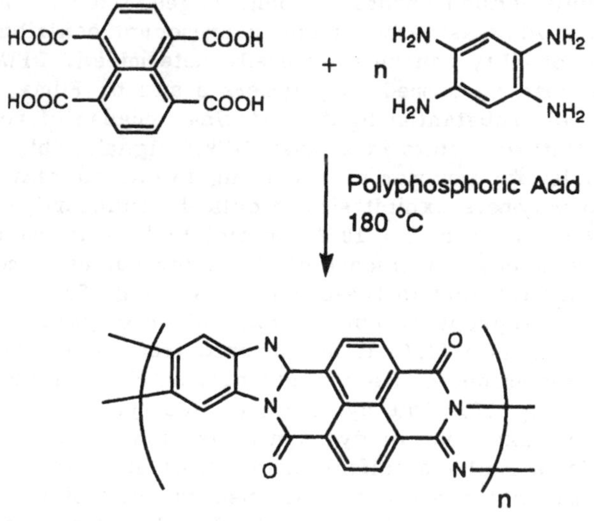

Figure 26. Preparation of the ladder polymer BBL.

X-ray and IR analyses revealed a compact interchain packing in
solution as molecular aggregates, and in the bulk after being precipitated out from solution [132].

Pristine BBL film is an insulator (ca 10^{-12} S/cm); however,
doping with fuming sulfuric acid produces highly conducting complexes [123,133]. The doped films retain their flexibility but swell
considerably. Unusually high dopant concentrations were observed
(6-8 dopants per polymer unit). The high dopant content and swelling
during doping were attributed to intercalation into the layered structure of the polymer. Films of BBL can be donor-doped both with
potassium vapor (300-500°C) and potassium naphthalide in tetrahydrofuran (THF). The latter process gives complexes with conductivity
as high as 1 S/cm.

Electron spin resonance (ESR) measurements were performed
on pristine and thermally annealed thin (20 microns) films of BBB
and BBL [134]. Both polymers gave a strongly anisotropic signal,
suggesting a highly planar polymer backbone (Fig. 27). The ESR
data imply the existence of two different types of spin species.
Thermally annealing the BBB and BBL films to high temperatures,
T_0, increases the spin susceptibility for $T_0 < 850$ K. However,
annealing to $T_0 > 900$ K results in a dramatic reduction in the spin
susceptibility and a dramatic increase in the conductivity (Fig. 28)
[135]. In fact, the room temperature in-plane conductivity of annealed
samples (900 K) exceeds $\sigma = 1\Omega$/cm. This is due to a phase transition
to a fully condensed, cross-linked aromatic structure.

The spin concentration can also be increased through photo-
generation at various excitation wavelengths as previously observed
in PQL ladders [136]. The results imply long-lived localized photogenerated polarons whose lifetime can be described by a two-component

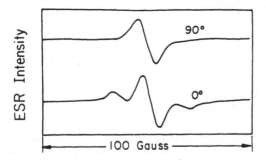

Figure 27. Typical ESR spectra of a BBB film showing anisotropic
signal. 0° = magnetic field parallel to the c-axis (perpendicular
to the film) 90° = perpendicular to the c-axis.

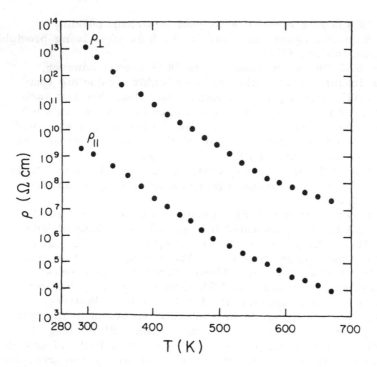

Figure 28. The temperature dependence of the resistivity for an annealed film.

exponential decay function. The two different decay processes may be explained as interchain and intrachain recombination of the localized polarons. Their lifetime dramatically increases with decreasing temperature and strongly depends on the wavelength of the exciting radiation. The similarities in the ESR properties for the photogenerated, thermally generated, and native spins indicates that the spin species all exhibit polaronic character. Third-order NLO studies are underway to correlate NLO properties to concentration of spins.

Current studies indicate that exposure of heat-treated BBB and BBL polymers to fluorine results in transparent and stable polymers with enhanced conductivities [137]. This is of interest because undoped ladder polymers are not attractive for device application due to their large light absorbance, particularly in the visible region. The polymers were fluorinated at room temperature at about

100 torr F_2; weight increase measurements indicate a weight uptake of approximately 22 (BBB) and 16 (BBL) fluorine atoms per monomer unit. Such treatment shifts the absorption edge from 2 eV in pristine BBL to 3.4 eV for fluorinated BBL. By careful thermal annealing before doping, it is possible to produce stable, transparent, and somewhat conducting polymers that are still partially conjugated. It is also possible to utilize laser annealing as a spatially selective heating technique, resulting in conducting patterns in an insulating BBB or BBL polymer matrix. Subsequent fluorination leads to conducting patterns embedded in a completely transparent polymer.

BBL can also be doped by ion implantation [138], using boron, argon, and krypton implantation. As the concentration of implanted impurities increases, the gold-yellow pristine BBL films gradually turn black and then silver-gray with metallic luster. BBL can be doped to conductivities as high as 224 S-cm at a dose of $4 \times 10^{16}/cm^2$ while retaining the mechanical properties of pristine films. Spatially selective implantation create regions of conducting lines in an insulating matrix, which suggests the potential of microelectronic device application for this ladder polymer.

Preliminary DFWM measurements indicate a third-order susceptibility of 2×10^{-9} esu at 532 nm ($\alpha = 18.3 \times 10^4$ cm^{-1}) and approximately 10^{-11} esu at 1064 nm [21,23].

Analogs of BBL (Fig. 29) have been synthesized by Marvel [139] and Arnold [140-142]. These polymers have low molecular weights and are thermally stable to 550°C.

Ladder-type polymeric tetraaza[14]annulene metal chelates (Fig. 30) were prepared from propynal aldehydes, tetraaminobenzene, and transition metal salts [143-145]. The structure of the products were determined by IR spectroscopy. The dark-colored polymers were insoluble in organic solvents, and their molecular weights are unknown. The polymers have high thermal stabilities. They decompose under nitrogen at ca. 700 K and have 10% weight loss at 1070 K.

For all of the chelates, the electrical conductivities are low; conductivities of ca. 10^{-9}-10^{-12} S/cm^{-1} have been observed. This can be understood from detailed x-ray diffraction investigations which showed that the delocalization of π-electrons is concentrated on the six-membered propan-1,3-diiminato-chelate ring. The intramolecular charge transfer is inhibited, and hence the charge carriers cannot be transported through the polymer chains.

We have conducted NLO measurements on polymer composites of [14]azaannulene-nickel complexes and have observed optical nonlinearity comparable to the metal phthalocyanines [79,23b]. A $\chi^{(3)}/\alpha$ value of 2.4×10^{-13} esu $\cdot$ cm has been observed for a composite (10% w/w) of the material in a polycarbonate matrix.

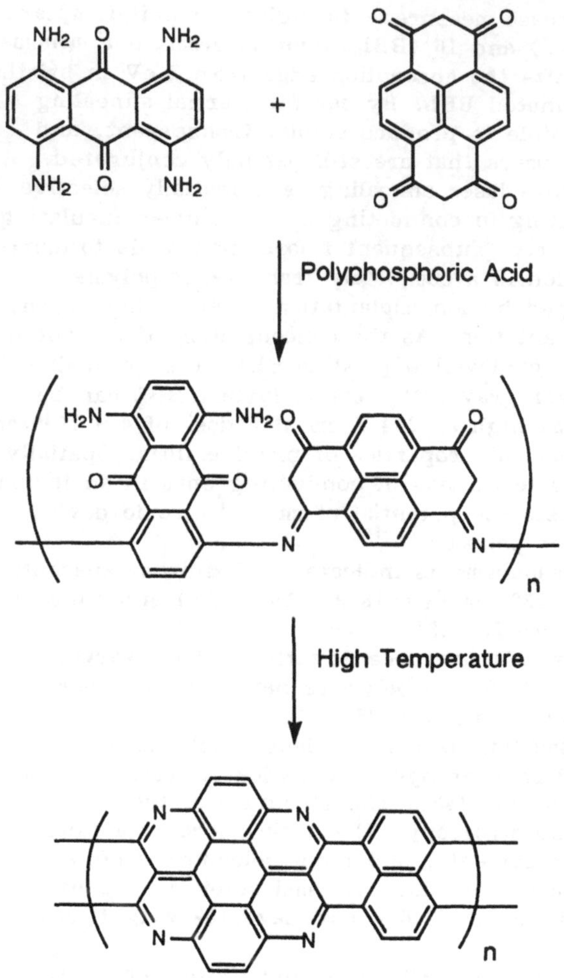

Figure 29. Preparation of BBL analog ladder polymers.

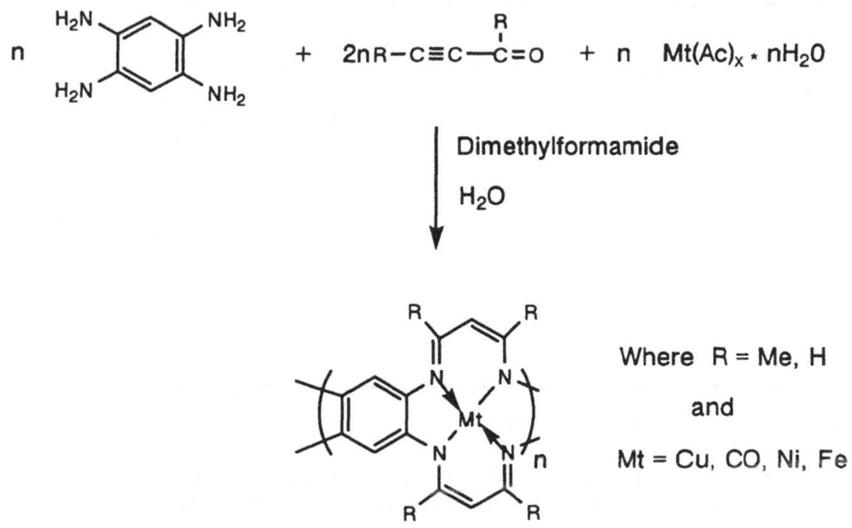

Figure 30. Preparation of polytetraazaannulene (5,7,12,14-tetramethyldibenzo[b,i][1,4,8,11]tetraazacyclotetradecahexenato-nickel(II)) complexes.

V. PROCESSING TO FABRICATE NONLINEAR DEVICES

Processing is necessary to improve a variety of characteristics ranging from optical clarity to mechanical stability to improved molecular order. For example, pure phthalocyanines are difficult to work with because of aggregation and associated light scattering problems. However, they can be fabricated as polymer composites with good optical clarity and useful nonlinearity (Fig. 31) [146a].

The same can be said for the fabrication of copolymer systems where the properties of the copolymer can be chosen to optimize the above properties. For example, if liquid crystalline phases can be generated, then oriented systems can be prepared by processing films from these phases. Some rigid-rod polymers already fall into this category [42,146b], and much greater control can obviously be obtained by working with copolymer systems.

A popular means of obtaining oriented materials is by the use of Langmuir-Blodgett fabrication methods [1-4,11], although most efforts along this line have been directed toward the generation of second-order materials. The difficulty encountered is that of realizing long-range order. Defects increase dramatically with increasing numbers of layers. A problem unique to application of this

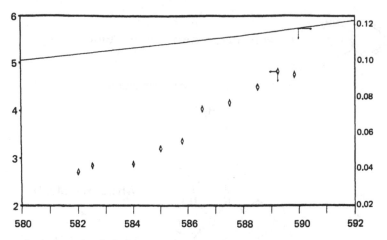

Figure 31. $\chi^{(3)}/\alpha$ ($\times 10^{-13}$ esu cm) and linear absorption (wavelength in nm) spectra of Pc–Cr(OH)(H$_2$O) in polycarbonate film.

technique to polymers is that high polymer viscosity inhibits the speed at which this fabrication technique can be effected. The most promising approach to the preparation of oriented polymers appears to be to use Langmuir-Blodgett methods to achieve monomer order followed by polymerization effected in the aligned film.

Smith et al. [147] proposed the production of oriented polymers using a reaction medium consisting of a mechanically coherent and stable gel fiber or film (a dilute network of carrier polymer filled with suitable solvent). Ingredients for the production of the final intractable polymer, such as catalyst, initiators, and/or monomers, are introduced into the gel, either through diffusion after the carrier gel is formed or prior to gel formation by dispersion or dissolution in the solution used to generate the gel. Subsequently, the ingredients are transformed into the final product by the proper chemical reactions

VI. CHARACTERIZATION OF OPTICAL NONLINEARITY

As mentioned in the introduction, a variety of techniques can be used to characterize optical nonlinearity. Not only do these different techniques measure different components of optical nonlinearity, but each experiment must be analyzed carefully to ensure meaningful interpretation of data and correlation of data between different experiments. To illustrate this point we now consider the analysis of a typical degenerate four-wave mixing experiment where the measured response depends upon the experimental configuration, upon molecular nonlinearities, and upon optical losses. Explicitly, we shall consider molecular optical nonlinearity characterized by electronic, thermal, and acoustic components. A typical DFWM experimental apparatus is illustrated in Fig. 32.

Laser pulses are split (by beam splitters BS1 and BS2) into three beams, which we call the "forward" beam F, the counter-propagating "backward" pump beam B, and the "probe" beam P (which impinges on the sample at a variable angle θ to beam F). Relative delays of the three incident pulses are achieved by mirrors DF and DP, which are moved on 1-m rails. The polarizations of all three incident beams are fixed by polarizers PF, PB, and PP. A signal pulse S is generated counter and phase conjugate to the probe beam P. The energy in each polarization component of beam S is measured by a calibrated photodiode PD2 after the analyzing polarizer PS. The incident and signal pulse energies can be determined by monitoring the energies at the calibrated photo-diodes and by knowing each beam splitter characteristic. The S pulse will depend upon the laser frequency, the characteristics (i.e., shape) of the laser pulses, the beam polarizations, and the relative pulse delays. The pulse width (FWHM) is determined from the behavior of the data as a function of beam delay (using the theory which we shall shortly present) and is verified using two-photon fluorescence.

A convenient formalism with which to analyze the DFWM experiment is quantum mechanical density matrix theory. This approach has the advantage of treating both diagonal (population) and off-

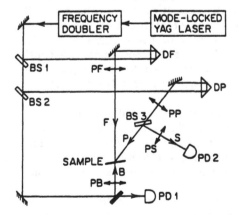

Figure 32. Typical DFWM experimental apparatus.

diagonal (dephasing) Hamiltonian terms. This approach also has
the advantage of permitting explicit treatment of the full details
of the experimental setup and permits as rigorous a treatment of
the molecular system as is possible given the complexity of the
molecular system. Quantum mechanics is applied to as much of the
problem as possible and the remaining aspects of the problem are
treated classically or phenomenologically. A detailed signal shape
is computed so that maximum information is extracted from the theoretical analysis. For example, use of a density matrix approach
[29] has shown that the temporal profile of the signal pulse is,
in general, different from that of the probe pulse, a result which
is inconsistent with the experiment treated as a diffraction process.
This will particularly be the case for pulse widths that are short
compared to molecular relaxation times (this is not typically the
case for the systems under consideration here). On the other hand,
pulse amplitudes are consistent with the diffraction description.
Our experience is that the density matrix approach is an excellent
bookkeeping procedure which ensures the treatment of vital terms.
In the following discussion, however, our attention is focused upon
illustration of the basics of theoretical analysis of the DFWM experiment; thus, we will not pursue the notational rigor of the density
matrix quantum mechanical Liouville equation. Quantum mechanics
is applied to as much of the problem as possible, and the remaining
aspects of the problem are treated classically or phenomenologically.

   In the following we review [32,148] the development of the
expression for pulse energy U_S detected by photodetector PD2 as
a function of the single pulse energies U_F, U_B, and U_P, measured
outside the sample. We obtain the signal expression as a function
of the following measured quantities: (1) the time delays t_F and
t_B of the F and B beam pulses with respect to the probe pulse P;
(2) the full-time width at half-maximum δt of the YAG laser pulse;
(3) the beam diameter D at half intensity of the congruent F, B,
and P pulses measured at the sample; and (4) the linear absorption
coefficient α and refractive index n of the sample at the pulse frequency ω; (5) the thickness of the sample L; and (6) the four
beam polarizations.

   We [32] make the following assumptions: (1) all beams are
directed near the forward or backward directions; (2) the pulse
width $c\delta t$ is much longer than the thickness L of the sample; and
(3) the three incident beams are negligibly affected by the nonlinear
interaction so that the phase conjugate reflectivity $R = U_S/U_p$ is
much less than unity.

   The real optical field in the sample medium can be expressed
as $\varepsilon_i(\underline{r},t) = \mathrm{Re}\, E_i(\underline{r},t)e^{-i\omega t}$, where E_i is the complex amplitude
which equals the sum of the amplitudes of the four beams $F_i + B_i +$

$P_i + S_i$. The three incident beam amplitudes are assumed to be given in the sample ($0 < z < L$) by

$$F_i = f_i F_0 \, u(t-t_F) \rho(x,y) e^{i\underline{k}_F\underline{r}-\alpha z/2} \tag{1}$$

$$B_i = b_i B_0 \, u(t-t_B) \rho(x,y) e^{i\underline{k}_B\underline{r}-\alpha(L-z)/2} \tag{2}$$

$$P_i = p_i P_0 \, u(t-t_F) \rho(x,y) e^{i\underline{k}_P\underline{r}-\alpha z/2} \tag{3}$$

where f_i, b_i, and p_i are normalized complex polarization vectors; $u(t)$ is the normalized temporal profile; the transverse profile is assumed to be Gaussian:

$$\rho(x,y) = \exp\left[-2(x^2+y^2)\frac{\ln 2}{D^2}\right] \tag{4}$$

with diameter D at half intensity. The F_0, B_0, and P_0 are the peak complex amplitudes on axis. The forward and backward beams are two opposite pump beams ($k_F = (n\omega/c)$, $\underline{k}_B = -\underline{k}_F$).

The three applied beams cause a nonlinear polarization density in the medium, which we write as Re $P_i^{NL}e^{-i\omega t}$. This polarization density radiates the output field S_i according to the classical dipole radiation law modified for attenuation α. In the first part of our analysis, we shall assume the nonlinear polarization density to be composed of electronic and thermal components. We shall first focus upon the former and then discuss the insights gained from analysis of the laser-induced bulk acoustic wave.

The fast "electronic" component of the optical nonlinearity can be expressed in terms of the conventional c tensor as

$$P_i^{NLE} = 6C_{ijkl}(-\omega,\omega,\omega,-\omega)F_jB_kP_l^* \tag{5}$$

For the moment, we also consider a thermal component given by

$$P_i^{NLT} = \chi^T F_i(t) \int_{-\infty}^{t} a \, ds \, P_j^*(s)B_j(s) \tag{6}$$

The complete nonlinear polarization density radiates a phase-matched signal field of amplitude S_i which can be calculated at the radiating sample face ($z = 0$) or in the far field ($Z \to -\infty$). In either case the signal pulse energy is given by

$$U_S = nc \int_{-\infty}^{\infty} \frac{dx \, dy \, dt \mid S_i(x,y,z,t)\mid^2}{8\pi} \tag{7}$$

A simple method to achieve the desired radiated pulse energy is to insert the temporal and spatial variations (Eqs. (1)-(3) in P_i^{NT} and compute the far-field radiated pattern from the classical dipole radiation formula as was done in Ref. 149. The result is most easily expressed for our purposes in terms of modifications of the cw expression for the phase conjugate reflectivity with plane pump waves. These modifications are expressed in terms of ξ-factors:

$$R \equiv \frac{U_S}{U_P} = \beta^2 l_c^2 I_F I_B \xi_D^2 \ \xi_{FBPS} \xi_{0\nu} \ \xi_\delta t^2 \ \xi_{BD}(t_F, t_B) \qquad (8)$$

The coupling coefficient β is defined in terms of the c coefficient and the normalized beam polarization vector as

$$\beta = \frac{196\pi^2 \omega C_{SFBP} I}{n^2 c^2} \qquad (9)$$

where

$$C_{SFBP} = C_{ijkl}(-\omega, \omega, \omega, -\omega) \ s_i^* f_j b_k p_l^* \qquad (10)$$

The effective interaction length l_c has the usual form

$$l_c = \frac{e^{-\alpha L/2}(1 - e^{-\alpha L})}{\alpha} \qquad (11)$$

which becomes the sample length in the limit of low absorption.

The nominal forward pump beam intensity in the equation for R is defined as the exterior pulse energy divided by the nominal beam area ($\pi D^2/4$) and by the pulse duration δt:

$$I_F = \frac{U_F}{\pi D^2 \delta t/4} \qquad (12)$$

and similarly for the incident backward pump intensity I_B. The interior forward incident F beam pulse energy is

$$\xi_F U_F = nc \iiint \frac{dx \ dy \ dt \ |\rho(x,y)u(t-t_F)F_0|^2}{8\pi} \qquad (13)$$

where ξ_F represents the reduction, by Fresnel reflection, of the interior intensity from the exterior intensity. There are similar definitions for the other beams. The factor ξ_{FBPS} is the product of four such factors for all beams and is approximately 0.77 for a polymer film on glass.

The factor ξ_D represents corrections to both I_F and I_B when U_S is not computed with flat pump beam profiles (the plane-wave case):

$$\xi_D = \frac{\pi D^2}{4} \mid \int dx \ dy \ |\rho(x,y)|^2 \tag{14}$$

If the pump beams are Gaussian, $\xi_D \to \ln 2$, which we will assume for these calculations.

The factor $\xi_{0\nu}$ gives a correction when the diameter D of the pump beams is not large compared to that of the probe beam. Suppose that the transverse probe beam profile is not $\rho(x,y)$ but, a different, $\rho_p(x,y)$. Then calculating U_S shows the overlap factor to be

$$\xi_{0\nu} = \int dx \ dy \ |\rho^2 \rho_p|^2 \int dx \ dy \ |\rho_p|^2 \tag{15}$$

This factor approaches unity for the broad pump profile ρ. In our case where ρ_p is approximately ρ in the Gaussian given by Eq. (4), then $\xi_{0\nu}$ approaches 1/3, the value we shall use in these calculations.

The factor $\xi_{\delta t}$ accounts for the fact that the peak pump pulse power is not quite $U_F/\delta t$ when the pulse envelope is not square. From the integral in Eq. (7)

$$\xi_{\delta t} = \frac{\delta t}{\int u^2 \ dt} \tag{16}$$

We assume the beam amplitude envelope function in Eqs. (1)-(3) is given by

$$u(t) = \text{sech } at \tag{17}$$

This gives results such as shown in Fig. 33. Obviously, reasonable reproduction of the experimental temporal profiles are obtained; moreover, the preceding equation allows for simple analytical results to be obtained for the several integrals above. Also, the above relationship fixes the time-scale parameter a in Eq. (6) in terms of which we have

$$\delta t \to \frac{2(\cosh^{-1} \sqrt{2})}{a} \approx \frac{1.763}{a} \tag{18}$$

and $\xi_{\delta t} \to 0.8813$. Note $\xi_{\delta t}$ would be 1 for a flat pulse and 0.9398 for a Gaussian temporal profile, indicating about a 6% uncertainty in the value of $\xi_{\delta t}$ in our experiment.

The beam delay factor has terms from both the fast and slow (transient) grating. We may write these as

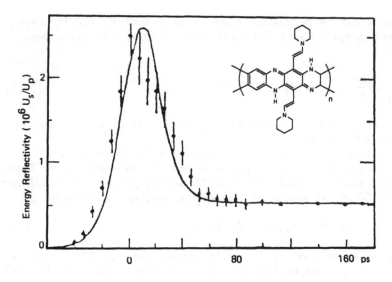

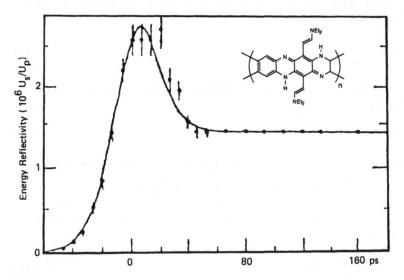

Figure 33.   Examples of DFWM signal pulse energy versus delay
time. Solid lines are from the theoretical calculations discussed
in the text. Uncertainty of the experimental points is indicated
by the vertical lines.

$$\xi_{BD} = \frac{\int dt \mid u(t)u(t-t_F)u(t-t_B) + g_T(t,t_F,t_B) \mid^2}{\int u^2 \ dt} \qquad (19)$$

In the limit of a no delayed slow response ($g_T = 0$) and no beam delays, ξ_{BD} approaches 1 for square pulses and approaches 0.5333 for the sech pulse of Eq. (17). The latter is 92% of the value for the Gaussian profile, indicating further possible systematic temporal profile error of approximately 8% in the same direction as the error in $\xi_{\delta t}$ of Eq. (8). Therefore, we take the probable limit of error in the factor $\xi_{\delta t}^2 \xi_{BD}$ to be ±20% when $g_T = 0$, with additional error for the separation of fast and slow components.

One may write the nonlinear susceptibility tensor element C_{SFBP} for isotropic media in terms of two independent elements, commonly called C_{1122} and C_{1221}, using the expression

$$C_{ijkl}(-\omega,\omega,\omega,-\omega) = C_{1122}(\delta_{ij}\delta_{kl} + \delta_{ik}\delta_{jl}) + C_{1221}\delta_{il}\ \delta_{jk} \qquad (20)$$

When all polarizations are parallel, the relevant combination is called C_{1111}, which is equal to $2C_{1122} + C_{1221}$. These C_{ijkl} coefficients of Maker and Terhume [150] are sometimes written as $\chi_{ijkl}^{(3)}$ with, less occasionally, a numerical factor.

To normalize our coefficients to others, it may be useful to use that fact that a linearly polarized electric field of amplitude E_ω driving a purely electronic nonlinearity at a low frequency ω creates a "third harmonic" polarization density at 3ω whose amplitude is $C_{1111}E_\omega^3$. In this case C_{1111} becomes the constant called $\chi^{(3)}$ most often used to describe third-harmonic generation. Of course, when the response is not purely electronic, this correspondence no longer exists. Caution must also be used when comparing our results with those of others utilizing DFWM. For example, Carter et al. [151] quote values of a quantity called $\chi^{(3)}$ for an isotropic polydiacetylene polymer which was obtained by DFWM of parallel-polarized, undelayed picosecond pulses. They expect their third-order susceptibility to approach the value obtained from harmonic measurements. However, a comparison of their Eq. (1) with our Eq. (8) shows that theirs equals $3C_{1111}$ or $6 \ln 2C_{1111}$, depending on whether one takes their "beam waist ω_0 cm" to be the beam radius at $1/e^2$ or at half-intensity points.

In addition to the responses we have already discussed we observe for some experimental configurations, a slowly oscillating tail (see Fig. 34). We attribute this latter response to the generation of thermal and acoustic grating in the polymer thin film. A laser-induced ultrasonic wave grating has been observed for molecular crystals [152] and other polymeric thin films [153]. A similar experiment has been used to study structural change during polymerization

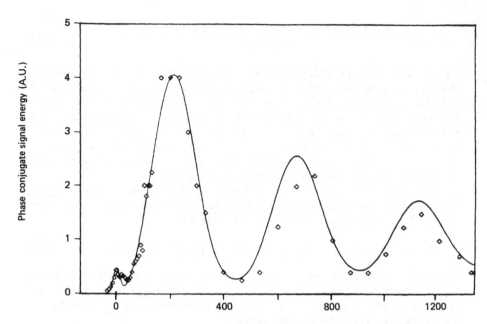

Figure 34. Energy of the phase conjugate signal versus the delay time of the forward pump pulse. The cross angle between the two forward beams is 22.6°. The solid line is derived from the theory presented in the text using the values of variables listed in Table 7. Experimental data for a ladder polymer sample are denoted by diamonds.

[154]. This effect can be used to measure sound velocity and damping constant and appears to be a useful route to improved insight into the organic solid state. As we shall shortly see, detailed analysis of this effect can provide insight into the signs of various components of the optical nonlinearity.

The physical interpretation of the laser-induced ultrasonic wave grating is as follows: When two coherent laser beams cross each other in an absorbing medium, an interference pattern is formed. If absorption of the laser energy results in heating of the medium, a spatially periodic temperature distribution is produced and the subsequent thermal expansion launches an ultrasonic standing wave whose wavelength and orientation match those of the intensity grating pattern. The propagation of this ultrasonic wave gives rise to a pressure density wave, thus producing an index of refraction variation that can Bragg scatter a third beam. However, right after the laser energy has been deposited, the medium has not yet had

Table 7. Constants Obtained from Fittings

Quantity[a]	Polymers		
	1	2	3
n	1.75	1.82	1.80
$\left[\dfrac{\partial n}{\partial \rho}\right]_T \rho_0$ [b]	0.99	1.12	1.16
$\left[\dfrac{\partial n}{\partial T}\right]_P$ [K^{-1}]	-7.28×10^{-5}	-2.79×10^{-5}	-5.34×10^{-5}
$\left[\dfrac{\partial n}{\partial T}\right]_\rho$ [K^{-1}]	-9.71×10^{-6}	-4.96×10^{-6}	-9.9×10^{-6}
ρ_0 [g/cm^3][c]	1.18	1.22	1.06
C_p [erg/(g-K)]	1.03×10^7	1.27×10^7	1.32×10^7
C_v [erg/(g-K)][c]	8.96×10^6	1.13×10^7	1.22×10^7
κ [erg/s-cm-K]	8.47×10^4	1.19×10^5	1.27×10^5
η [poise]	1.14	0.76	0.63
V_s [cm/s]	3.25×10^5	2.97×10^5	2.86×10^5
$\left[\dfrac{\partial \rho}{\partial p}\right]_T /\rho_0$ [atm^{-1}]	9.23×10^{-6}	1.02×10^{-5}	1.32×10^{-5}
$\left[\dfrac{\partial \rho}{\partial T}\right]_p /\rho_0$ [K^{-1}]	-5.53×10^{-5}	-2.05×10^{-5}	-3.75×10^{-5}
C_{1212} [esu]	-2.2×10^{-10}	-1.9×10^{-10}	-2.8×10^{-10}

[a]See text for notation.
[b]Estimate from Clausius-Mosotti relation.
[c]Measured quantity by other methods.

time to expand. The density remains constant, the index change is then due to the intrinsic variation in the index of refraction with the temperature at constant density. As the sound wave is produced, the pressure density wave gives rise to an oscillating refractive index variation. Finally, when the sound wave damps out, the usual thermal diffusion mode becomes dominant. Since the index variations are proportional to the temperature change induced by the crossing beams, this phenomena has been termed *forced Rayleigh scattering* [155].

In the following paragraphs, we review a hydrodynamical calculation [147], including the effect of sound waves as well as thermal
diffusion on the phase conjugate reflectivity. Then we will study
the time evolution of thermal-induced density and temperature variation in the medium by analytically solving the Navier-Stokes equation
and thermal diffusion equation, and use our theoretical model to
describe the phase conjugate signal including both the fast electronic
and the thermally induced acoustic part.

There are two gratings which can be written in the medium,
the grating written by the forward pump beam and the probe beam
is called the Δk grating, and the grating written by the backward
pump beam and the probe is called the 2k grating. We will consider
these two gratings separately.

Assume two writing beams form an interference pattern in the
medium. The intensity variation is

$$\Delta I(r,t) = \Delta I(r,t=0)u\ (t) \tag{21}$$

where

$$\Delta I(r,t=0) = \begin{cases} \dfrac{nc}{8\pi}\ \text{Re}\ F_i(\underline{r},t=0)\ P_i{}^*\ (\underline{r},t=0) & \text{for}\ \Delta k\ \text{grating} \\[1.5em] \dfrac{nc}{8\pi}\ \text{Re}\ B_i(\underline{r},t=0)\ P_i{}^*\ (\underline{r},t-0) & \text{for}\ 2k\ \text{grating} \end{cases} \tag{22}$$

Assuming the light absorbed by the medium is totally converted
to heat, and the intensity grating gives rise to a spatially modulated
heat production rate per unit volume:

$$h(\underline{r},t) = \alpha\ \Delta I(\underline{r},t) \tag{23}$$

This thermal effect creates a thermal nonlinear polarization
density in the medium which can be expressed as

$$P_i{}^{NLT}(\underline{r},t) = \frac{n\ \Delta n(\underline{r},t)}{2\pi}\ \varepsilon_i(\underline{r},t) \tag{24}$$

Here $\Delta n(\underline{r},t)$ is the thermally induced refractive index variation.
If we assume that it has the same spatial dependence of the intensity
variation $\Delta I(\underline{r},t)$ but different time dependence, $\Delta n(\underline{r},t)$ can be
be expressed as

$$\Delta n(\underline{r},t) = n_2{}^T(t)\ \Delta I(\underline{r},t = 0) \tag{25a}$$

Of course, the fast-responding electronic component will be expressed as before.

In order to study the dynamics of the index grating, we must relate the refractive index variation to a measurable quantity which is the phase-conjugated reflectivity with R expressed as before by Eq. (8).

The slow responding part $g_T(t,t_B$, or $t_F)$ which represents the time dependence of the thermally induced index variation can be expressed as

$$g_T(t,t_d) = \frac{u(t-t_d)n c n_2{}^T(t)}{24\pi C_{SFBP}} \qquad (25b)$$

where t_d is the delay time, $t_d = t_B$ for the Δk grating, and $t_d = t_F$ for the 2k grating.

To study the time evolution of the index grating $\Delta n(\underline{r},t)$ we first express it in terms of the density and temperature variations in the medium:

$$\Delta n(\underline{r},t) = \left[\frac{\partial n}{\partial \rho}\right]_T \Delta\rho\,(\underline{r},t) + \left[\frac{\partial n}{\partial T}\right]_\rho \Delta T(\underline{r},t) \qquad (26)$$

where $\Delta\rho\,(\underline{r},t)$ and $\Delta T(\underline{r},t)$ are the local density and temperature variations, respectively. Since the polymers of interest here are highly absorptive at the operating laser wavelength, we can assume that the acoustic wave is created by thermal expansion and neglect the electrostriction term. Also, as we have arranged beam polarizations so as not to excite any transverse wave, we will consider only the longitudinal wave.

As the basic equations to be solved we choose the equation of continuity, the Navier-Stokes equation, and the energy transport equation.

$$\frac{-\partial^2\Delta\rho}{\partial t^2} + \left(\frac{V_S^2}{\gamma}\right)\nabla^2\,\Delta\rho + \frac{\eta\nabla^2\,\Delta\rho}{\rho_0} + \left(\frac{V_S^2\beta\rho_0}{\gamma}\right)\nabla^2\Delta T = 0 \qquad (27)$$

$$\frac{\partial\,\Delta T}{\partial t} - \frac{\lambda}{\rho_0 C_V}\,\nabla^2\,\Delta T - \frac{\gamma-1}{\beta\rho_0}\frac{\partial\,\Delta\rho}{\partial t} = \frac{h}{\rho_0 C_V} \qquad (28)$$

Here ρ_0 is the equilibrium density $\eta = \kappa + 4\,\zeta/3$; κ, ζ are the bulk and shear viscosity, respectively; $\beta = -(1/\rho_0)(\partial_\rho/\partial T)p$ is the thermal cubic expansion coefficient, λ is the heat conductivity coefficient, $\gamma = C_p/C_V$ is the ratio of the specific heat at constant pressure and at constant volume, and h is the heat production rate resulting from the absorption of the laser energy.

If we again assume $\Delta\rho(\underline{r},t)$ and $\Delta T(\underline{r},t)$ have the same spatial dependence of the intensity variation $\Delta I(\underline{r},t)$, then by trying solutions with the form

$$\Delta\rho(t) = \Delta\rho(t)\ \Delta l(\underline{r},t=0) \tag{29}$$

$$\Delta T(t) = \Delta T(t)\ \Delta l(\underline{r},t=0) \tag{30}$$

we can obtain the exact solution for $\Delta n(t)$ and $\Delta T(t)$. Using Eq. (26), we can express the refractive index variation as

$$\Delta n(\underline{r},t) = \frac{\alpha}{\rho_0 C_v}\ \Delta I(\underline{r},t=0) \int_{-\infty}^{t} G(q,t-t')u^2(t')\ dt \tag{31}$$

where $G(q,t-t')$ is the Green's function:

$$G = \frac{r_s\alpha_1^2 - \mu\alpha_1 + \nu}{(\alpha_1-\alpha_2)(\alpha_1-\alpha_3)}\ e^{-\alpha_1 t} + \frac{r_s\alpha_2^2 - \mu\alpha_2 + \nu}{(\alpha_2-\alpha_1)(\alpha_2-\alpha_3)}\ e^{-\alpha_2 t}$$
$$+ \frac{r_s\alpha_3^2 - \mu\alpha_3 + \nu}{(\alpha_3-\alpha_1)(\alpha_3-\alpha_2)}\ e^{-\alpha_3 t} \tag{32}$$

In the above equation, $r_s = (\partial n/\partial T)_\rho$, $\mu = r_s \eta q^2/\rho_0$, $\nu = (\partial p/\partial\rho)_T\ r_p q^2$, and:

$$r_P = \left[\frac{\partial n}{\partial T}\right]_p = r_s + \left[\frac{\partial n}{\partial\rho}\right]_T \left[\frac{\partial\rho}{\partial T}\right]_p$$

Here q is the amplitude of the grating wave vector. Constants α_1, α_2, α_3 are the solutions of the cubic dispersion relation:

$$[\omega^3 + \omega^2(\Gamma_B + \gamma\Gamma_R) + \omega(\omega_0^2 + \gamma\Gamma_B\Gamma_R) + \omega_0^2\ \Gamma_R] =$$
$$(\omega + \alpha_1)(\omega + \alpha_2)(\omega + \alpha_3) = 0 \tag{33}$$

where $\Gamma_B = \eta q^2/\rho_0$ is the Brillouin bandwidth, $\Gamma_R = \lambda q^2/\rho_0 C_n$ is the Rayleigh bandwidth, and $\omega_0 = V_s q$ is the acoustic frequency. Under the approximation $\omega_0 \gg \Gamma_B$ or Γ_R, we obtain solutions of the cubic equation:

$$\alpha_1 = -i\omega_0 + \Gamma \tag{34}$$

$$\alpha_2 = i\omega_0 + \Gamma \tag{35}$$

$$\alpha_3 = \Gamma_R \tag{36}$$

with $\Gamma = [\Gamma_B + (\gamma-1)\Gamma_R]/2$. Notice that α_1 and α_2 are complex conjugated, representing two opposite propagating sound waves which form a standing damped acoustic wave which gives rise to the damped oscillating nature of the index change $\Delta n(\underline{r}, t)$. With this result, the Green's function becomes

$$G(q, t-t') = (r_s - r_p) e^{-\Gamma(t-t')} \cos \omega_0(t-t') + r_p \, e^{-\Gamma_R(t-t')} \qquad (37)$$

We can now calculate the slowly responding part g_T and the beam delay factor and thus predict the time evolution of the phase-conjugated signal (the experimental observable). If we assume $u(t) = \exp(-a^2 t^2/2)$ and $a \gg \omega_0 \gg \Gamma, \Gamma_R$, we obtain

$$\xi_{BD}(t_d) = \frac{1}{\sqrt{3}} \, \exp\left(\frac{-2a^2 t_d^2}{3}\right)$$

$$+ \frac{\sqrt{2} \, C \, \exp(-a^2 t_d^2/2)}{1 + \exp(-1.1 a t_d)} \, [\exp(-\Gamma_R t_d)$$

$$- r_0 \, \exp(-\Gamma_B t_d) \cos(\omega_0 t_d)]$$

$$+ \frac{C_2 [\exp(-\Gamma_R t_d) - r_0 \, \exp(-\Gamma_B t_d) \cos(\omega_0 t_d)]^2}{[1 + \exp(-2.1 a t_d)]^2} \qquad (38)$$

where

$$C = \frac{\sqrt{\pi} \, r_p \alpha n c}{24\pi^2 \, C_{SFBP} \rho_0 C_p a} \qquad (39)$$

$$r_0 = \frac{1 - \gamma r_s}{r_p} \, \exp \frac{-\omega_0^2}{2a^2} \qquad (40)$$

In this calculation, we used the approximation $\mathrm{erfc}(x) \simeq 2/(1+e^{2.5x})$; the error of this approximation is less than 1%. The first term in the above expression for the beam delay factor represents the fast-responding part, the second is the cross term between the fast part and the acoustic part which is small for a Δk grating, and the third term is the damped oscillating acoustic part. Notice that when t_B is small there will be a contribution from the instantaneous thermal grating due to the nonzero r_s. It takes half an oscillation period to reach the first acoustic peak. Here we also assume that the beam diffracted from the thermal induced acoustic grating is in phase with that from the fast-responding grating.

A typical experimental curve is given in Fig. 34 with theoretical curves computed for the parameters listed in Table 7 (these are best-fit values).

That the first peak appears at half period of oscillation after the zero delay clearly justifies the neglect of the electrostriction term. We also see that the pulse width is much shorter than the oscillation period and the decay time for small q grating. Since $r_s \neq 0$, the signal does not fall to zero at very short delay time. These data give us a unique way to measure r_s. The acoustic decay time is about 3-4 ns for the Dk grating. The thermal diffusion decay time is only measurable at Δk grating, and from that measurement we can obtain the thermal conductivity. The amount of information one can get out of this simple experiment is quite sizable; with a knowledge of density ρ, heat capacity C_p, and the estimate of $[\partial n/\partial p]_T$ from the Clausius-Mosotti relation, one can obtain most of the thermodynamic constants of interest (see Table 7).

The cross term between the fast-responding (electronic) and the acoustic part is important because it can determine the phase difference between two processes. Clearly the cross term becomes larger in the 2k grating case. Data fitting shows that these two processes are in phase. Unlike other experiments, from this observation we believe that both the fast-responding and acoustic parts are due to a phase-grating mechanism (change in the real part of the refractive index). The fact that they are in phase indicates that C_{SFBP} has the same sign as r_p, which is normally negative. This result is similar to the measurement of third-order susceptibility reported for polydiacetylene and is quite different [155a] from that reported for linear polyenes.

VII. FINAL COMMENTS

Rather than simply summarizing the preceding sections, we shall attempt to address issues and review data that have come to our attention since the first draft of this manuscript was submitted for editorial consideration.

The first point that we address is the assertion that the magnitude of optical nonlinearity in organic materials is always moderate to small and the response time for this nonlinearity is fast. Materials with long-lived states involved in the excitation/relaxation cycle can exhibit large third-order susceptibilities characterized by slow response times as is expected from an analysis of the evolution of state populations. For example, Tompkin et al. [156] have observed acridine orange and acridine yellow dyes dissolved in lead-tin fluorophosphate glass to exhibit third-order susceptibilities on the order of 0.1 esu characterized by response times of approximately 1 ms. As might be expected, this observation is lattice-dependent consistent with the dependence of triplet state lifetimes on lattice.

It is important to realize that many potential mechanisms for optical nonlinearity exist for organic and organometallic materials. While general trends such as the relationship between state lifetimes and optical nonlinearity can be discussed for particular models, the range of possibilities relevant to device considerations simply cannot be deduced from available data at this time. The preceding observation together with our earlier comment about misconceptions regarding the ease of processing of electroactive polymers suggests perhaps that we should be more careful in drawing distinctions between organic and inorganic systems in terms of their relative advantages and disadvantages.

With regard to simple but intriguingly effective theoretical modeling, we mention the coupled anharmonic oscillator (CAO) model calculations of Prasad, Perrin, and Samoc [44]. They have used this model to calculate band gap, polarizability, and hyperpolarizability as a function of the number of repeat units for oligomers of thiophene and benzene. In this simple model, the coupling constant tensor K models the effect of electron delocalization while the anharmonic coefficient b models structural factors such as substituent effects. The coupling tensor can also be expressed as

$$K = -k(\omega_0)^2$$

where ω_0 is the resonance frequency of a single unit and $0 < k < 0.5$. With the CAO model, a graph of the ratio of second hyperpolarizability γ to the number of repeat units N plotted against N will obviously depend upon k. For $k = 0$, no coupling exists and γ/n will be independent of N. For small values of k, saturation (a maximum in the graph) of γ/N versus N occurs quickly (e.g., by $N = 10$). For $k = 0.5$, saturation is observed only at very large N. For π-electron systems such as those considered by Prasad and co-workers, the coupling tensor attempts to model the through space electron coupling arising from the β overlap integrals (or hopping integrals in the language of physicists). It is reasonable to assume that the CAO model will work best for systems with strong π-electron coupling within repeat units and an interunit coupling which does not vary as one moves from one repeat unit to the next. In structural terms, this would correspond to a constant dihedral angle between, for example, adjacent phenyl rings for the phenyl oligomers. It should be kept in mind that steric interactions determine dihedral angle and thus substituents may produce significant variations in k. Another important point to keep in mind is that K depends only upon a single characteristic frequency ω_0 which in turn corresponds to a two-state model. While the two-state model has proven useful in the consideration of second-order susceptibilities, the data on

third-order susceptibilities are insufficient to draw such conclusions.
In particular, for heteroatomic systems one might anticipate that
nonbonding as well as π states might contribute to nonlinear optical
processes. It is, of course, consideration of these detailed features
of the electronic system that argue for quantum mechanical calculations. On the other hand, it seems clear from available data that
π-electrons dominate nonlinear processes and that many features
of the total electronic problem such as sigma states can be safely
neglected in many cases.

Fortunately, Prasad and co-workers [157,158] have carried
out quantum mechanical (QM) calculations on the same oligomeric
systems examined employing the CAO model. They have, moreover,
effected an interesting correlation of the CAO model results with
those from their QM calculations and QM calculations of other workers
[14,159-161]. The QM results can frequently be fit to a power law
expression of the form

$$\alpha \text{ or } \gamma = A + B(N - \delta)^c$$

where A and δ take into account end group effects and adjustment
for effective conjugation length. A systematic study [158] of the
dependence of $<\alpha>$ and $<\gamma>$ on chain length employing ab initio
methods yielded a value for the exponent value c in the range
1.3-1.5 for polarizability and 3.2-3.4 for second hyperpolarizability.
In contrast, a free-electron model [159] predicted 3 and 5 for α
and γ, respectively, while a PPP model [160] yielded 2.3 and 5.3.
The CAO model of Prasad and co-workers yielded an α of 1.7 for
thiophene oligomers and values for γ of 4 and 3.2 for thiophene
and benzene oligomers, respectively. The CAO model also yields
ratios of longitudinal to transverse polarizability and hyperpolarizability which are in good agreement with QM ratios. The agreement
of CAO results with experimental values is evident from Table 8.

It is not reasonable to expect the CAO model to work for every
oligomeric series; however, when it is found to work for several
members of a series it has obvious predictive utility and provides
a simple and convenient scheme for rationalizing existing data with
a limited set of parameters.

The recent study [48] of second hyperpolarizability for aryl
vinyl derivatives of ferrocenes is another excellent example of a
study providing insight into trends and providing a correlation
between theory and experiment. For these systems, second hyperpolarizability increases strongly with the length of the conjugated
π-electron systems (analogous to the systems discussed above) and
the effective conjugation is determined predominantly by the length
of the aryl vinyl system. The contribution from the ferrocenyl group

Table 8. Experimental and Calculated Values of the Band Gap, Orientationally Averaged Polarizabilities, Orientationally Averaged Second Hyperpolarizabilities for the Thiophene and Benzene Oligomers[a]

N	E_g expt (eV)	E_g cal (eV)	(α) expt (cm^{-3})	(α) cal (cm^{-3})	(γ) expt (esu)	(γ) cal (esu)
1 thiophene	5.39	5.39	9.8×10^{-24}	9.8×10^{-24} [b]	4.1×10^{-36}	4.1×10^{-36} [b]
2	4.12	4.15	2.5×10^{-23}	2.4×10^{-23}	2.2×10^{-35}	2.1×10^{-35}
3	3.55	3.55	4.5×10^{-23}	4.4×10^{-23}	1.6×10^{-34}	1.5×10^{-34}
4	3.19	3.22	1.0×10^{-22}	1.0×10^{-22}	8.0×10^{-34}	7.5×10^{-34}
5	3.00	3.00	1.9×10^{-22}	2.1×10^{-22}	2.6×10^{-33}	2.4×10^{-33}
6	2.90	2.91	4.5×10^{-22}	4.5×10^{-22}	1.0×10^{-32}	6.9×10^{-33}
1 benzene	6.11	6.11			5.5×10^{-36}	5.5×10^{-36} [b]
2	5.04	5.00			2.9×10^{-35}	2.7×10^{-35}
3	4.49	4.51			8.5×10^{-35}	8.7×10^{-35}
5[c]	3.76	4.10			2.1×10^{-34}	4.3×10^{-34}

[a]The laser wavelengths used for the measurement of polarizabilities and second hyperpolarizabilities are 632.8 and 602.0 nm, respectively.
[b]This value has been taken to normalize all the parameters.
[c]This molecule is pentaphenyl with a $OC_{10}H_{21}$ substituent group on the center phenyl ring.
Source: Ref. 44.

is less significant. Thus, optical nonlinearity appears little affected
either by d-π coupling or by d-d resonances. Qualitative agreement
is obtained with the semiempirical quantum mechanical calculations
of Waite and Papadopoulos [47] although the experimental value
for ferrocene is about a factor of 4 greater than the theoretical
value. This discrepancy could arise either from resonance contributions to the experimental value (measured at 602 nm) or from limitations of the theoretical method.

Another important advance is the theoretical semiempirical calculations of second hyperpolarizability for bisthiazole model compounds
[162] which consider length, substituent, and conformational effects.
Experimental studies now underway should permit an important
correlation of effects in this important series of heteroaromatic materials.

The study of Alfano and co-workers [163] of thiophene-based
polymers provides important insight into resonance and nonresonance
contributions to optical nonlinearity. They have investigated the
wavelength dependence of third-order susceptibility over the range
532 to 1064 nm. Values in the range 532 to 640 are observed to
increase as one progresses to higher energy consistent with resonant
contributions dominating the optical nonlinearity. At 532 nm, third-
order susceptibility values ranging between 6 and 11 $\times$ 10^{-9} esu
are observed. At 1064 nm, the third-order susceptibility is on the
order of 3 $\times$ 10^{-11} esu. This is very similar to behavior that we
have observed for ladder polymers, phenylpolyenes, and a variety
of organic and organometallic dye copolymer systems (see synthetic
section of this chapter). A representative example is the copolymer
data shown in Fig. 25.

The effect of electron donating and withdrawing substituents
on optical nonlinearity has been both theoretically predicted and
experimentally observed as is evident from a reading of this chapter.
More quantitative insight into this phenomenon must await further
data. What still remains a question is the effect of doping (or the
population of intragap states). Both chemical and photoinduced
population and depopulation of polaron and bipolaron states need
to be investigated; hopefully, such studies will appear in the near
future.

Finally, in the accompanying appendix, we provide a tabulation
of some representative third-order susceptibility and hyperpolarizability data. This is done simply to provide the reader access to
the literature and no attempt is made at standardizing values.

ACKNOWLEDGMENTS

This work was supported in part by Air Force Office of Scientific
Research contracts F49620-87-C-0100 and F49620-88-C-0071. Acknowl-

edgment is also made to the National Science Foundation (Grant DMR-88-15508). We wish to thank Dr. Donald Ulrich and Professor Paras Prasad for helpful discussions.

APPENDIX. Third-Order Susceptibility Data for Selected Materials

Material	$\chi^{(3)}$ (esu)	λ(eV)	Reference
Polyacetylene	0.5×10^{-10}	0.65-1.5	34
Polyacetylene	5.0×10^{-9}	2	3
	4×10^{-10}	1.17	
Soluble polyacetylene	0.9×10^{-9}	2.33	46
Polydiacetylene	1.6×10^{-10}	0.47	6
Polytoluene sulfonate	8.5×10^{-10}	0.66	
(PDA-PTS)	9.0×10^{-9}	On resonance (1.91)	40
	5×10^{-10}	Off resonance	
PDA-TCDU	$0.37-0.7\times10^{-10}$	0.47-0.66	6
2d-PDA, 2j-PDA	7×10^{-12}	2.33	164
PDA-4BCMU	4×10^{-10}	2.07	42
PDA-mBCMU monomer	1.6×10^{-13}	1.17	31
Yellow	1.4×10^{-12}	1.17	
Red (4BCMU)	1.3×10^{-11}	1.17	
Blue (3BCMU)	9×10^{-10}	1.17	
PMMA/PDA-4BOMV	3×10^{-10}	On resonance	37
	3×10^{-11}	Off resonance	
PDA-LB films	3×10^{-12}	1.17	165
Derivatized PDA LB-film (EFISH)	1.3×10^{-12}	Off resonance	38
blue form	7.0×10^{-12}	On resonance	
Blue form LB-film (THG)	6×10^{-11}	On resonance	
Red 4BCMV (EFISH)	9×10^{-12}	On resonance	
Blue/red 3BCMV(EFISH)	7×10^{-12}	On resonance	

(continued)

(Appendix, Continued)

Material	$\chi^{(3)}$ (esu)	λ(eV)	Reference
Derivatized PDA	4×10^{-9} 4×10^{-11}	On resonance Off resonance	40
Vapor-deposited PDA	1.4×10^{-11} 6.5×10^{-12}	0.65 0.61	166
Derivatized PDA	1.1×10^{-11}	0.65	167
PDA-pDCH	1.2×10^{-10}	0.89	168
PBT (poly-*p*-phenylene benzobisthiazole)	7.2×10^{-12}		11,174
LARC-TPI	1×10^{-12}		15
PPV (poly-*p*- phenylenevinylene)	7.8×10^{-12} 5×10^{-12}	0.67 0.58	169
Dimethoxy-PPV	2.4×10^{-9}		15
Oriented-PPV	4×10^{-10}		15
Modified polythiophene	$0.05\text{-}1.81\times10^{-7}$	2.33	170
Polydithieno(3,2-b 2',3'-b)thiophene (pDTT)	$0.1\text{-}1.15\times10^{-8}$	1.94-2.33	174
Polythiophene	4×10^{-10}	2.07	157a
Polycondensed PT	6.5×10^{-9} $>3\times10^{-9}$	2.07 1.17	163
Thiophene oligomers	$0.005\text{-}1.0\times10^{-10}$	2.06	44,157b
Polysilanes	1.5×10^{-12} $7.2\text{-}10\times10^{-12}$ $1.3\text{-}1.9\times10^{-12}$	1.17 1.17 0.65	175 176
Polygermanes	6.5×10^{-12} 1.4×10^{-12}	1.17 0.65	177
Pendant Azo dye	1.3×10^{-12}	0.61	178
Polyacenequinones	1.1×10^{-11}	2.38	179
Fluoro-Al PC Phethalocyanine	5×10^{-11}	1.17	37

Material	$\chi^{(3)}$ (esu)	λ(eV)	Reference
Metal-free tetra-kiscumylphenoxy PC LB-film	$3-10\times10^{-9}$	2.07	174
BETD-TTF	5.0×10^{-8}	1.91	180
4-Dimethylamino-4'-nitrostilbene (DEANS)	$(6+4i)\times10^{-12}$	1.55-2.07	177
COT/COD copolymers %COT			
8	3.3×10^{-13}		73
15	6.0×10^{-13}		
27	1.3×10^{-13}		
32	1.6×10^{-13}		
Aniline	32×10^{-12}		182
Nitrobenzene	3.32×10^{-12}		
m-Nitroaniline	3.32×10^{-12}		
o-Nitroaniline	4.83×10^{-12}		
p-Nitroaniline	1.97×10^{-11}		
Polydiacetylene-PTS	1.6×10^{-10} [a]		
Polydiacetylene-TCDU	3.7×10^{-11} [a]		6
GaAs	1.2×10^{-11} [b]		171
InSb	5×10^{-11} [b]		
B-carotene	$<1.4\times10^{-13}$		181
Nigrosine	1.8×10^{-12}		
DTIC	5.7×10^{-13}		
DNTPC	7.3×10^{-13}		
BDN	1.2×10^{-12}		
A9860	1.3×10^{-12}		
IR5	1.5×10^{-12}		
SFO1	9.0×10^{-12}		
Arylvinyl derivatives of ferrocenes	$\gamma=1.6-155\times10^{-36}$	at 2.06 eV	48

[a] At 2.62 μm.
[b] Far from the band gap.

REFERENCES

1. J. Messier et al., eds., *Nonlinear Optical Effects in Organic Polymers*, Kluwer Academic, Boston, 1989.
2. D. S. Chemla and J. Zyss, eds., *Nonlinear Optical Properties of Organic Molecules and Crystals*, Vol. 2, Academic Press, New York, 1987.
3. A. J. Heeger, J. Orenstein, and D. R. Ulrich, eds., *Nonlinear Optical Properties of Polymers*, Mats Res. Soc. Symp. Proc., Vol. 109, Materials Research Society, Pittsburgh, 1988.
4a. R. A. Hann and D. Bloor, eds., *Organic Materials for Nonlinear Optics*, Royal Society of Chemistry, 1989.
4b. D. Jerome and L. G. Caron, eds., *Low Dimension Conductivity and Superconductivity*, NATO ASI Series, New York, 1987.
5. K. C. Rustagi and J. Ducuing, *Opt. Comm.*, *10*:258 (1974).
6. C. Sauteret, J. P. Hermann, R. Frey, F. Pradere, J. Ducuing, R. H. Baughman, and R. R. Chance, *Phys. Rev. Lett.*, *36*:956 (1976).
7. G. P. Agrawal, C. Cojan, and C. Flytzanis, *Phys. Rev. B*, *17*:7765 (1978).
8. G. P. Agrawal and C. Flytzanis, *Chem. Phys. Lett.*, *44*:366 (1976).
9. G. P. Agrawal and C. Flytzanis, *ACS Symp. Ser.*, *233* (1983).
10. D. N. Beratan, J. N. Onuchic, and J. W. Perry, *J. Phys. Chem.*, *91*:2696 (1987).
11. P. N. Prasad and D. R. Ulrich, eds., *Nonlinear Optical and Electroactive Polymers*, Plenum Press, New York, 1987.
12. J. R. Heflin, K. Y. Wong, O. Zamani-Khamiri, and A. F. Garito, *Phys. Rev. B*, *38*:1573 (1988).
13. J. W. Wu, J. R. Heflin, R. A. Norwood, K. Y. Wong, O. Zamani-Khamiri, A. F. Garito, P. Kalyanaraman, and J. Sounik, *J. Opt. Soc. Am.*, *6*:707 (1989).
14. C. P. de Melo and R. Silbey, *Chem. Phys. Lett.*, *140*:537 (1987).
15. M. Druy, *Proc. First Int. Symp. on Nonlinear Optical Polymers for Soldier Survivability*, to be published.
16. P. N. Prasad, *Nonlinear Optical Properties of Polymers*, A. J. Heeger et al., eds., Materials Research Society, Pittsburgh, 1988, p. 271.
17. A. O. Patel, Y. Shenoue, F. Wudl, and A. J. Heeger, *J. Am. Chem. Soc.*, *109*:1858 (1987).
18. R. L. Elsenbaumer, K. Y. Jen, G. G. Miller, and L. W. Shacklette, *Synth. Met.*, *18*:277 (1987).
19. M. Sato, S. Tanaka, and K. Kaeriyama, *Synth. Met. 18*:229 (1987).
20. L. R. Dalton, *Nonlinear Optical and Electroactive Polymers*, P. N. Prasad and D. R. Ulrich, eds., Plenum Press, New York, 1987, p. 243.

21. L. R. Dalton, *Nonlinear Optical Properties of Polymers*, A. J. Heeger et al., eds., Materials Research Society, Pittsburgh, 1988, p. 301.
22. L. R. Dalton, *SPIE Proc.*, *878*:102 (1988).
23a. L. P. Yu and L. R. Dalton, *Synth. Met.*, *29*:E463 (1989).
23b. L. P. Yu, Ph.D. Dissertation, University of Southern California, 1989.
23c. L. P. Yu and L. R. Dalton, *Macromol.*, *233439* (1990).
24. L. R. Dalton, *Nonlinear Optical Effects in Organic Polymers*, J. Messier et al., eds., Kluwer Academic, Boston, 1989, p. 123.
25. R. W. Lenz, C.-C. Han, J. Stenger-Smith, and F. E. Karasz, *J. Polym. Sci. A*, *26*:3241 (1988).
26. S. Antoun, F. E. Karasz, and R. W. Lenz, *J. Polym. Sci. A*, *26*:1809 (1988).
27. T. Granier, E. L. Thomas, D. R. Gagnon, F. E. Karasz, and R. W. Lenz, *J. Polym. Sci. B*, *24*:2793 (1986).
28. L. P. Yu and L. R. Dalton, *J. Am. Chem. Soc.*, *111*:8699 (1989).
29. R. F. Loring and S. Mukamel, *J. Chem. Phys.*, *83*:4353 (1985).
30. B. H. Robinson and L. R. Dalton, *Chem. Phys.*, *36*:207 (1979).
31. J. M. Nunzi and D. Grec, *J. Appl. Phys.*, *62*:2198 (1987).
32. X. F. Cao, J. P. Jiang, D. P. Bloch, R. W. Hellwarth, L. P. Yu, and L. R. Dalton, *J. Appl. Phys.*, *65*:5012 (1989).
33. F. Kajzar and J. Messier, *Phys. Rev. A*, *32*:2352 (1985).
34. F. Kajzar, S. Eternad, G. L. Baker, and J. Messier, *Solid State Commun.*, *63*:1113 (1987).
35. F. Kajzar and J. Messier, *Polym. J.*, *19*:275 (1987).
36. P. P. Ho, *J. Opt. Soc. Am. B*, *4*:1025 (1987). P. P. Ho, N. L. Yang, T. Jimbo, Q. Z. Wang, and R. R. Alfano, *J. Opt. Soc. Am. B*, *4*:1025 (1987).
37. Z. Z. Ho, C. Y. Ju, and W. M. Heaterington III, *J. Appl. Phys.*, *62*:716 (1987).
38. P. A. Chollet, F. Kajzar, and J. Messier, *Synth. Met.*, *18*:459 (1987).
39. F. Kajzar, I. Ledoux, and J. Zyss, *Phys. Rev. A*, *36*:2210 (1987).
40. G. M. Carter, Y. J. Chen, and S. K. Tripathy, *Opt. Eng.*, *24*:609 (1985).
41. C. C. Frazier, E. A. Chauchard, M. P. Cockerham, and P. L. Porter, *Mat. Res. Soc. Symp. Proc.*, *109*:323 (1988).
42. D. N. Rao, P. Chopra, S. K. Ghoshal, J. Swiatkiewicz, and P. N. Prasad, *J. Chem. Phys.*, *84*:7049 (1986).
43. A. F. Garito, J. R. Heflin, K. Y. Wong, and O. Zamani-Khamiri, *Organic Materials for Nonlinear Optics*, R. A. Hann and D. Bloor, eds., Royal Society of Chemistry, 1989.
44. P. N. Prasad, E. Perrin, and M. Samoc, *J. Chem. Phys.*, *91*:2360 (1989).
45. A. F. Garito, J. K. Heflin, K. Y. Wong, and O. Zamani-Khamiri, *Photoresponsive Materials*, S. Tazuka, ed., 1988.

46. R. Dorsinville, L. Yang, R. R. Alfano, R. Tubino, and S. Destri, *Solid State Comm.*, *68*:875 (1988).

47. J. Waite and M. G. Papadopoulous, Z. Naturforsch. A, *Phys. Sci.*, *42*:749 (1987).

48. S. Ghosl, M. Samoc, P. N. Prasad, and J. J. Tufariello, *J. Phys. Chem.*, *94*:2847 (1990).

49. C. S. Winter, S. N. Oliver, and J. D. Rush, *Opt. Commun.*, *45*:69 (1988).

50. C. W. Spangler, T. J. Hall, K. O. Havelka, M. Badr, M. R. McLean, and L. R. Dalton, *SPIE Proc.*, 1147 (1989).

51. J. Kao and A. C. Lilly, Jr., *J. Am. Chem. Soc.*, *109*:4149 (1987).

52. J. L. Bredas, B. Themans, and J. M. Andre, *J. Chem. Phys.*, *78*:6137 (1983).

53. J. L. Bredas, R. R. Chance, R. Silbey, G. Nicholas, and P. Durand, *J. Chem. Phys.*, *77*:371 (1982).

54. Z. Iqbal, D. M. Ivory, J. Marti, H. L. Bredas, and R. H. Baughman, *Mol. Cryst. and Liq. Cryst.*, *118*:103 (1985).

55a. M. A. Sato, S. Tanaka, and K. K. Kaeriyama, *Synth. Met.*, *17*:229 (1987).

55b. G. L. Baker and F. S. Bates, *Macromol.*, *17*:2619 (1984).

56. T. Ito, H. Shirakawa, and S. J. Ikeda, *J. Polym. Sci.*, *Polym. Chem. Ed.*, *12*:11 (1974).

57. R. H. Friend, D. C. Bott, D. D. C. Bradley, C. K. Chai, W. J. Feast, P. J. S. Foot, J. R. M. Giles, M. E. Norton, C. M. Pereira, and P. D. Townsend, *Phil. Trans. R. Soc. London A*, *314*:37 (1985).

58. T. M. Swager, D. A. Dougherty, and R. H. Grubbs, *J. Am. Chem. Soc.*, *110*:2973 (1988).

59. F. L. Klavetter and R. H. Grubbs, *Synth. Met.*, *26*:311 (1988).

60. K. Knoll, S. A. Krouse, and R. R. Schrock, *J. Am. Chem. Soc.*, *110*:4424 (1988).

61. M. Aldissi, M. Hou, and J. Farrell, *Synth. Met.*, *17*:229 (1987).

62. H. Naarmann and N. Theophilou, *Synth. Met.*, *22*:1 (1989).

63. N. Theophilou, D. B. Swanson, A. G. MacDiarmid, A. Chakrabory, H. H. S. Javadi, R. P. McCall, S. P. Treat, F. Zuo, and A. J. Epstein, *Synth. Met.*, *28*:D35 (1989).

64. K. Akagi, M. Suezaki, H. Shirakawa, H. Kyotani, M. Shimomura, and Y. Tanabe, *Synth. Met.*, *28*:D1 (1989).

65. Y. W. Park, C. O. Yoon, C. H. Lee, H. Shirakawa, Y. Suezaki, and K. Akagi, *Synth. Met.* *28*:D27 (1989).

66. E. J. Ginsburg, C. B. Gorman, S. R. Marder, and R. H. Grubbs, Submitted for publication.

67. R. L. Elsenbaumer, K. Y. Jen, and C. C. Han, *Polym. Prepr.*, 144 (1989). M. Feldhues, G. Kampf, H. Litterer, T. Mecklenburg,

and P. Wegener, *Synth. Met.*, *28*:C487 (1989). K. Y. Jen,
R. Oboodi, and R. L. Elsenbaumer, *Polym. Mat. for Sci. Eng.*,
53:79 (1985). M. Sato, S. Tanaka, and K. Kaeriyama, *J. Chem.
Soc. Chem. Comm.*, 873 (1986). J. E. Osterholm, J. Laakso,
and P. Nyholm, *Synth. Met.*, *28*:C435 (1989). J. Roncali,
A. Yassar, and F. Garnier, *J. Chem. Soc. Chem. Comm.*, 581
(1988). K. Y. Jen, G. G. Miller, and R. L. Elsenbaumer,
J. Chem. Comm., 1346 (1989). J. Roncali, R. Garreau,
A. Yassar, P. Marque, F. Garnier, and M. Lemaire, *J. Phys.
Chem.*, *91*:6706 (1987). R. Souto-Maior and F.Wudl, *Synth.
Met. 28*:C281 (1989). S. Hota, S. D. D. V. Rughoooputy,
A. J. Heeger, and F. Wudl, *Macromol.*, *20*:212 (1989).
68. M. Inoue, R. E. Navarro, and M. B. Inoue, *Synth. Meth.*,
    30:199 (1989).
69. T. Inagaki and T. A. Akotheim, *Synth. Met. 28*:C245 (1989).
70a. M. Schott and G. Wegner, *Nonlinear Optical Properties of
    Organic Molecules and Crystals*, Vol. 2, D. S. Chemla and
    J. Zyss, eds., Academic Press, New York, 1987, p. 1.
70b. G. M. Carter, Y. J. Chen, M. F. Rubner, D. J. Sandman,
    M. K. Thakur, and S. K. Tripathy, *Nonlinear Optical Properties
    of Organic Molecules and Crystals*, Vol. 2, D. S. Chemla and
    J. Zyss, eds., Academic Press, New York, 1987, p. 85.
71. J. R. Reynolds, P. A. Poropatic, and R. L. Toyooka, *Synth.
    Met.*, *17*:95 (1987).
72a. F. L. Klavetter and R. H. Grubbs, *Synth. Met.*, *28*:699 (1989).
72b. F. L. Klavetter and R. H. Grubbs, *Synth. Met.*, *28*:D105
    (1989).
73. S. R. Marder, J. W. Perry, F. L. Klavetter, and R. H. Grubbs,
    Chem. Mat., *1*:171 (1989).
74. L. P. Yu and L. R. Dalton, *J. P. S. Polym. Lett.* (in press).
75a. A. Ray, G. E. Asturias, D. L. Kershner, A. F. Richter,
    A. G. MacDiamid, and A. J. Epstein, *Synth. Met.*, *29*:E141
    (1989).
75b. T. C. Chung, J. H. Kaufuran, A. J. Heeger, and F. Wudl,
    Phys. Rev. B, *30(2)*:702 (1984).
76. D. W. Polis, C. L. Young, M. R. McLean, and L. R. Dalton,
    Macromol., *23*:3231 (1990).
77. C. L. Young, D. W. Polis, A. N. Bain, L. S. Sapochak, and
    L. R. Dalton, *Macromol.*, *23*:3236 (1990).
78. D. W. Polis, D. J. Vachon, and L. R. Dalton, *Mat. Res. Soc.
    Symp. Proc.*, *134*:679 (1989).
79a. L. R. Dalton, L. P. Yu, L. Sapochak, and M. Chen, *Liq.
    Cryst. Molecul. Cryst.* (in press).
79b. L. P. Yu, M. Chen, and L. R. Dalton, *Polymer* (in press).
79c. M. R. McLean, D. W. Polis, C. L. Young, and L. R. Dalton,
    Pacific Polym. Preprints, *1*:205 (1989).

80. D. W. Polis and L. R. Dalton, unpublished results.
81. L. S. Sapochak, D. W. Polis, A. N. Bain, P. Bryson, L. R. Dalton, and D. W. Spangler, *Mat. Res. Soc. Symp. Proc.* (in press, 1990).
82. J. Capistran et al., *Polymer Preprints*, *25*:282 (1984); D. Gagnon, *Polymer Preprints*, *25*:284 (1984); F. Karasz, *Mol. Cryst. Liq. Cryst.*, *118*:327 (1985).
83. C. W. Spangler, T. J. Hall, L. S. Sapochak, and P. K. Liu, *Polymer*, *30*:1166 (1989).
84a. L. R. Dalton, J. Thomson, and H. S. Nalwa, *Polymer*, *28*:543 (1987).
84b. A. Heeger, D. Moses, and M. Sinclair, *Synth. Met.*, *17*:343 (1987).
84c. M. Sinclair et al., *Solid State Comm.*, *61*:221 (1987).
85. C. W. Spangler, E. Nickel, and T. Hall, *Polym. Preprints*, *28*:219 (1987). C. W. Spangler, L. S. Sapochak, and B. Gates, *Organic Materials for Nonlinear Optics*, R. A. Hann and D. Bloor, eds., Royal Society of Chemistry, 1989.
86. A. Seher and T. Seifen, *Austri Chemittel*, *54*:544 (1952).
87. F. Bohlmann and E. Inhoffen, *Ber*, *89*:1276 (1956).
88. A. W. Snow, *Nature*, *292*:40 (1981).
89. P. Teyssie and A. C. Korn-Giard, *J. Polym. Sci.*, A-2:2849 (1964).
90. N. Kobayashi, M. Mikitoshi, H. Ohno, E. Tsuchida, H. Matsuda, N. Nakanishi, and M. Kato, *New Polym. Mater.*, *1*:3 (1987).
91. A. N. Nesmeianov, M. I. Rybiskaya, and G. L. Slonimskii, *Vysokomol. Soedin*, *2*:526 (1960).
92. C. S. Marvel and C. L. Levesque, *J. Am. Chem. Soc.*, *60*:280 (1938).
93. T. Kitamura, K. Hasumi, and N. Fujisawa, *Jpn Kokai Tokkyo Koho*, *52*:179 JP 509 (87, 178, 509) (1987).
94. M. L. Kaplan, P. H. Schmidt, C-H Chen, and W. M. Walsh, Jr., *Appl. Phys. Lett.*, *36*:867 (1980).
95. Z. Iqbal, D. M. Ivory, J. Marti, J. L. Bredas, and R. H. Baughman, *Mol. Cryst. Liq. Cryst.*, *118*:103 (1985).
96. P. H. Schmidt, D. C. Joy, M. L. Kaplan, and W. L. Feldmann, *Appl. Phys. Lett.*, *40*:93 (1982).
97. S. R. Forest, M. L. Kaplan, P. H. Schmidt, T. Venkatesan, and A. J. Lovinger, *Appl. Phys. Lett.*, *41*:708 (1982).
98. M. Murakami and S. Yoshimura, *Mol. Cryst. Liq. Cryst.*, *118*:95 (1985).
99. Z. Iqbal et al., *Synth. Met.*, *15*:161 (1986).
100. R. C. Houtz, *Textile Res. J.*, *20*:786 (1950).
101. W. G. Vosburgh, *Textile Res. J.*, *30*:882 (1960).
102. N. Grassie and I. C. McNeil, *J. Chem. Soc.*, 3929 (1956), *J. Polym. Sci.*, *27*:20 (1958), *J. Polym. Sci.*, *33*:171 (1958), *J. Polym. Sci.*, *39*:211 (1959).

103. N. Grassie and R. Mcguchan, *Eur. Polym. J.*, 7:1357, 1503 (1971); 8:243, 257 (1972).
104. P. J. Goodhew and A. J. Clarke, Bailey, *Mat. Sci. Eng.*, 17:3 (1975).
105. G. Arey, S. K. Chadda, and R. C. Poller, *Eur. Polym. J.*, 19:313 (1983).
106. G. Arey, S. K. Chadda, and R. C. Poller, *J. Polym. Sci. Polym. Chem. Ed.*, 20:2249 (1982).
107. A. Brokman, M. Wager, and G. Marson, *Polymer*, 21:1114 (1980).
108. W. Watt, *Conf. on Ind. Carbons and Graphite*, Soc. Chem. Ind. London (1970).
109. J. E. Bailey and A. J. Clarke, *Nature*, 234:529 (1971).
110. J. Wallach and J. Manassen, *J. Polym. Sci. A-1*, 7:1983 (1966).
111. C. Carlini and J. C. W. Chien, *J. Polym. Sci. Polym. Chem. Ed.*, 22:2749 (1984).
112. J. Z. Ruan and M. H. Litt, *Macromol.*, 20:285 (1987).
113. J. Z. Ruan and M. H. Litt, *Macromol.*, 20:299 (1987).
114. J. Z. Ruan and M. H. Litt, *Macromol.*, 21:876 (1988).
115. J. Z. Ruan and M. H. Litt, *Macromol.*, 21:882 (1988).
116. J. K. Stille and M. E. Freeburger, *J. Polym. Sci. A*, 1:161 (1968).
117. J. K. Stille and E. Mainen, *J. Polym. Sci. Polym. Lett.*, 4:39 (1966).
118. J. K. Stille, E. L. Mainen, M. E. Freeburger, and F. M. Harris, *Polym. Prepr.*, 8:244 (1967).
119. J. K.Stille and E. L. Mainen, *Macromol.*, 1:36 (1968).
120. O.-K. Kim, *J. Polym. Sci. Polym. Lett. Ed.*, 23:137 (1985).
121. K. Chiba, T. Ohsaka, Y. Ohnuki, and N. Oyama, *J. Electro. Anal. Chem.*, 218:117 (1987).
122. S. Kunimura, T. Ohsaka, and N. Oyama, *Macromol.*, 21:894 (1988).
123. Oh-Kil Kim, *Mol. Cryst. Liq. Cryst.*, 105:161 (1984).
124. D. N. Berntar, J. N. Onuchi, and J. W. Perry, *J. Phys. Chem.*, 91:2696 (1980).
125. R. L. Van Deusen, *J. Polym. Sci. B*, 4:211 (1966).
126. F. E. Arnold and R. L. Van Deusen, *Macromol.*, 2:497 (1969).
127. F. E. Arnold and R. L. Van Deusen, *J. Appl. Polym. Sci.*, 15:2035 (1971).
128. R. L. Van Deusen, O. K. Goins, and A. J. Sicree, *J. Polym. Sci. A-1*, 6:1777 (1968).
129. V. L. Bell and G. F. Pezdirtz, *J. Polym. Sci. Polym. Lett.*, 3:977 (1968).
130. J. G. Colson, *J. Polym. Sci. A*, 4:59 (1966).
131. F. Davans and C. S. Marvel, *J. Polym. Sci. A*, 3:3549 (1965).

132. G. C. Berry, *J. Polym. Sci. Polym. Symp.*, *65*:143 (1978).
133. R. Liepins and M. Aldissi, *Mol. Cryst. Liq. Cryst.*, *105*:151 (1984).
134. I. Belaish, C. Rettori, D. Davidov, M. R. McLean, L. R. Dalton, and H. Nalwa, *Mat. Res. Soc. Symp. Proc.*, *134*:689 (1988).
135. F. Coter, I. Belaish, D. Davidov, L. R. Dalton, E. Ehren-freund, M. R. McLean, and H. S. Nalwa, *Synth. Met. 29*:E471 (1989).
136. I. Belaish, C. Rettori, D. Davidov, L. P. Yu, M. R. McLean, and L. R. Dalton, *Synth. Met.*, *33*:341 (1989).
137. I. Belaish, D. Davidov, H. Selig, M. R. McLean, and L. R. Dalton, *Angew. Chem. Adv. Mater.*, *101*:1601 (1989).
138. S. A. Jenekhe and S. J. Tibbeths, *J. Polym. Sci. Polym. Phys. Ed.*, *26*:201 (1988).
139. F. Davan and C. S. Marvel, *J. Polym. Sci. A*, *3*:3549 (1965).
140. F. E. Arnold and R. L. Van Deusen, *J. Polym. Sci.*, *B-6*:815 (1968).
141. F. E. Arnold and R. L. Van Deusen, AFML-Tr-68-1, 1968.
142. F. E. Arnold, *J. Polym. Sci. A-1*, *8*:2079 (1970).
143. R. Muller and D. Worhle, *Makromol. Chem.*, *176*:2775 (1975).
144. R. Muller and D. Worhle, *Makromol. Chem.*, *177*:2241 (1976).
145. R. Muller and D. Worhle, *Makromol. Chem.*, *179*:2161 (1976).
146a. L. P. Yu, R. Vac, L. R. Dalton, and R. W. Hellwarth, *SPIE Proc.*, 1147 (1989).
146b. I. Belaish, D. Davidov, C. Rettori, M. R. McLean, L. R. Dalton, and L. P. Yu, *Mat. Res. Soc. Symp. Proc.*, *Multi-Functional Materials*, D. R. Ulrich, A. J. Buckley, F. E. Karasz, and G. Gallagher-Daggit, eds., Materials Research Society, Pittsburgh, 1990.
147. P. Smith, A. J. Heeger, F. Wudl, and J. Chiang, *Nonlinear Optical Properties of Polymers*, A. J. Heeger et al., eds., Materials Research Society, Pittsburgh, 1988, p. 283.
148. X. F. Cao, J. P. Jiang, R. W. Hellwarth, L. P. Yu, and L. R. Dalton, Technical Digest, Conference on Lasers and Electro-Optics (CLEO), Optical Society of America, Washington, D.C., 1989, paper TUGG28; *J. Appl. Phys.*, to be published.
149. R. W. Hellwarth, *J. Opt. Soc. Am.*, *67*:1 (1977).
150. P. W. Maker and R. W. Terhume, *Phys. Rev.*, *137*:A801 (1965).
151. G. M. Carter, M. K. Thakur, Y. J. Chen, and J. V. Hrynie-wicz, *Appl. Phys. Lett.*, *47*:457 (1985).
152. J. R. Salcedo and A. E. Siegman, *IEEE J. Quantum Electronics*, *QE-15*:250 (1979).
153. D. Narayana Rao, R. Burzynski, X. Mi, and P. N. Prasad, *Appl. Phys. Lett.*, *48*:387 (1986).

154. D. Blanchard, R. Casalegno, M. Pierre, and H. P. Trommsdorff, *J. Phys.*, *46*:C7, 517 (1985).
155. R. C. Desai, M. D. Levenson, and J. A. Barker, *Phys. Rev. A*, 27:1968 (1983).
155a. A. F. Garito, J. R. Heflin, K. Y. Knong, and O. Zamini-Khamiri, *Mat. Res. Soc. Symp. Proc.*, *91*:109 (1988).
156. W. R. Tompkin, R. W. Boyd, D. W. Hall, and P. A. Tick, *J. Opt. Soc. Am. B*, *4*:1030 (1987).
157a. P. N. Prasad, J. Swiatkiewicz, and J. Pfleger, *Mol. Cryst. Liq. Cryst.*, *160*:53 (1988).
157b. M. T. Zhao, B. P. Singh, and P. N. Prasad, *J. Chem. Phys.*, 89:5535 (1988).
158. P. Chopra, H. F. King, and P. N. Prasad, to be published.
159. P. Chopra, L. Carlacci, H. F. King, and P. N. Prasad, *J. Phys. Chem.*, in press.
160. M. P. Boggadd and B. J. Orr, *Int. Rev. Sci. Phys. Chem. Ser.*, 2:149.
161. E. F. McIntyre and H. F. Hameka, *J. Chem. Phys.*, *68*:(1978). G. J. B. Hurst, M. Dupuis, and E. Clementi, *J. Chem. Phys.*, 89:385 (1988).
162. I. J. Goldfarb and J. Medrano, Low Dimension Conductivity and Superconductivity, D. Jerome and L. G. Caron, eds., NATO ASI Series, New York, 1987.
163. L. Yang, R. Dorsinville, Q. Z. Wang, W. K. Zou, P. P. Ho, N. L. Yang, R. R. Alfano, R. Zamboni, R. Danieli, G. Ruani, and C. Taliani, *J. Opt. Soc. Amer. B*, *6*:753 (1989).
164. W. M. Dennis, W. Blau, and D. J. Bradley, *Appl. Phys. Lett.*, 47:200 (1985).
165. G. Berkovic, Y. R. Shen, P. R. Prasad, *J. Chem. Phys.*, 87:1182 (1987).
166. S. Tomaru, K. Kubodera, T. Kurihara, and S. Zembutsu, *Japan J. Appl. Phys.*, *26*:L1957 (1987).
167. S. Tomaru, K. Kubodera, S. Zembutsu, K. Takeda, and M. Hasegawa, *Electron. Let.*, *23*:595 (1987).
168. J. LeMoigne, A. Thierry, P. A. Chollet, F. Kajzar, and J. Messier, *J. Chem. Phys.*, 88:6647 (1988).
169. T. Kaino, K. I. Kubodera, S. Tomaru, T. Kurihara, S. Saito, T. Tsutsui, and S. Tokito, *Electron Lett.*, *23*:1095 (1987).
170. S. A. Jenekhe, S. K. Lo, and S. R. Flom, *Appl. Phys. Lett.*, 54:2524 (1989).
171. S. K. Kurtz, *Treatise of Quantum Electronics*, Vol. 1, Part A, H. Rabin and C. L. Tary, eds., Academic Press, New York, 1975, p. 210.
172. P. N. Prasad, *Nonlinear Optical Effects in Organic Polymers*, J. Messier et al., eds., Kluwer Academic, Boston, 1989, p. 351.
173. S. Etemad, G. L. Baker, D. Jaye, F. Fajzar, and J. Messier, *SPIE Proc.*, *682*:44 (1986).

174. D. N. Rao, J. Swiatkiewicz, P. Chopra, S. K. Ghoshal, and P. N. Prasad, *Appl. Phys. Lett.*, *48*:1187 (1986).
175. F. Kajzar, J. Messier, and C. Rosilio, *J. Appl. Phys.*, *60*:3040 (1986).
176. J. C. Baumert, G. G. Bjorkund, D. H. Jurich, M. C. Jund, H. Looser, R. D. Miller, J. Rabolt, J. D. Swalen, and R. J. Twieg, Technical Digest, Conference on Lasers and Electro-Optics (CLEO), Vol. 7, Anaheim, CA, 1988.
177. H. Uchiki and T. Kobayashi, *J. Appl. Phys.*, *64*:2625 (1988).
178. S. Matsumoto, K. I. Kubodera, T. Kurihara, and T. Kaino, *Appl. Phys. Lett.*, *51*:1 (1987).
179. P. F. Barbara, *SPIE Proc.*, *878*:65 (1988).
180. P. G. Huggard, W. Blau, and D. Schweitzer, *Appl. Phys. Lett.*, *51*:2183 (1987).
181. W. Blau, *Phys. Tech.*, *18*:250 (1987).
182. B. F. Levine, *Chem. Phys. Lett.*, *37*:516 (1976).
183. A. F. Garito, *Symposium on Rigid Rod Polymers, Mat. Res. Soc. Symp. Proc.*, in press.

4

Processable Electronically Conducting Polymers

John R. Reynolds and Martin Pomerantz / The University of
Texas at Arlington, Arlington, Texas

I. INTRODUCTION

It has now been over 15 years since the discovery that strong,
flexible, freestanding films of polyacetylene could be prepared via
Ziegler-Natta polymerization on glass surfaces [1]. Even in light
of the highly elevated conductivity attained for polyacetylene when
redox doped [2], the complete intractability of the material, coupled
with its air instability, prevents its acceptance as a useful material.

A significant body of work by chemists, physicists, and materials scientists now exists which provides insights for the basic structural requirements for electronically conducting polymers. Numerous reviews [3] have been written that detail the need for conjugation, the ability to inject charge onto the polymer chain using redox doping techniques, changes in the optical properties during doping, and the numerous potential applications that exist for conducting polymers. As research has progressed, the electronic properties of these polymers have steadily improved. Generally, polyacetylene has led the way as the prototype system. Originally, conductivities of 200-750 S cm^{-1} were found for films synthesized using a Ti(OBu)$_4$/AlEt$_3$ initiator system. These films have become commonly known as Shirakawa polyacetylene [2]. Improved syntheses, yielding polymers containing fewer defects in their structure, have now led to polyacetylenes having conductivities [4] of 20,000-100,000 S cm^{-1}. These latter conductivities are truly metallic and, when adjusted for the low density of the polymer relative to metals, the conductivity of polyacetylene is approximately the same as that of copper. In addition to polyacetylene, improved syntheses of other conjugated polymers have led to highly conducting materials whose conductivity is a strong function of synthetic conditions. Both poly(p-phenylene vinylene) (PPV) and poly(3-methylthiophene) [5] exhibit conductivities in excess of 2000 S cm^{-1} when properly prepared. An additional class of soluble and processable conjugated polymers is produced via the solid-state polymerization of 1,4-disubstituted-1,3-diynes commonly called diacetylenes. The optical properties of these polymers have been thoroughly investigated, as have been the topochemical polymerization mechanism and their electronic properties, and fully reviewed [6]. Attempts to attain highly conductive polydiacetylenes have, in general, been unsuccessful, and thus this class of polymers will not be addressed in detail here.

As is the case with the electronic properties, physical properties are a strong function of the macromolecular structure of conducting polymers. Just as many attempts have been made to improve the electronic properties, so too has considerable effort gone into improving physical properties. These physical properties include processability, stability, and mechanical integrity. Conducting polymers are now available that are soluble and fusible, rather thermally stable, ambient atmosphere stable, and exhibit good mechanical strength. The inducement of processability into these typically intractable polymers has been a major goal of synthetic chemists and is the focus of this manuscript. The techniques employed range from the preparation of processable precursor polymers which are subsequently converted to an intractable conjugated polymer, to the formation of solutions of the polymer in the conducting form, to the use of flexible substituents, copolymerization, and blend and composite formation. The processability now available allows the materials to be used much like common polymers. Films can be extruded, fibers can be spun, and parts can be molded. These material property advances can be traced to the numerous syntheses and polymer preparations developed. It is impossible to acknowledge every contribution to such a broad area but we hope to provide a framework that will stimulate further research.

II. POLYACETYLENE

As the simplest conjugated polymer, polyacetylene [1,$(CH)_x$] has served as a model for developing both the electronic and physical

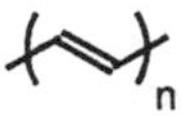

1

properties of electronically conducting polymers. As mentioned earlier, the complete intractability of $(CH)_x$ can be attributed to the extremely rigid conjugated backbone and strong interchain forces which give rise to crystallinity. Although films can be prepared by the direct polymerization of acetylene, they tend to be highly porous and fibrillar.

The metathesis polymerization of 7,8-bis(trifluoromethyl)tri-cyclo[4.2.2.0^{2,5}]deca-3,7,9-triene (2) by WCl_6 and Me_4Sn shown in Eq. (1) leads to a high-molecular-weight, soluble, precursor

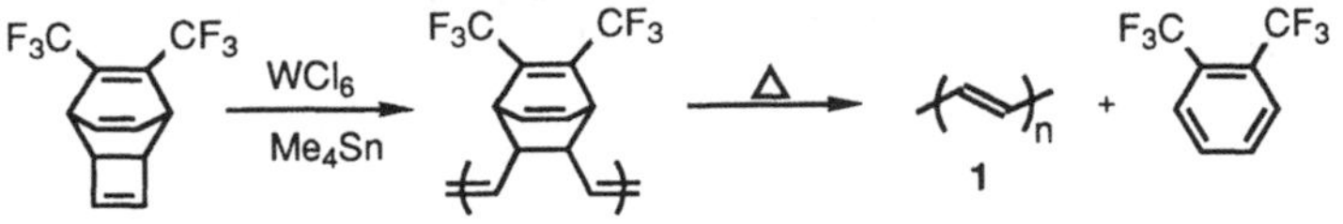

(1)

polymer (3) that can subsequently be thermally converted to $(CH)_x$ [7]. The concept of using soluble precursor polymers is now used extensively for the preparation of a variety of conjugated polymers, as will be made evident throughout this chapter. In addition to the bis(trifluoromethyl)benzene precursor polymers, which contain a readily thermally eliminated group, other precursors having groups which can be eliminated and with a range of stabilities have been prepared. These polymers are shown in structures 4, 5, 6, and 7. The thermal instability of polymers 3, 4, and 5 originally precluded any significant characterization in the solution phase. Precursor polymer 6,

on the other hand, could be purified by precipitation and was found
to have a hydrodynamic volume determined by gel permeation chromatography (GPC) comparable to polystyrene with a number average
molecular weight (M_n) of 40,000 and a molecular weight distribution

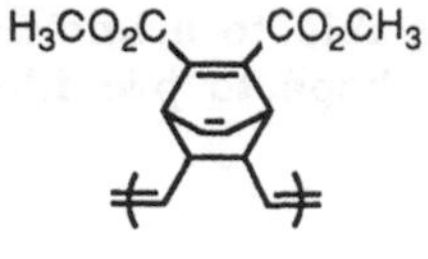

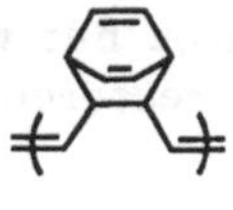

(polydispersity) of $M_w/M_n > 4$. Further discussion of GPC analyses
of the soluble polymers, unless otherwise noted, will be relative
to polystyrene standards, and so the molecular weights should be
taken as approximations only. One of the major benefits of the
soluble precursor method is that, after solution casting and careful
thermal elimination, the polyacetylene film formed is continuous
and space-filling with densities greater than 1 g cm^{-1}. The polyacetylene thus obtained is essentially amorphous.

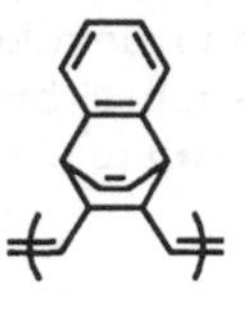

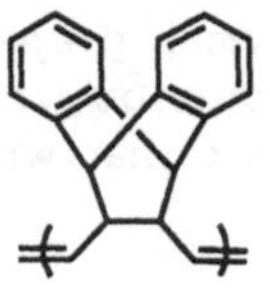

A trade-off of the thermal instability of the polymers with ease
of synthesis motivated Feast and Winter to investigate these soluble
precursor syntheses in greater detail [7b]. Their ability to isolate
and polymerize 3,6-bis(trifluoromethyl)pentacyclo[6.2.0.0^{2,4}.0^{3,6}.0^{5,7}]-
dec-9-ene (8) as shown in Eq. (2) yielded the soluble precursor
polymer 9, which was quite stable at room temperature and easily

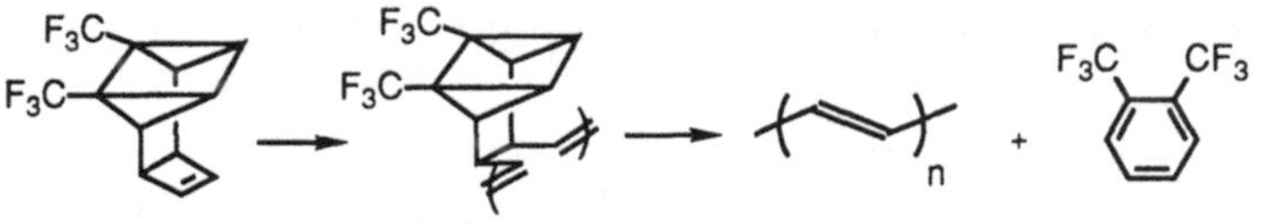

$$(2)$$

converted to (CH)$_x$ at 75°C. Unfortunately, the strain energy in
this polymer, which is stable only because its thermal reversion
to 3 is symmetry-forbidden, caused the conversion reaction to be
extremely exothermic and often explosive. Conversion of this polymer
to (CH)$_x$ is recommended in thin films only.

The conversion of polymer 3 to (CH)$_x$, which has become known
as the Durham route, and the properties of the resultant materials
have been extensively studied. The prepolymer can be significantly
oriented and, upon thermal conversion, this orientation is retained
and a highly ordered and crystalline (CH)$_x$ is obtained which can
be doped to high conductivity [8,9].

A drawback to the Durham precursor method for the synthesis of $(CH)_x$ is the necessary expulsion of a relatively large molecule during the conversion reaction. In addition to representing a relatively large fraction of the overall mass of the precursor polymer, removing these molecules from the bulk of the polymer is difficult, limiting the process to thin films. This has been overcome by the inclusion of strained rings into polymers which can be converted into double bonds. This is exemplified by the work of Grubbs et al. [10], who investigated the ring-opening metathesis polymerization (ROMP) of benzvalene (10) by a tungsten alkylidene initiator system as outlined in Eq. (3). The fact that no molecule is eliminated during conversion suggests this method may be employed in the formation

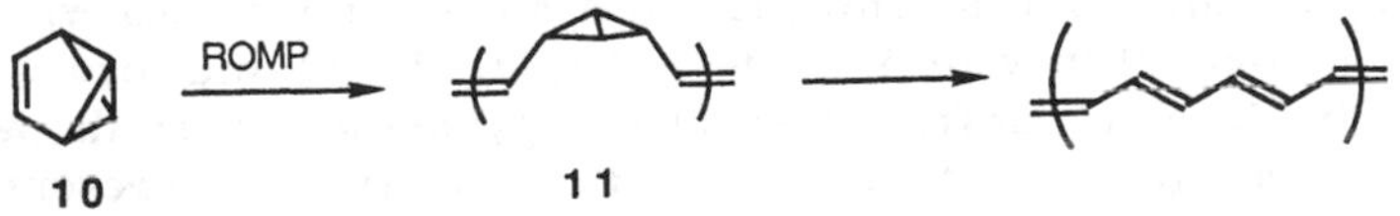

$$(3)$$

10   **11**

of relatively thick $(CH)_x$ samples. The isomerization of bicyclobutane rings to 1,3-dienes can typically be accomplished thermally, photochemically, and by transition metals. Interestingly, polybenzvalene (11, PBV) cannot be successfully converted into $(CH)_x$ using either thermal or photochemical methods. Solutions of $HgCl_2$, $HgBr_2$, and Ag^+ salts in tetrahydrofuran (THF) were found to transform films of polybenzvalene into "shiny silvery materials resembling polyacetylene in appearance." IR and NMR spectroscopic analyses showed these films to be $(CH)_x$ with a high density (19%) of saturated defects and ca. 40:60 cis:trans double-bond content. Films of $(CH)_x$ prepared in this manner are strong and flexible with a relatively low extent of crystallinity. Swager and Grubbs [11] extended the homopolymerization of benzvalene in a study of a series of tungsten alkylidenes and concurrently examined the ability of benzvalene and norbornene to form block copolymers having the general structure shown in 12. The homopolymer 11 was found to be quite unstable, undergoing spontaneous detonation with mechanical stress or rapid heating and cross-linking in the solid or gel state. The latter property prevented purification by reprecipitation and dissolu-

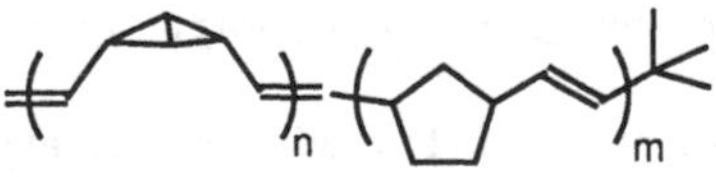

12

tion, forcing film casting directly from the polymerization reaction mixture. The ability to carry out living polymerizations of norbornene

allowed formation of the block copolymers which were more stable
and were more soluble than the PBV homopolymer.

Conversion of polybenzvalene (11) to conjugated polymers was
accomplished using a variety of catalysts, including $AgBF_4$, ZnI_2,
$HgBr_2$, $HgCl_2$, and $Rh(COD)Cl_2$ (COD = 1,5-cyclooctadiene). In
addition to the fully conjugated $(CH)_x$, a cross-conjugated polymer
13 containing an exocyclic double bond was expected. It was found

13

that $HgCl_2$ performed best and yielded a silvery black film which
exhibited a conductivity of 1 S cm^{-1} when I_2 doped and also was
mechanically durable. Infrared spectroscopy showed a small peak
at 895 cm^{-1} attributable to the vinylidene moiety, the presence
of residual saturation, and a more highly disordered structure when
compared to Shirakawa $(CH)_x$. This is also evident in the electronic
spectra of the polymer as the $(CH)_x$ formed in this manner has a
band gap (absorption onset) at ~1.9 eV (650 nm) and an absorption
maximum at 2.8 eV (440 nm) (compared to a band gap of 1.4 eV
(885 nm) for Shirakawa $(CH)_x$).

As is the case with Durham $(CH)_x$, orientation can be obtained
in this system by stretching the precursor polymer. Polybenzvalene
(11) was cast onto a polyethylene support and stretched to orientation draw ratios, l/l_0, of 2.3 and 6.0. After conversion and doping
with I_2, these polymers exhibited conductivities of 13 and 49 S
cm^{-1}, respectively. Scanning electron microscopy (SEM) showed
this form of $(CH)_x$ to have a fibrillar morphology, and the crystallinity of the polymer was increased dramatically with stretching.

Examination of the block copolymers, prior to conversion, by
differential scanning calorimetry (DSC) leads to no discernible T_g
for the polynorbornene fraction and suggests the material exists
as a single phase. Some phase separation occurs upon subsequent
conversion of 11 to $(CH)_x$ with the polynorbornene T_g evident at
38°C. The morphologies of these block copolymers are quite smooth
and continuous. Elevated conductivities are still found after saturation with I_2 with oriented samples of 50:50 block copolymers reaching
ca. 0.4 S cm^{-1}.

The direct formation of copolymers of $(CH)_x$ with a variety of
monomers and carrier polymers has also been employed in the preparation of soluble polyacetylenes. The controllable alteration of the
electrical conductivity and physical properties of a random acetylene
copolymer was demonstrated by Chien et al. [12] with the synthesis
of poly(acetylene-co-methylacetylene) (14). Although these copolymers

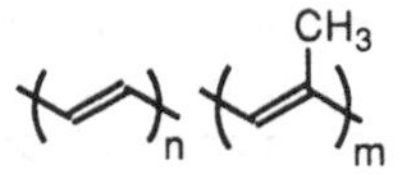

14

were not soluble, doped conductivities ranging from 10^{-3} to 36 S cm^{-1} were measured and, as the methylacetylene content in the copolymer was increased, the copolymer exhibited a greater propensity to swell in solvent.

Retention of long domains of unsubstituted $(CH)_x$ chains in soluble materials was accomplished with the application of block and graft copolymer methods [13-23]. A number of synthetic techniques have been utilized to graft $(CH)_x$ chains onto carrier polymers. In general these fall into two classes, where either a growing $(CH)_x$ chain in solution is grafted onto a solubilized carrier polymer, or the $(CH)_x$ chain is polymerized off of the carrier polymer as a side chain. As an example of the former method, Bates and Baker [13,16,17] polymerized acetylene using the $Ti(OC_4H_9)_4/Al(C_2H_5)_3$ system in the presence of modified polyisoprene or polystyrene. They suggest that nucleophilic attack of growing $(CH)_x$ chains on electrophilic sites (previously incorporated) of the carrier polymer leads to termination of the $(CH)_x$ and grafting. Destri et al. [19], on the other hand, utilized an initial reaction between carrier polymer and initiator to subsequently polymerize $(CH)_x$ side chains onto polybutadiene as illustrated in Scheme 1. Using this method, estimations of the extent of grafting were obtained from the relative concentration of 1,2- and 1,4-diene linkages and initiator/vinyl group ratio. In addition, this method allows control of the length of the polyene through monomer concentration and cis:trans ratios through temperature.

Scheme 1

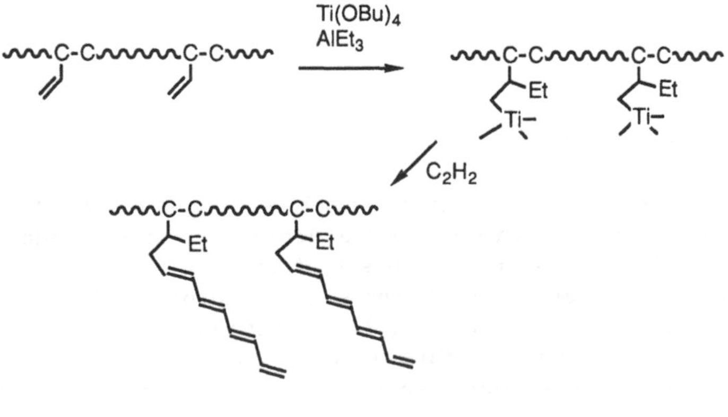

Block copolymers have been prepared by carrying out a living polymerization of one monomer and, after consumption of this monomer, switching to a more reactive second monomer. The number of copolymers possible is broadened by the ability to change the nature of the active site after synthesis of the first block, and subsequently use a different *mechanism* for preparation of the second block. This has been utilized in the preparation of acetylene block copolymers via both anionic to Ziegler-Natta [15,18,20,21] and anionic to metathesis [22,23] methods. This is illustrated in Scheme 2 for the formation of poly(acetylene-b-styrene). An elaborate double-labeling experiment, utilizing ^{14}C and tritium, was employed to show that indeed copolymer was forming and not $(CH)_x$ homopolymer [20]. These $PS/(CH)_x$ block copolymers, which were initially soluble, could be cast into films. Upon subsequent I_2 oxidation however, conductivities $<10^{-6}$ S cm^{-1} were obtained. In an extension of this work, Aldissi and Bishop [21] examined the relationship between the copolymer composition and electrical conductivity. At low levels ($\leq 20\%$) of PS incorporation, conductivities above 1 S cm^{-1} were obtained which quickly fell off to 10^{-5} S cm^{-1} when the PS level was raised to 40%.

Scheme 2

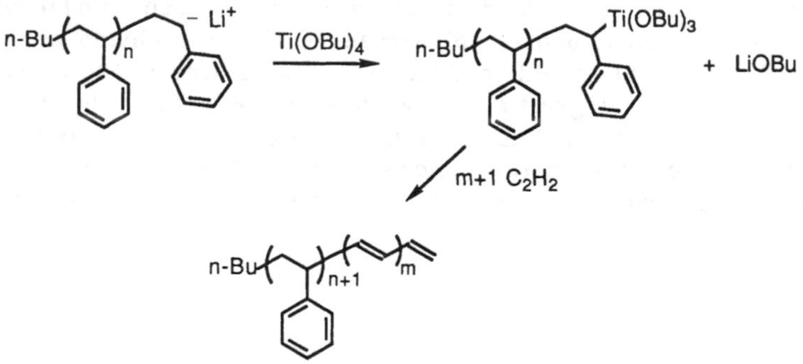

III. POLY(p-PHENYLENE)

Processable precursor routes have also been developed for the synthesis of the fully conjugated and completely non-processable poly(p-phenylene) (PPP). Previous syntheses of this polymer generally yielded poorly characterized oligomers with approximately 10-15 phenylene units in each chain [24-27]. Attempts to improve molecular weight, purity and processability were stimulated by the fact that these oligomers could be oxidized by AsF$_5$ to form complexes with conductivities as high as 200 S cm^{-1} [28].

The ability to bacterially oxidize benzene, using the micro-organism *Pseudomonas putida*, to form 5,6-*cis*-dihydroxycyclohexa-1,3-diene in useful quantities has led researchers at ICI to polymerize it as a PPP precursor as outlined in Eq. (4) [29,30]. The fact that ca. 250 g of the diol was utilized in the conversion to ester in this preparation indicates that relatively large scales are possible in these syntheses. Polymerization of a variety of esters (diacetate, dipivalate, dibenzoate, etc.) was accomplished using common free-radical methods (benzoyl peroxide, 90°C) to yield conversions to 15

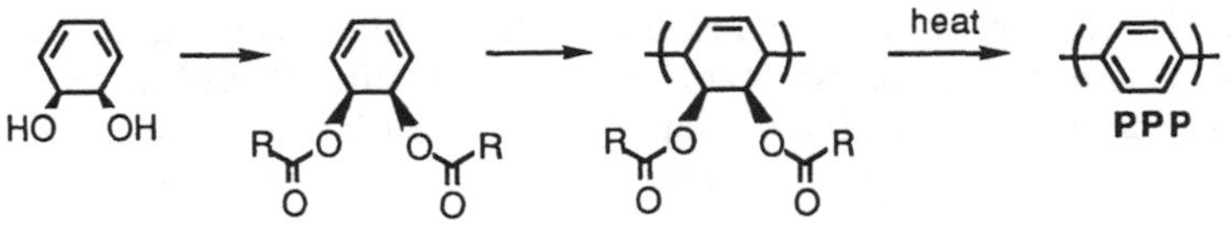

greater than 80% in less than 10 min. Using the diacetate as an example, degrees of polymerization of ca. 50-200 and polydispersities M_w/M_n) of 2-4 were obtained for 15 ($R = CH_3$). These cyclohexene ring containing polymers were processable to films and fibers and could be subsequently thermally converted at 300°C to form PPP. These temperatures could be reduced to 220-260°C by catalyzing the reaction using strong tertiary amine bases and alkali metal salts. Extents of aromatization could be controlled by varying the time and temperature used during conversion with essentially complete aromatization possible.

Using common synthetic organic methods, McKean and Stille developed a nonenzymatic preparation of the *cis*-dibenzoate and *cis*-dipivalate esters [31]. Polymerizations were carried out using free-radical methods. A close examination of the structure showed that the polymer (R = Ph and *t*-Bu) was not completely 1,4-linked, as desired for full conjugation in the PPP, but contained ca. 10-15% of 1,2-linkages (16). These 1,2-defects should decrease charge

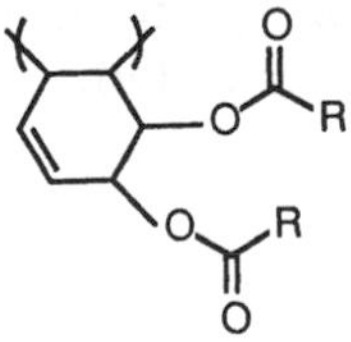

16

mobility along the polymer chain in the conducting complex and would also decrease order by preventing crystallization.

Both groups [29-31] examined the electrical properties of their PPPs upon doping with AsF_5. Although both polymers became highly conducting, the PPP prepared by Ballard exhibited a conductivity of 100 S cm^{-1} while that prepared by Stille was somewhat lower at ca. 1 S cm^{-1}. It should be pointed out that these were initial studies and did not indicate the upper limits possible for the conductivity of these polymers.

Although the precursor polymers described above have high molecular weights and are processable, the PPPs produced are completely intractable. This has been overcome by the synthesis of PPPs that are dialkyl substituted and quite soluble [32,33]. The use of alkyl substitution to improve processability has been used extensively for polyheterocycles and will be discussed in detail later.

The Grignard coupling reaction of alkyl substituted 1,4-dibromobenzenes was initially employed, as outlined in Eq. (5), where R = C_6H_{13} or C_8H_{17}. These side chains were chosen since

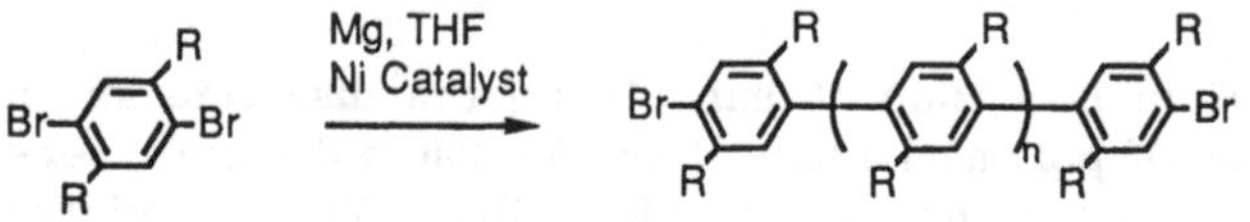

$$(5)$$

they were expected to be long enough to induce solubility but not so long that the side chains would crystallize in the solid state. Although these polymerizations were found to produce only para-linked phenylenes, and were thus highly regiospecific, the chain lengths obtained were found to be quite small. Chromatographic analysis yielded an average degree of polymerization (DP) of about 8 with the largest chains containing about 20 repeat units. This low DP was attributed to the loss of chain end functionality during polymerization and difficulty in maintaining stoichiometric balance.

To overcome these problems, an A-B step-growth polymerization system was investigated [33] as shown in Eq. (6). An NMR spectral analysis of the dihexylsubstituted PPP (17) showed essentially complete para linkages and no detectable end groups, suggesting a relatively high molecular weight. Vapor phase osmometry showed the polymer to have a DP of 28. No characterization of the dopability or electrical properties for these soluble PPPs was reported.

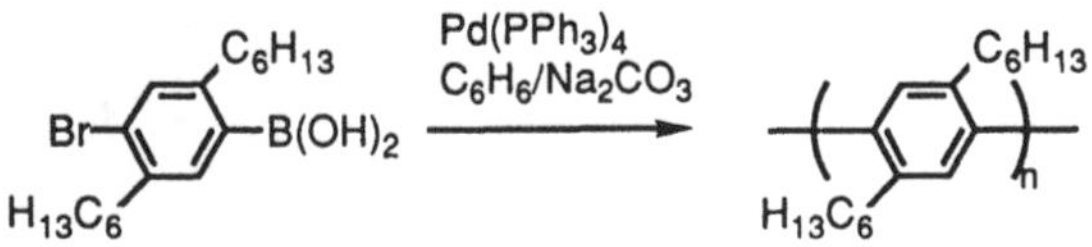

$$(6)$$

17

In addition to the purely para linked polyphenylenes, substituted fluorenes have been used to prepare [34] melt-processable polymers which can be oxidatively doped to conductivities as high as 10^{-3} S cm^{-1}. Polymerization of a series of alkyl-substituted fluorenes 18

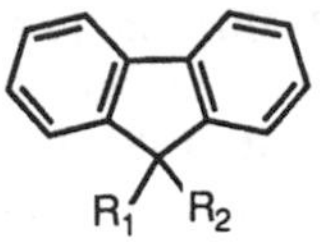

$R_1 = H; R_2 = C_6H_{13}, C_{16}H_{33}$

$R_1 = R_2 = C_6H_{13}, C_8H_{17}, C_{10}H_{21},$
$\quad\quad\quad C_{12}H_{25}, C_{16}H_{33}, C_{22}H_{45}$

18

by FeCl$_3$ led to polymers which are expected to contain a conjugated backbone equivalent in structure to poly(p-phenylene). The annulated ring system locks the polymer into a stable conformation, and thus the optical band gaps of the mono- and disubstituted species are at approximately the same energy (3.0-3.1 eV). This is in strong contrast to the poly(3-alkylthiophenes), as will be detailed later, where 3,4-disubstitution induces an increase in the band gap via steric interactions. The melting points of the polymers were found to be a strong function of the length of the alkyl group and the extent of substitution. For example, poly(9-hexylfluorene) melts at 140°C which was reduced to 100°C for poly(9,9'-hexylfluorene). It should be pointed out that these melting points were made by visual inspection, and no results on the molecular weights of the polymers were available. As the length of the side chain increases, the observed melting point decreases. In the case of the bis-dodecyl substituted polymer, the melting point is reported to be less than 25°C.

IV. POLY(ARYLENE VINYLENES): PROCESSABLE PRECURSOR POLYMERS

A. Poly(p-Phenylene Vinylene)

As previously discussed, one of the important ways to process otherwise intractable materials is via the so-called precursor polymer route. In this, a processable polymer is prepared, processed and then converted in one step to the dopable and ultimately electronically conducting, but intractable, polymer. One area where this has been exploited and has worked extremely well is in the poly(arylene vinylenes) (19) which include both benzenoid and heteroaromatic species as the Ar group.

$+Ar-CH=CH+_n$

19

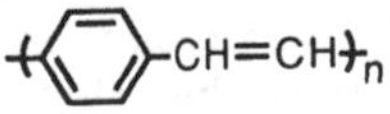

$+\langle\bigcirc\rangle-CH=CH+_n$

PPV

Since the preparation of poly(p-phenylene vinylene) (19, Ar = p-C_6H_4; PPV) using either a Wittig condensation or a dehydrohaloge-nation reaction [35] gave oligomers, a different synthetic methodology was examined. In 1984 Karasz and Lenz [36,37] and Murase [38] and their co-workers reported the use of a procedure previously devised by Kanbe [39] and by Wessling and Zimmerman [40] using a soluble precursor polymer for the preparation of PPV which is still being used and has been extended to other poly(arylene viny-lenes). The method, shown in Scheme 3, involves preparation of a bis-sulfonium salt such as 20 [p-phenylenedimethylenebis(dimethyl-sulfonium chloride)] for the parent PPV (or the diethylsulfonium chloride) followed by a sodium hydroxide induced elimination-polymerization reaction, at 0°C, to produce an aqueous solution of the polyelectrolyte, precursor polymer 21 (poly[p-phenylenedi-methylene-α-(dimethylsulfonium chloride)]). This polymer 21 could be processed into films, foams, and fibers [36-40]. Heating a cast film of 21 at 200°C or 300°C for more than 2 h resulted in a yellow, freestanding film of PPV; however, rapid heating at 220°C provided a flexible foam structure. This is not surprising since, at 200°C two gaseous molecules [$(CH_3)_2S$ and HCl] are being eliminated from each monomer unit. These polymers were doped with AsF_5 or H_2SO_4 vapors to yield PPV having conductivities of 10 and 100 S cm^{-1},

Scheme 3

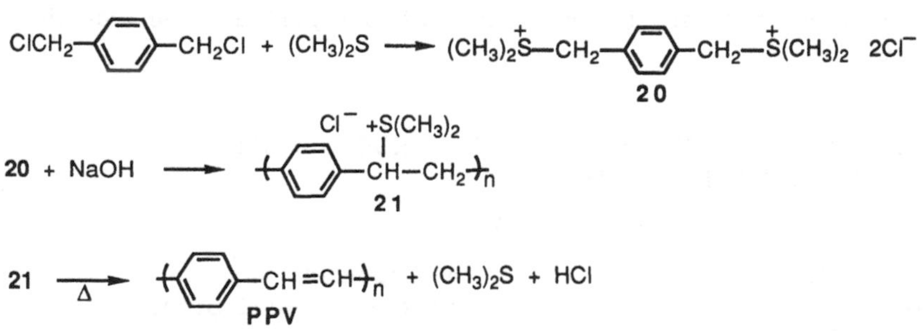

respectively. Further, films could be stretch oriented and conductivi-ties after AsF_5 and SO_3 doping reached 2780 and 685 S cm^{-1}, respec-tively. This was considerably higher than had been achieved with other methods [35], where the polymers were made by condensation reactions and were in fact oligomers. A major problem in the conver-sion of 21 to PPV is incomplete loss of all of the sulfur and chlorine producing a polymer with sp^3 hybridized carbon atoms in the main chain. This reduces the overall conjugation. The optimum conditions for the preparation of 21 and, in turn, for the conversion of 21 to PPV with the minimum amount of sulfur, have been determined by heating above 380°C [41].

Molecular weights for polymer 21 have recently been obtained by low-angle laser light scattering and were reported to be 990,000 and 500,000 for the weight average and number average respectively, with a polydispersity of 2.0 [41].

In a recent paper [42], Karasz and co-workers made an extensive study of the molecular weight of the precursor polymer 22 as the fluoroborate salt and also converted it to a neutral polymer, 23, by

BF_4^- $\overset{+}{S}$
$+\!\!\langle\bigcirc\rangle\!\!-CH_2-CH\!+\!)_n$

22

S
$+\!\!\langle\bigcirc\rangle\!\!-CH_2-CH\!+\!)_n$

23

reaction with thiophenoxide. Using GPC with polystyrene standards and a nonaqueous solvent [22 is soluble in nonaqueous solvents such as dimethylformamide (DMF) and acetonitrile] number average molecular weights for 22 were about 5.5×10^6 with a polydispersity of 2.1. For polymer 23 M_n was about 2×10^5, M_w about 9×10^5 with a polydispersity of about 6. These numbers represent the average of three runs on each polymer and are in qualitative agreement with the low-angle light scattering results for this system of $M_n \approx 5.5 \times 10^5$ and $M_w \approx 1.1 \times 10^6$. Further, ultracentrifugation of polymer 21 gave $M_n = 5.0 \times 10^5$ and $M_w = 9.9 \times 10^5$. Thus all techniques provide similar results and indicate high degrees of polymerization [42].

Since precursor polymer 21 could be cast into films and these converted to films of PPV which could then be doped, one could get better conductivities, up to 100 S cm^{-1}, than from the material from the polycondensation reactions. Improvement in the conductivity requires alignment of the polymer molecules, often achieved by stretching a film or fiber. Indeed, as indicated above, 21 (or the diethylsulfonium derivative) could be stretch oriented at elevated temperatures by uniaxial stretching with a draw ratio of up to 15 [37,38,42-44]. This increased the conductivity in the direction of stretching enormously, to well over 10^3 S cm^{-1} with AsF$_5$ [38,43] dopant and up to 5000 S cm^{-1} with H$_2$SO$_4$ dopant [44].

The UV-Vis transition edge, or the band gap, of PPV has been variously reported to be 2.7 eV [45a] and also 2.4 eV [41,45b], intermediate between those of polyacetylene and poly(*p*-phenylene).

X-ray diffraction, electron diffraction, and infrared dichroism studies have shown that the PPV molecules are highly oriented along the stretch axis, and studies have been made of the increase of molecular orientation along the stretch axis as a function of draw ratio [41,43,46]. Further, in stretched precursor polymer 21 the extent of crystallinity of the film increases with extent of conversion to PPV [47].

When the anion in 21 is exchanged for an anion which can produce a dopant molecule, such as $AsF_6^- \to AsF_5 + F^-$, the thermal elimination gives $(CH_3)_2S$, HF in the case of the AsF_6^- anion, and AsF_5 which then dopes the polymer. This has been referred to as *incipient doping* [48].

The effect of altering the sulfonium structure was studied by examining the diethyl, tetramethylene, and pentamethylene xylene sulfonium salt derivatives, 24, 25, and 26, respectively, and the precursor polymers derived from them, 27, 28, and 29, respectively. These were subsequently compared to the dimethyl sulfonium derivatives 20 and 21. Better yields of PPV with somewhat higher conductivity were obtained in all three cases, with the cyclic sulfonium precursor polymers, 28 and 29, being the best [49]. Indeed elemental analysis showed very little residual sulfur and chlorine compared to the larger amounts observed with the dialkyl sulfonium polymers after thermal elimination [49].

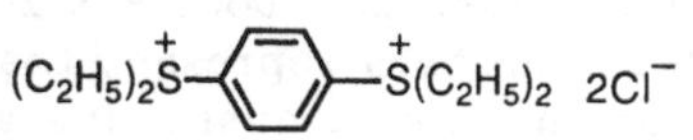

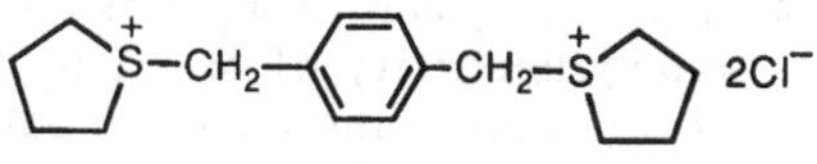

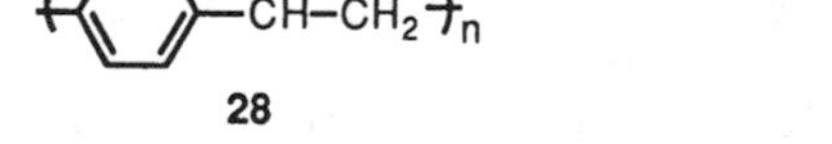

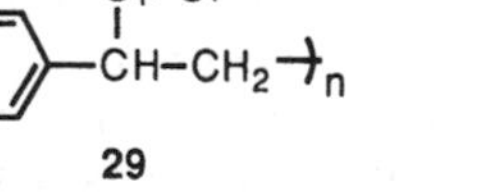

It is interesting to note that a comparison of properties of PPV formed from precursor polymer 21 with that formed from 28 has revealed that the UV-Vis spectra of the undoped polymers are slightly different [50]. The PPV from 28 shows a slightly longer wavelength absorption maximum corresponding to longer conjugation lengths. This is consistent with the observation of considerably less sulfur and chlorine, and hence fewer sp^3 hybridized carbon atoms, and therefore greater conjugation lengths in the polymers. The UV-Vis spectrum of 28 shows phonon sidebands attributed to a narrower distribution of crystallite sizes [50].

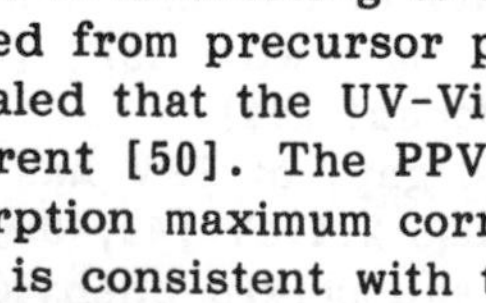

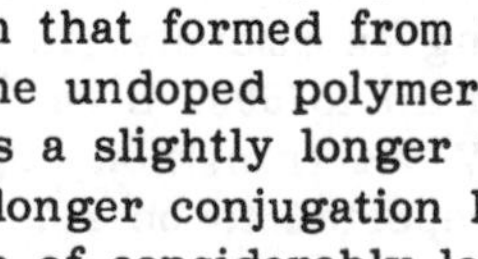

When PPV was examined by DSC, no glass or melt transition was observed between -196 and 500°C. Decomposition began at about 550°C in a nitrogen atmosphere [41]. The thermogravimetric analyses (TGA) of the precursor polymer 21 (and also 27-29) showed weight loss at about 50°C with rapid loss at 100°C to ca. 130°C. There was a second region of slower weight loss up to about 230-240°C with little weight loss, of what is now PPV, up to about 600°C.

Electrochemical oxidation-reduction of PPV in a PPV, PPV$^+$/LiAsF$_6$, propylene carbonate/Li/Li$^+$ system was investigated by cyclic voltammetry [51]. Two broad peaks, one anodic and one cathodic, are observed at around 3.85 V. With a sweep rate of 2 mV/s the anodic and cathodic peaks occur at 3.90 V and 3.80 V, respectively. In the absence of impurities such as O_2, N_2, and H_2O PPV can be cycled between oxidized conducting and neutral insulating states hundreds of times over several months without decomposition [51].

Mechanical testing showed that a drawn film of PPV was significantly stronger and stiffer than an undrawn film. This PPV, formed from the diethylsulfonium precursor polymer 27, drawn fourfold at 90-120°C, and heated 30 min at 300°C, gave an elastic modulus of 9221 MPa, a tensile strength of 274 MPa, and an elongation of 7.0% compared with an elastic modulus of ~1800 MPa, a tensile strength of 71 MPa, and an elongation of 16% for the film which had not been drawn [52].

B. Poly(Naphthylene Vinylenes)

There are a number of other poly(arylene vinylenes), where the arylene group is a benzenoid aromatic moiety, which have been studied, but much less has been done with these systems than with PPV itself. One of the earliest of this type to be prepared was poly(2,6-naphthylene vinylene) (32) also prepared by the precursor polymer route, shown in Scheme 4, from the bis-sulfonium salt 30 via the precursor polymer polyelectrolyte 31 [36]. No properties were reported, however. Another naphthalene derivative, using 1,4-linkages, namely 33, was also prepared by this same methodology [53,54]. Films of the precursor polymer 34 were prepared and thermally eliminated at 300 and 350°C to produce 33. The transition edge (band gap) was at 615 nm or 2.02 eV, which is about 0.5 eV lower than for PPV. This is due to the more effective resonance interaction of the naphthylene with the vinylene unit than the phenylene has with the vinylene unit as shown in Scheme 5.

Electrical conductivity, however, was considerably less than that of PPV. With AsF$_5$ as dopant a maximum conductivity (for the as-produced film) was 0.032 S cm^{-1} [53].

Scheme 4

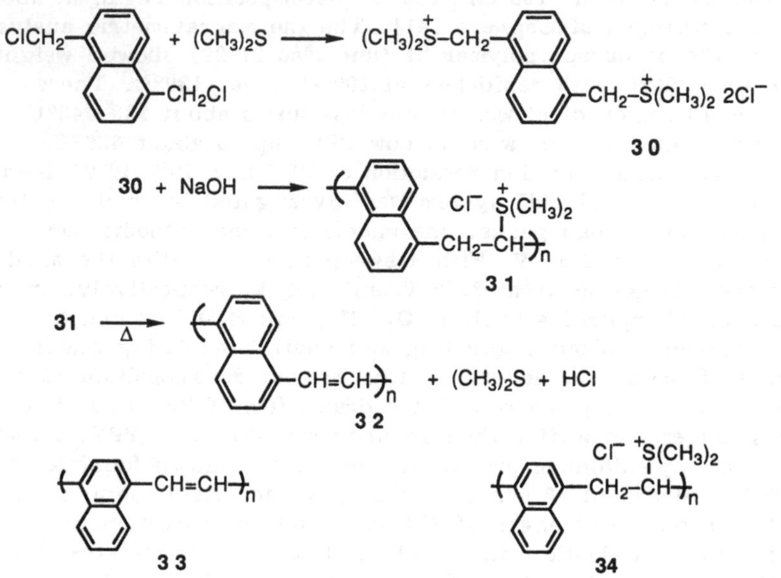

30

31

32

33   **34**

Scheme 5

33

compare with

P P V

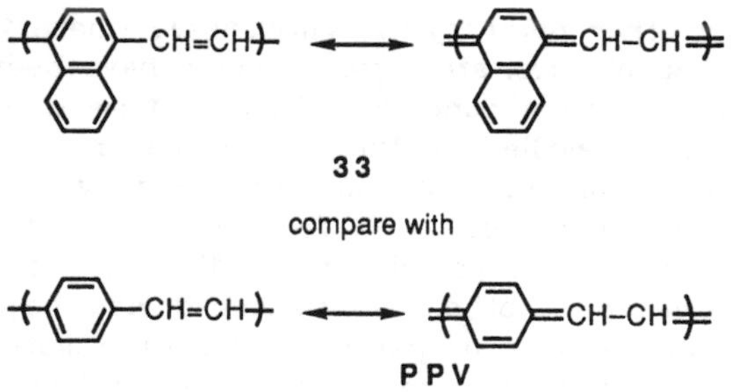

C. Substituted and Soluble Poly(p-Phenylene Vinylenes) from Processable Precursor Polymers

The earliest report of substituted poly(p-phenylene vinylenes) made by the precursor polymer route employing sulfonium salts was by Wessling and Zimmerman [40]. They reported on the preparation of methyl substituted (on the p-phenylene ring) PPV using the same precursor polyelectrolyte sulfonium polymer procedure they used for the parent PPV. The systems prepared contained 2,5-dimethyl- and 2,3,5,6-tetramethyl-p-phenylene groups. More recent studies of the 2,5-dimethyl derivative showed fairly low conductivities after doping with I_2 or SO_3, 2×10^{-4} S cm^{-1} and 10^{-4} S cm^{-1}, respectively [44], and with AsF$_5$, 3.2×10^{-2} S cm^{-1} [54].

An interesting derivative of PPV that is in fact soluble as the conjugated polymer is poly(p-phenylene-1,2-diphenylvinylene) (35) [55]. It was prepared by a dehalogenation polycondensation reaction of 1,4-bis(phenyldichloromethyl)benzene (36) with chromium(II) acetate and gave material, which after fractionation, showed a number average molecule weight of up to about 80,000 by membrane osmometry. The UV-Vis-near-IR spectra [55,56] showed the long wavelength λ_{max} to be at 350-360 nm with the absorption edge (band gap) at around 425-430 nm (2.9 eV), and this is the same as a cast film

35   36

or in solution. It is clear from the spectra and the fact that this material is soluble that there is considerable steric distortion in the backbone away from planarity, and hence there is less conjugation. The glass transition temperature (T_g) is reported to be 137°C while TGA (argon atmosphere) shows the polymer to be stable to 550°C then rapidly loses weight. The un-cross-linked polymer 35 does, however, produce a stable highly viscous fluid at about 300°C.

Upon electrochemical oxidation, 35 shows two new Vis-near-IR absorptions at 667 nm and 2500 nm. The electrochemical oxidation is reversible and shows an anodic peak at 1.24 V and a cathodic peak at 1.10 V using CH$_2$Cl$_2$ solvent, Et$_3$BzN SbCl$_6$ electrolyte, and a platinum disk electrode with Ag/AgCl as the reference electrode.

Poly(2,5-dimethoxy-1,4-phenylene vinylene) (37) has been made
by the dehydrohalogenation reaction [35], and this low-molecular-
weight oligomer showed a conductivity of 4.1×10^{-4} S cm^{-1} when
doped with AsF$_5$. More recently, Lenz and Karasz [54], Murase
[44], and Elsenbaumer [57] and their co-workers prepared 37 by
the precursor polymer route from the sulfonium polymers 38 and
39, respectively. The incorporation of these electron-donating substituents serve to stabilize the doped cationic form of the polymer and
thus lower the ionization potential of the species. Red films of 37
were made by heating polymer 38 to 200°C. Either cast films of 38
were heated or 38 was hot-pressed into a film. The films were tough
and transparent and doping with I$_2$, FeCl$_3$, SO$_3$, AsF$_5$, H$_2$SO$_4$ or
electrochemically gave a conducting polymer with four-probe conductivities as high as 500 S cm^{-1} [44,54,57]. It should be noted that,
while I$_2$ is a rather poor dopant for PPV itself, it is a good dopant
for 37, which is a reflection of the greater ease of oxidizing this
polymer compared to PPV. Indeed electrochemical studies suggest

37   38

39

the oxidation potential of 37 is comparable to that of polyacetylene
[57]. It has also recently been demonstrated that 37 could be doped
by protonic acids (pK$_a$ < 2) [58] to give conductivities up to about
70 S cm^{-1}.

The absorption edge of 37 is 610 nm, corresponding to a band
gap of 2.0 eV, which is in agreement with the band gap obtained
from electrochemical studies of the redox properties of the polymer
[57].

When the poly(2,5-dialkoxy-1,4-phenylene vinylene) has butoxy
(40) [58] or hexyloxy (41) [59] groups as substituents, not only
are the polyelectrolyte precursor polymers from which they are
made soluble and processable but, because of the long-chain alkyl

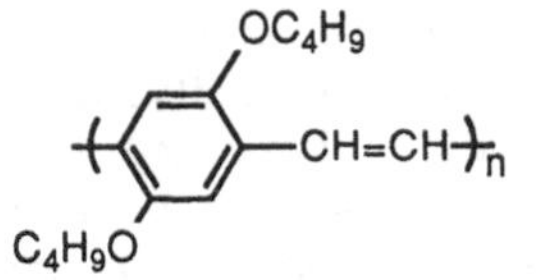

substitution, so are the conjugated polymers 40 and 41 themselves. The dimethylsulfonium precursor polymer to 41, containing some unsaturation, was gumlike and could be stretched up to 20 times its length. Heating this precursor polymer to 220°C gave 41 as a red soluble material. Prestretched precursor polymer gave 41 which resisted further stretching. The elimination of dimethylsulfide could be carried out in refluxing (214°C) 1,2,4-trichlorobenzene to produce an orange-red solution of 41 from which films could be cast.

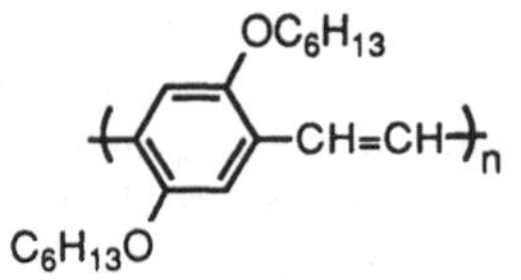

The UV-Vis spectrum was similar to that of the dimethoxy polymer 37 and a band gap (transition edge) of 2.08 eV (595 nm) virtually identical to that of 37 was reported [59]. As with many other soluble conjugated polymers, 41 shows both solvatochromism (color change depending on the solvent) and thermochromism (color change with temperature) in 1,2,4-trichlorobenzene solution. When acetonitrile is added to the 1,2,4-trichlorobenzene solution λ_{max} shifts to longer wavelength and in trichlorobenzene as the temperature is increased λ_{max} shifts to shorter wavelength. The reasons for these effects is not clear. When unstretched and stretch aligned (draw ratio of 7) films were doped with I_2 the maximum conductivities were 3-4 and 200 S cm^{-1}, respectively [59].

Another route to polymer 40 has been developed by the initial replacement of the sulfonium moiety (Scheme 6) by butoxy to produce

Scheme 6

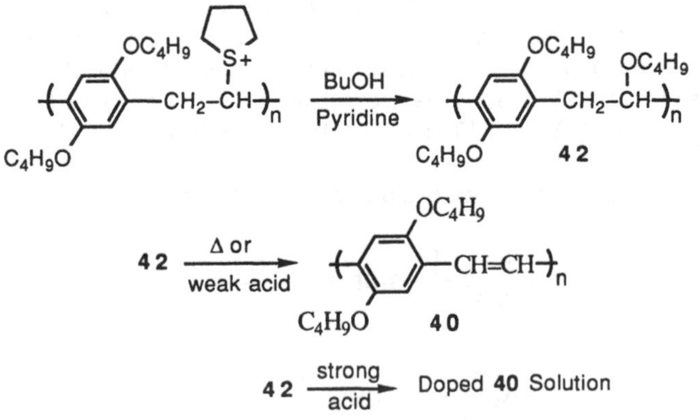

neutral polymer 42, which is now soluble in organic solvents [58].
This neutral polymer 42 can either be eliminated thermally or with
a weak acid. When a strong acid was used, the polymer 40 was
produced in the doped state [58]. Polymer 40 is said to be freely
soluble in solvents such as chloroform, tetrahydrofuran, chloro-
benzene, nitrobenzene, and toluene [58].

D. Copolymers of PPV Prepared from Processable Precursor Polymers

A number of copolymers of PPV have been reported, made from
sulfonium precursor polymers. These include polymers 43 where
$R = CH_3$, C_2H_5, and C_4H_9 [60] and 44 [61]. The copolymers 43
were prepared by copolymerizing the two monomeric bis-sulfonium

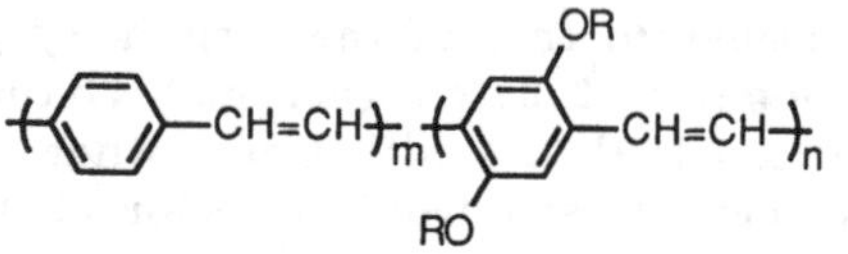

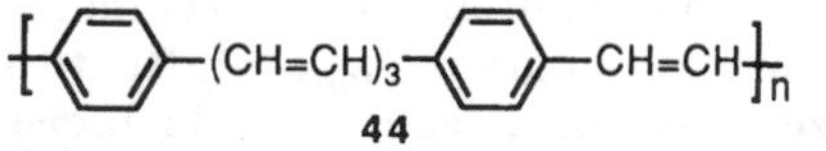

salts followed by thermal elimination at 220°C. The precursor polymers
could be stretch oriented before heating. The maximum conductivity
for unstretched copolymer (several were studied) was 28 S cm^{-1}
compared to dimethoxy PPV homopolymer, which was 51 S cm^{-1}.
The stretch-aligned systems, with draw ratios of up to 13, showed
maximum conductivities of over 1500 S cm^{-1} using I_2 as dopant.
A comparison with PPV homopolymer cannot be made except to note
that it does not dope well with I_2. TGA of the precursor copolymers
showed sulfide elimination between 50°C and about 200°C, and then
all were stable to over 430°C. This latter represents the stability
of the copolymers 43. UV-Vis spectra were similar to the parent
PPV although λ_{max} for PPV is 440 nm and for dimethoxy PPV homopolymer it is 408 nm.

Copolymer 44 was prepared from monomer 45 via precursor

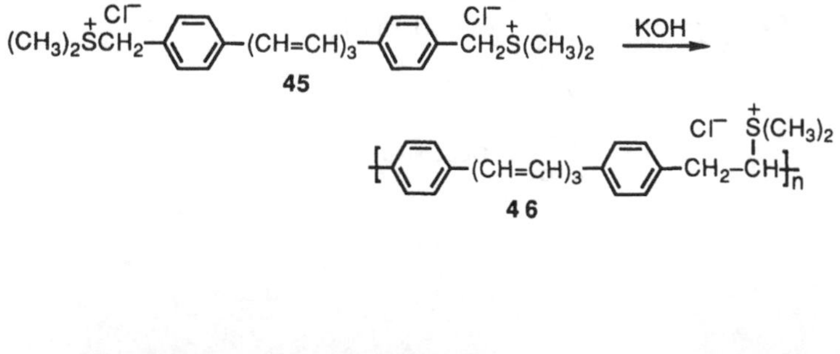

polymer 46 [61]. Doping of 44 with I_2 gave a maximum conductivity of 7×10^{-4} S cm^{-1}. Sulfuric acid doping, however, gave conductivities of up to 0.2 S cm^{-1}. Copolymer 44 is similar to PPV in that H_2SO_4 is a much better doping molecule. X-ray diffraction indicated the polymer was highly crystalline.

Finally, there is a report on the preparation of copolymer 47, by a Wittig reaction, which can be considered a soluble alternating PPV copolymer [62]. The molecular weight of the polymer, which

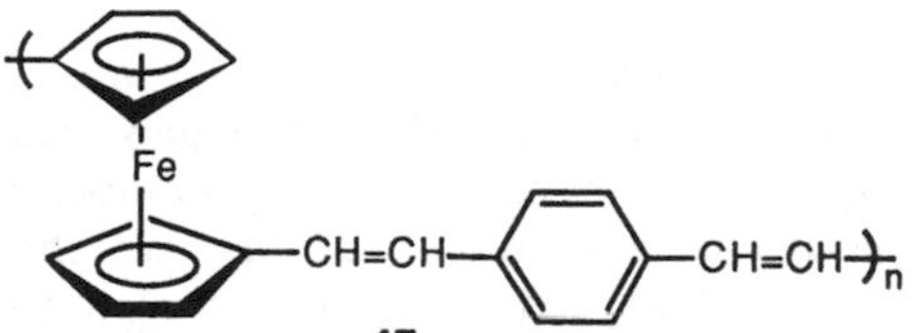

47

was soluble in DMF, was quite low, ca. 3000. There was also an insoluble component which probably had a higher molecular weight. Pressed pellet conductivities of I_2, Br_2, or AsF_5 doped samples was no higher than 5×10^{-5} S cm^{-1}. Interestingly, I_2 was the best dopant, the doped material was stable in air, and the conductivity actually increased on long exposure to air.

E. Poly(Heteroarylene Vinylenes) from Processable Precursor Polymers

Poly(thienylene vinylene) (48) has been prepared as a film or a foam beginning with bis-sulfonium salts 49 [63] and 50 [64,65] via processable precursor polymers. However, while 49 gave 48

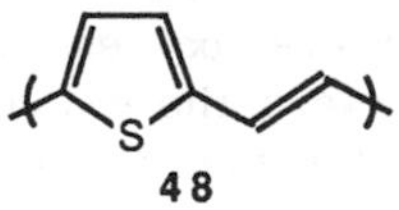

48 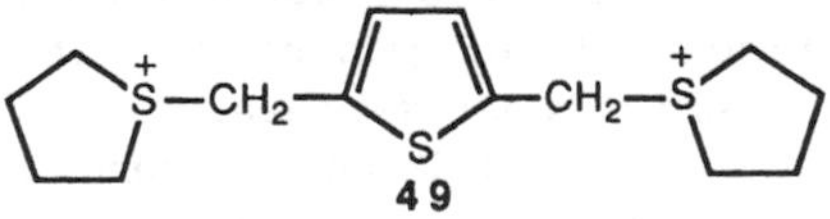   49

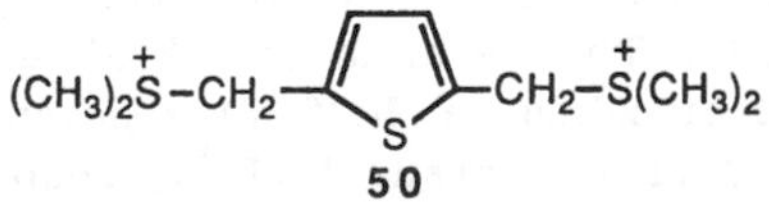

50

directly [63], 50 gave the organic soluble precursor polymer 51

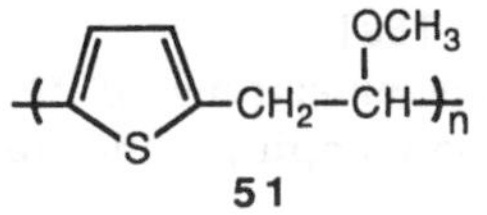

51

when the elimination from 50 was attempted in aqueous methanol.
This in turn, upon heating, produced poly(thienylene vinylene)
(48). One report [63b] says that 48 can be formed from 50; however, the molecular weight is lower than the polymer prepared from
49. Poly(thienylene vinylene) (48) has a lustrous golden color and
is stable in air for long periods of time [63-65]. Doping of cast
films of 48 with a number of oxidants including I_2 and $FeCl_3$ gave
four-probe conductivities of up to 230 S cm^{-1} [63-65]. Uniaxially
stretched film gave over a 10-fold increase in conductivity to values
approaching 2700 S cm^{-1}.

The UV-Vis spectrum of 48 showed a maximum at 530 nm [66]
(previously reported to be 540 nm [65]) and the band gap was
reported to be 1.8 eV from optical spectroscopy. Elsenbaumer et al.
report that the absorption maximum is at about 600 nm and the
band gap is 1.74 eV [63b]. A band gap of 1.64 eV has also been
reported [67]. Cyclic voltammetry of 48 showed a broad oxidation
wave with a maximum at 3.61 V and a reversible reduction wave
at 3.34 V (versus Li/Li$^+$) [63a]. A later publication gives the oxidation and reduction maxima as 3.56 V and 3.51 V, respectively [67].

An interesting system prepared recently contains an alkoxy
group on the 3-position of poly(thienylene vinylene). Both poly(3-
methoxylthienylene vinylene) (52) and poly(3-ethoxythienylene
vinylene) (53) have been studied [68,69]. They were not prepared

52   53

by the processable precursor polymer route but by a Grignard
coupling reaction. The oxygen renders the monomers too reactive
for the precursor polymer synthesis route but lowers the oxidation
potential of the polymers 52 and 53 compared to the parent 48.
For comparison, the band gap, based on optical spectroscopy, of
52 is reported to be 1.55 eV (800 nm) [63b] while that of 53 is
1.50 eV (825 nm) [68,69]. This is at least 0.2 eV less than the
band gap in 48. The electrochemical half-wave potentials for 52
and 53 are 3.23 V and 3.11 V (versus Li/Li$^+$), respectively, compared with 3.54 V for 48 [67-69]. The band gap of 53, based on
electrochemical n-doping and p-doping, was found to be somewhat
lower, 1.32 V [67-69]. Conductivities (pressed pellet) of $FeCl_3$ doped
samples were only as high as about 2 S cm^{-1} [68].

Perhaps the most interesting aspect of the chemistry of the
poly(3-alkoxythienylene vinylene) is that, upon doping, the absorp-

tion band in the visible region of the spectrum decreases significantly while the absorption of the doped state in the infrared region of the spectrum increases. This causes thin films on glass substrates to appear much less colored (faint blue-gray) in the doped state [68,69].

Finally, a poly(furylene vinylene) (54) synthesis has been reported via the processable precursor polymer route. It was prepared from precursor sulfonium polymer 55 in the usual way [63b,69,70]. The properties are similar to those of poly(thienylene vinylene) (48), with a band gap, from optical spectroscopy, of 1.76 eV (700 nm) [63b,69,70]. Upon doping, with, for example, I_2 or $FeCl_3$, conductivities of cast films up to 36 S cm^{-1} have been reported [63b,69,70].

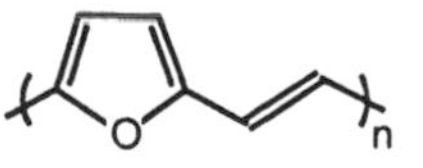

54 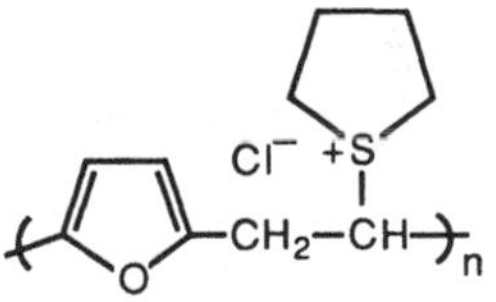   55

There are reports of copolymers of thienylene vinylene and either phenylene vinylene, 56a, or 2,5-dimethoxyphenylene vinylene, 56b [71], and of thienylene vinylene and furylene vinylene 57 [70]. These have been prepared by copolymerization of mixtures of sulfonium salts which provided processable precursor copolymers 58 [71] and 59 [70].

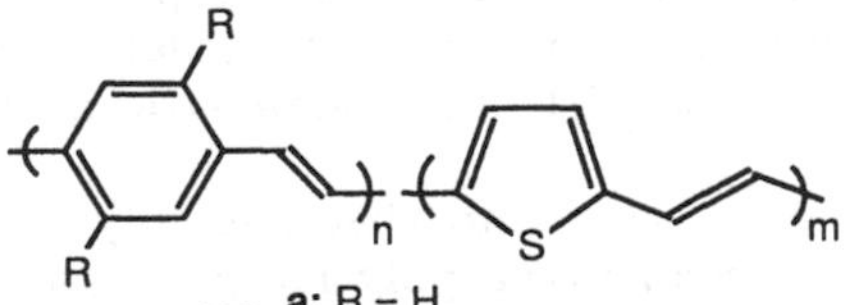

56 a: R = H
     b: R = OCH$_3$

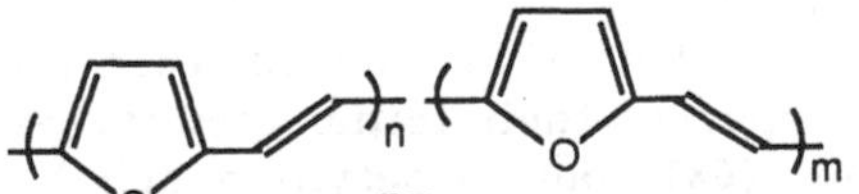

57

Iodine doping of stretched films of 56 gave fairly high electrical conductivities with the conductivities increasing with the amount of the thienylene vinylene in the copolymer. Copolymer 56a showed a maximum conductivity of 20 S cm^{-1} with a draw ratio of 5 (42% thiophene units) and for 56b it was 120 S cm^{-1} with a draw ratio of 2 (11% thiophene units) [71]. The I_2 doped samples were found to be stable in air for at least several months. Copolymer 57 was reported to have properties intermediate between 48 and 54 [70].

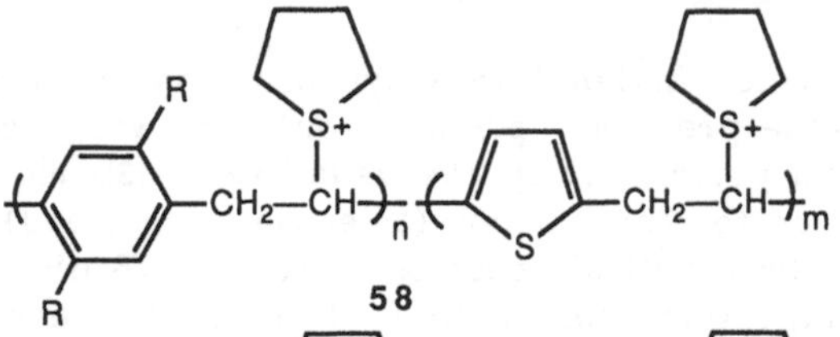

58

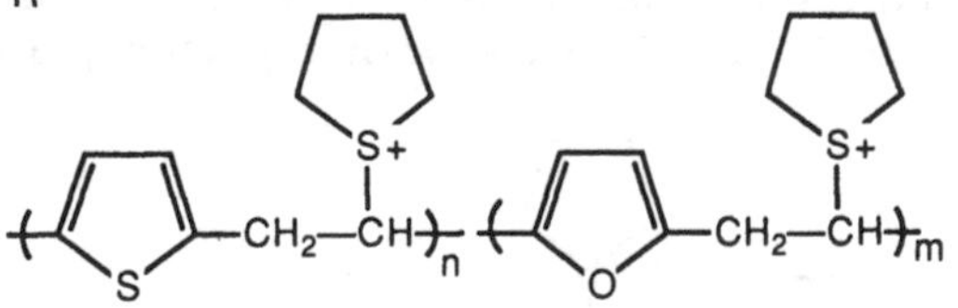

59

V. PROCESSABLE POLYHETEROCYCLES

A. Poly(3-Alkylthiophenes)

One of the best ways to render an otherwise intractable conjugated polymer processable is to attach long chains of atoms to the backbone. Unfortunately, this frequently causes steric interference with the ability of these polymers to remain planar, and so properties requiring this planarity, such as relatively low band gap and high conductivity of the doped system, are usually compromised. Indeed the steric problems usually get worse with increasing size of these chains.

In the case of polythiophene it was found that not only does a long straight-chain substituent render the polymer processable, both soluble and fusible, but there is only a minimal effect on the properties derived from the conjugation of the backbone.

The first reports of the preparation of soluble poly(3-alkylthiophenes) were in 1986 when Sato [72], Elsenbaumer [73,74], Yoshino [75], and co-workers reported their initial results. Sato prepared poly(3-hexylthiophene) (63), poly(3-octylthiophene) (64), poly(3-dodecylthiophene) (66), poly(3-octadecylthiophene) (67), and poly(3-eicosylthiophene) (68) by electrochemical coupling of the appropriate monomers. Elsenbaumer et al. prepared poly(3-methylthiophene) (60), poly(3-ethylthiophene) (61), and poly(3-butylthiophene) (62) along with the octyl derivative 64 by a chemical coupling route involving the nickel-catalyzed coupling of the mono-

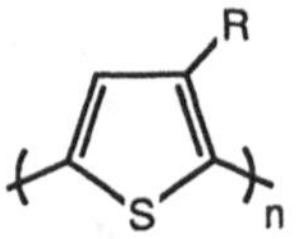

60: R = CH$_3$ 63: R = C$_6$H$_{13}$ 67: R = C$_{18}$H$_{37}$

61: R = C$_2$H$_5$ 64: R = C$_8$H$_{17}$ 68: R = C$_{20}$H$_{41}$

62: R = C$_4$H$_9$ 65: R = C$_{10}$H$_{21}$ 69 R = C$_{22}$H$_{45}$

66: R = C$_{12}$H$_{25}$

Grignard reagent derived from 3-alkyl-2,4-diiodothiophene as shown in Eq. (7). Yoshino and co-workers used a polymerization method

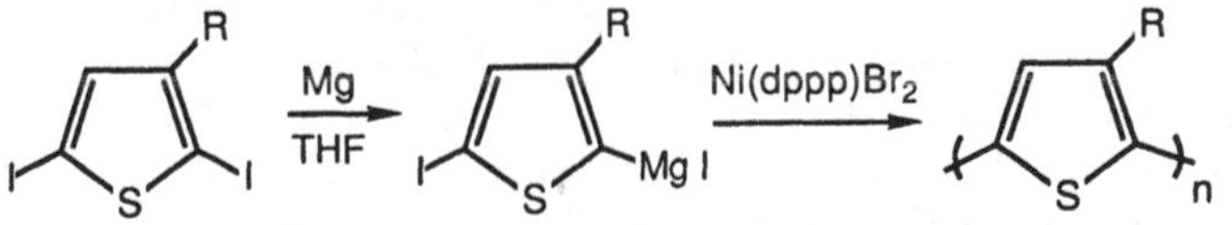

$$(7)$$

which involved chemical oxidation of the monomers with transition metal halides such as FeCl$_3$, MoCl$_5$, and RuCl$_3$ [Eq. (8)]. Several

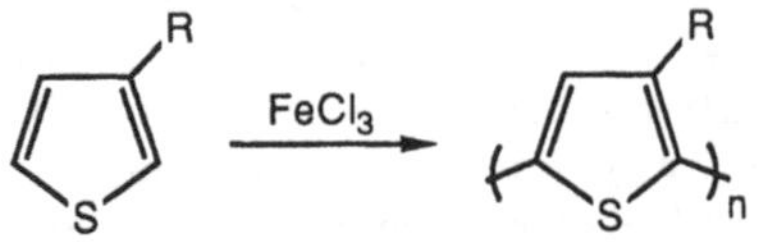

$$(8)$$

polymers were prepared, but only poly(3-hexylthiophene) (63) was discussed in detail and the 3-octyl polymer 64 was mentioned. More recently, this and other groups have used this Lewis-acid-induced polymerization of the 3-alkylthiophene monomers, generally using FeCl$_3$ [76-79], as an excellent method for polymer preparation. It should be pointed out that this is a modification of the procedure first reported in 1984 by Yoshino et al. [80].

The room temperature conductivity of these polymers as cast films compares quite favorably with those of polythiophene films produced either chemically or electrochemically. Thus, for example, iodine doped poly(3-butylthiophene) (62), prepared by Grignard coupling [73,74], showed a conductivity of 4 S cm^{-1} and polymers 63, 64, 66, 67, and 68 which were electrochemically prepared and doped showed conductivities of 95, 78, 67, 17, and 11 S cm^{-1} respectively [72]. The FeCl$_3$ produced polymers 63 and 64 had conductivities [75] of over 10 and 11 S cm^{-1} [76] when I$_2$ was used as the dopant. This is to be compared with chemically produced (FeCl$_3$ oxidant) polythiophene doped with I$_2$ which showed a conductivity of 14 S cm^{-1} [80], and electrochemically produced and doped polythiophene which was somewhat higher at 190 S cm^{-1} (PF$_6$$^-$ anion) [81,82].

A further comparison of electrochemically prepared films showed that poly(3-methylthiophene) (60) had a somewhat higher conductivity than polythiophene when doped electrochemically (500 S cm^{-1}; PF$_6$$^-$ or AsF$_6$$^-$ anion) [81,82] while poly(3-butylthiophene) (62) was 110 S cm^{-1} (PF$_6$$^-$ anion) [81]. Other reports of 60 showed electrochemical doping gave a conductivity of 120 S cm^{-1} (ClO$_4$$^-$ anion) while for iodine doping it was 5 S cm^{-1} [83]. A later report gives a conductivity of up to 750 S cm^{-1} for electrochemically doped 60 (ClO$_4$$^-$ anion) [84].

There are a number of other, more recent reports of conductivity of the soluble polymers but none are substantially different from these values and are generally within the 1-100 S cm^{-1} range. All of this, of course, shows that there is a very small effect on conductivity due to alkyl substitution in polythiophene, which suggests little twisting of the backbone out of planarity [85].

When branched-chain substituents were incorporated into the poly(3-alkylthiophenes), the conductivity of the electrochemically doped films decreased substantially compared to their linear analogs. When the chain branching was moved away from the ring the conductivity increased, but was still below the analogous straight-chain system [86]. It should also be mentioned that Lemaire et al. [86] report a dependence of the conductivity on the chain length in the poly(3-alkylthiophenes) prepared electrochemically in a similar way. In these polymers the conductivity changes from 450 S cm^{-1} for the methyl system 60 to 2 S cm^{-1} for the octadecyl system 67 (PF$_6$$^-$ dopant ion) [86]. All values are in reasonable agreement with the other measurements reported above and all polymers show good conductivities.

A study of the effect of film thickness on the conductivity of poly(3-dodecylthiophene) (66) was reported recently [87] and, for thicknesses below 6 μm, the conductivity increases from about 70 S cm^{-1} to over 100 S cm^{-1} at a thickness of 3 μm.

When cast films of poly(3-hexylthiophene) (63) were stretched up to a draw ratio of 5 the conductivity went from 27 S cm^{-1} for unstretched to 200 S cm^{-1} for the iodine-doped, stretched film [88]. It should be noted that whereas polythiophene itself is only slightly stretchable (up to a draw ratio of 1.5 and a 30% conductivity increase upon doping) the poly(3-hexylthiophene) can be stretched about five times [88].

Another interesting feature of conducting polymers is their UV-Vis spectra and band gaps. As was pointed out, the alkyl groups have only a minimal effect on the planarity of the backbone in the

poly(3-alkylthiophenes) and so it is not surprising that the band
gaps and absorption maxima are only slightly different from that
in the parent. This has been incorporated in a recent review of
the optical properties of conducting polymers [3k]. The band gap
of the parent polythiophene prepared electrochemically is 2.0 eV,
while for the 3-methyl and 3-ethyl polymers (60 and 61 respectively)
it is 1.96 and 2.06 eV [82,89]. UV-Vis spectra of the soluble poly(3-
alkylthiophenes) have been obtained both as films and in solution,
and the Vis-near-IR spectra of the doped polyemrs have likewise
been obtained. The spectra of films are similar to those above;
all band gaps are around 2 eV [3k,89]. Further, the as synthesized
and cast films are also essentially identical to the above polymers
both in the neutral state and after doping [3k].

In the doped state the Vis-near-IR spectra of the films are
consistent with the appearance of the bipolaron states [3k,81,82].
Further, electron spin resonance studies of doped [with $(NO)^+ (PF_6)^-$]
poly(3-hexylthiophene) (63) solutions in chloroform coupled with
spectroscopic data have indicated that the spinless bipolaron state
is the lowest-energy charge storage configuration in dilute solution
[90a]. Polarons are formed either because of an odd number of
charges on the polymer or, in somewhat more concentrated solutions,
as a result of interchain interactions. Gu et al., however, conclude
from ESR and spectral measurements that only polaron and no bipolaron states are supported when they used benzonitrile or methylene
chloride as solvent [91]. Recently, however, it has been shown
that whether polarons or bipolarons are favored in solution depends
on the solvent polarity and polymer concentration [90b].

When the spectra are examined in solution several interesting
phenomena emerge. First there is a rather large hypsochromic shift
from film to solution [75]. Thus, for example, the maximum in the
absorption of poly(3-hexylthiophene) (63) moves from around 488
nm as a film [84] to 435 nm in THF solution [90,91] and 430 nm
in chloroform solution [92]. Further, the color of the solution can
vary with solvent. For example at 30°C the color of a solution of
poly(3-docosylthiophene) (69) is red, orange and yellow in methylene
chloride, chloroform, and carbon tetrachloride respectively [93a].
Poly(3-hexylthiophene) (63) also changes colors from yellow in a
good solvent to magenta when a poor solvent is added [92]. A more
recent report [94] confirms the observations on 63 and reports
an absorption maximum at 443 nm in chloroform with a bathochromic
shift as methanol (a poor solvent), with λ_{max} = 506 nm when there
is only 10% chloroform.

Other detailed studies of the solvatochromism [93b] have shown
that there is a correlation of the UV-Vis maximum with Taft's solvatochromic π^* parameters with 63 but not with 69. The conclusion was

that, for poly(3-hexylthiophene), (63) the solvatochromism is the
result of normal solute-solvent interactions where the alkyl groups
and the backbone both interact with the solvent. However, for
poly(3-docosylthiophene) (69) the large alkyl groups shield the
backbone chain from the solvent and so the interaction of solvent
is primarily with the side chains [93b].

Further, in addition to this solvatochromism, solutions of poly(3-
alkylthiophenes) are also thermochromic [92,93a]. Thus, 63 goes
from yellow to magenta when the temperature of a solution in 2,5-
dimethyltetrahydrofuran [92] is lowered while 69 in chloroform solution is reported to go from red at 5°C to orange at 30°C to yellow
at 50°C [93a]. A very similar phenomenon has been observed with
films of poly(3-alkylthiophenes). Thus, for example, the absorption
maximum of poly(3-hexylthiophene) (63) moves from 515 nm at room
temperature to 419 nm at 190°C [94]. Similar observations have
been made with other poly(3-alkylthiophenes) [95]. In addition to
the bathochromic shift in λ_{max} as the temperature is lowered and
the solvent is changed to a poorer one, two new peaks appear on
the long-wavelength side of the band at 607 nm and 560 nm [92,94].

The explanation for the change in absorption maximum from
long to short wavelength is that it is the result of reduction in
the conjugation length and a change from a more rodlike to a more
coil-like conformation [92-95]. By bending or twisting the chain,
defects are introduced which shorten the average conjugation length
and thus shift the maximum absorption to shorter wavelength. This
is what occurs at higher temperature with more thermal energy.
The shift to shorter wavelengths as the solvent is changed to a
"poorer" one may be a solvent effect or an effect due to aggregation
of polymer chains (which could also occur in the films). This has
yet to be clarified [96].

It has been suggested that since the thermochromism of the
600-nm band is independent of temperature all spectral changes
in the thermochromism and solvatochromism are due to changes in
single chains and not to interchain interactions and aggregation
[92], although there must be some aggregation at lower temperatures
and in the poorer solvents [92]. Inganäs et al. have reported slightly
different results and suggest a somewhat different interpretation
[94]. They report that the 607- and 560-nm bands do not change
position with temperature but do disappear with increasing temperature. However, they disappear at different rates and the band
at 607 nm disappears last as the temperature is raised. Thus the
two shoulders do not come from the same phase. These phases could
be on the same polymer chains but this still requires considerable
clarification [94].

It is interesting to note that not only does the absorption maximum shift to shorter wavelength as the temperature is increased but also there is a more rapid decrease in wavelength with temperature around the melting point [96], indicating a more rapid loss of conjugation length at this point. In addition, x-ray photoelectron spectra shows two peaks for the carbon 1s core level at 3.4 and 5.3 eV higher than the main peak growing in as the temperature is raised [96]. Once again this is consistent with shorter conjugation lengths at higher temperatures.

These ideas of shortening conjugation lengths (more rotational defects along the chain) with temperature giving rise to these observations has been supported by quantum mechanical valence effective Hamiltonian calculations which suggest fully coplanar rings along the polymer chain at low temperature (-60°C) and conjugation lengths of only a few rings at high temperatures (190°C) for poly(3-hexylthiophene) (63) [97]. This has been further reinforced by photoelectron spectroscopy [97,98].

An interesting observation is that the thermochromic behavior is nearly reversible when the poly(3-alkylthiophenes) are heated to <200°C. After many days at room temperature the UV-Vis spectrum has not completely returned to what it was before heating although it is close [94]. When heated above 200°C the poly(3-alkylthiophene) spectra show considerable irreversibility.

Recent x-ray studies of poly(3-butylthiophene) (62) and poly(3-hexylthiophene) (63) have revealed partial crystallinity with periodicities of 12.7 Å and 16.8 Å, respectively, and a lamellar structure with an interlayer spacing of 3.8 Å in 63 [99]. There is also considerable interweaving of the side chains from the neighboring polymer chains. The x-ray data on thermally cycled 63 has shown that the crystalline portion of the polymer disappears at the higher temperatures (>160°C) which corresponds to the melting transition. However, the persistence of a broad x-ray peak, even at elevated temperatures, is assigned to retention of appreciable nematic alignment of the polymer chains [99]. Indeed, the data suggests two coexisting regions, crystalline and nematic. In the intermediate temperature range there is hysteresis and the ordering is a slow process during the cooling cycle.

Winokur et al. observed a single low-energy shoulder on the UV-Vis spectrum of a thin film of poly(3-hexylthiophene) (63) which diminished on heating and irreversible changes were noted using infrared spectroscopy after heating to 300°C [99], consistent with the earlier observations.

When poly(3-dodecylthiophene) (66) film was subjected to pressures up to 8 kbar at 116°C the UV-Vis spectrum absorption maximum shifted to longer wavelength. There was also a shift, but smaller, at 32°C [100a]. Similar observations have been made with poly(3-hexylthiophene) (63) in solution; however, at high pressures (up to 8 kbar) a toluene solution of 63 showed two new peaks at longer wavelength at the expense of the main absorption [100b]. It was also noted that the conductivity of undoped 66 film increased with pressure at 70°C [100a]. These results suggest greater order and fewer defects along the polymer chain under pressure.

When the conductivity of the undoped polymer films, poly(3-dodecylthiophene) (66) and poly(3-docosylthiophene) (69) were examined as a function of temperature there was a marked decrease in the conductivity around the melting point. Cooling reversed the trend with some hysteresis [95,100a,101]. This, too, is explained on the basis of shorter conjugation lengths at the higher temperatures, above the melting point [95,100a,101]. This effect cannot be seen in an iodine doped film of 66. The conductivity ($\sim$10 S cm^{-1}) decreases very slightly with decreasing temperature [102], but temperatures near the melting point were not examined. This is presumably because doped poly(3-alkylthiophenes) rapidly degrade at elevated temperatures [103].

The UV-Vis absorbances of films of poly(3-decylthiophene) (65) shifts to shorter wavelength upon elongation (draw ratio = 2). This is explained as being due to increased twisting between thiophene rings upon stretching [104].

There have been a number of optical emission studies of the poly(3-alkylthiophenes) both in the solid and in solution as a function of temperature. In solution the emission quantum yield increases with increasing temperature [93a,105]. It has been reported to be fluorescence emission with a lifetime of about 600 ps [105]. The explanation for the increase in luminescence with temperature, which is opposite to that usually observed, is that the number of torsional defects increases in the chains with temperature, and this results in more localized excitations and hence more fluorescence [93a,105]. The same trend is observed in films, increased emission with temperature, up to the melting point. Above the melting temperature the emission intensity decreases with increasing temperature [106]. Here, since the number of defects does not increase very much with increasing temperature, the more typical nonradiative processes, which increase with temperature, take over and give rise to decreasing luminescence [106].

Since these species are all soluble, molecular weights can be determined. The majority of those reported have been obtained by gel permeation (size exclusion) chromatography using polystyrene standards. As noted earlier, these are actually comparisons of hydrodynamic volume, not molecular weight, but the relative values allow comparison of various polymers. Electrochemical polymerization of 3-hexylthiophene provided a weight average molecular weight (M_w) for 63 of about 48,000 (DP ~300) with the peak maximum at about 20,000 and a polydispersity of approximately 2 [89]. This was about an order of magnitude larger than that reported by Elsenbaumer et al. [73]. Hotta and co-workers have reported an M_w of about 250,000 also for 63 but, with the rather high polydispersity of 5.5, the number average molecular weight (M_n) was about 45,000. The peak maximum was at a molecular weight of 110,000 and the 250,000 M_w corresponds to a relative degree of polymerization about 1500 [88]. These samples were prepared either by dehydrohalogenation of 2-halo-3-hexylthiophene or by metal halide polymerization of 3-hexylthiophene. Inganäs, Salaneck, Österholm, and their co-workers have reported M_w = 19,000 and M_n = 8300 for 63 prepared by nickel-catalyzed coupling of the mono-Grignard reagent of 2,5-diiodo-3-hexylthiophene [94]. The poly(3-octylthiophene) (64) prepared by this method gave M_w = 12,500 and M_n = 6250 while 64 produced by FeCl$_3$ polymerization of 3-octylthiophene provided M_w = 178,000 and M_n = 44,000 [77]. Yoshino and co-workers reported molecular weights for FeCl$_3$ prepared polymers 64, 66 and 69. M_n and M_w were, respectively, 23,400 and 147,000 for 64, 20,600 and 109,000 for 66, and 28,600 and 135,000 for 69 [76].

Recently in our laboratories [79,107] we have examined poly(3-butyl-), poly(3-hexyl-), poly(3-octyl-), poly(3-decyl-), and poly(3-dodecylthiophene) (62-66) by GPC using polystyrene standards and have compared the molecular weights obtained with those obtained using a multiangle laser light scattering detector (MALLS). These polymers were prepared by the FeCl$_3$ polymerization of the 3-alkylthiophene in the presence of oxygen (dry air was bubbled through the reaction mixture) [79]. Using a refractive index detector (the results are somewhat different using a UV detector at 254 nm) and polystyrene standards, M_w for 62-66 were, respectively, 68,000, 175,000, 98,000, 127,000, and 121,000. Our preliminary results using a MALLS GPC detector was that the light scattering molecular weights were considerably greater (by factors of 2-5) than when polystyrene was used as a GPC standard. Our results for these samples (62-66) were, respectively, for M_w, 340,000, 398,000, 204,000, 390,000, and 227,000; for M_n (polydispersity in parentheses), 205,000 (1.7), 297,000 (1.3), 136,000 (1.5), 187,000 (2.1), and 99,000 (2.5), and for the position of the peak maximum, 280,000, 320,000, 190,000, 260,000, and 88,000. The peak maxima based on polystyrene standards and using the RI detector for 62-66 were 44,000, 80,000, 48,000, 73,000, and 53,000, respectively. It should be noted that using light scattering, which provides absolute molecular weights, M_n is less accurate than M_w since the lower-molecular-weight species scatter less than the higher ones.

A study of the molecular weights of 66 produced electrochemically as a function of current density has shown that the molecular weight increases with current density and that cross-linking takes place at high current density [87]. In addition, membrane osmometry has demonstrated that 62 prepared by $FeCl_3$ polymerization had a molecular weight of about 70,000 [78].

A number of thermal studies of poly(3-alkylthiophenes) have been carried out. The following are typical. Thermomechanical studies seem to indicate a glass transition for the hexyl derivative 63 of 33°C [88]. The octyl derivative 64 shows good thermal stability by thermogravimetric analysis under nitrogen to well over 300°C and in oxygen decomposition begins around 250°C [77]. With residual iron in the polymer (from the preparation) decomposition in oxygen begins at lower temperatures [77]. Differential scanning calorimetry of 64 revealed a glass transition of -20°C. For 62 T_g has been reported to be 48°C [74].

The melting points of the various poly(3-alkylthiophenes) have been obtained by DSC and have been reported [77,102]. There is a steady decrease in T_m with chain length of the 3-alkyl substituent to the point where the docosyl derivative 69 melts well below 100°C [77,102].

TGA of a film of doped poly(3-methylthiophene) showed decomposition beginning at about 100°C. Using I_2 as the dopant there was a rapid weight loss between 100 and 200°C, while with other dopants the rapid weight loss was at somewhat higher temperatures, between 200 and 300°C for $NOSbF_6$ [73]. A recent study [103,108] has shown that electrochemically or $FeCl_3$-doped thin films of poly(3-hexylthiophene) (63) and poly(3-decylthiophene) (65) rapidly degrade and lose conductivity at elevated temperatures (110°C). Also in the presence of water or oxygen the conductivity rapidly degrades. In all cases this is due to dedoping of the polymer as shown by optical spectroscopy. There is no evidence that oxygen or water attack the polymer. Thermal undoping has also been observed in solution but, unlike the thin-film case, the polymer is redoped on cooling [103,108]. Optical and x-ray photoelectron spectroscopy have shown that this undoping of $FeCl_3$-doped polymer is accompanied by conversion of $FeCl_4^-$ to $FeCl_2$. A tentative model which assumes that the oxidation potential of the polymer increases with increasing twists at elevated temperatures has been postulated [103,108].

Proton and ^{13}C NMR spectra of a number of poly(3-alkylthiophenes) have been reported. The hexyl polymer 63 is reported to show six ^{13}C lines corresponding to each side chain carbon and four-ring carbons [88]. In the 3-decyl polymer 65 we have observed the four thiophene ^{13}C NMR peaks at essentially the same chemical shifts as reported for the hexyl polymer 63 [88], and see only five broadened peaks for the 10 carbons on the side chain [79] at approximately the chemical shifts reported for 63. We have further observed that the monomer ^{13}C NMR spectrum (of 3-decylthiophene) shows all 10 separate ^{13}C side-chain resonances, and the main difference in the spectra of the monomer and polymer is that the two upfield monomer aryl carbons move downfield by about 10 ppm in the polymer [79]. A proton NMR study of several poly(3-alkylthiophenes) has revealed two different α-CH$_2$ groups (also β-CH$_2$), indicating there are different thiophene ring conformations in the polymer chains [109].

Cyclic voltammetry on an electrochemically prepared film of poly(3-dodecylthiophene) (66; DP about 90) has revealed differences in films prepared in different ways. The as-grown film showed two anodic and two cathodic peaks, while a film cast from an oxidized polymer solution showed only one anodic and one cathodic peak. Two peaks would suggest two different conjugation lengths or the ability to separate different charged states in the polymer [110].

Small-angle neutron scattering (SANS) studies of both doped and undoped poly(3-butylthiophene) (62; two samples of molecular weight 49,000; polydispersity 1.5 and 22,000; polydispersity 5) in nitrobenzene-d$_5$ solution have recently been reported [111]. At room temperature a strong-chain-chain-attractive interaction is observed. Even at the lowest concentration which could be studied, 0.1 mg/mL, there were still chain attractions observed, albeit weaker. At 65°C there were also some residual chain-chain interactions; the statistical molecule length was 50-60 Å and suggested a fairly flexible coil. The lateral width was 11-12 Å corresponding to fairly extended butyl groups. Fully extended butyl groups would provide a value of 15 Å. With the 62 of lower molecular weight 22,000, there were negligible chain interactions at 70°C even up to concentrations of 2 mg/mL. Doped solutions of this latter polymer 62, doped by adding NOSbF$_6$ to the polymer solution at room temperature, showed behavior strongly dependent on polymer concentration and that the conformation was very different from that of the neutral polymer. At low concentration and a high doping level a rod conformation was observed with a length greater than 850 Å. When a solution of polymer, at 80°C, was doped and studied at different doping levels the scattering behavior depended on the doping level. The higher the doping level the more rigid the chain. It was also observed that the doped polymers do not behave like common polyelectrolytes [111].

Morphological studies using scanning electron microscopy have shown that when poly(3-octylthiophene) (64) powder was hot pressed the film had a clear uniform surface [76]. For poly(3-butylthiophene) (62) prepared by FeCl$_3$ polymerization, the films cast from THF

solution by slow solvent evaporation were dense and showed some
fibrillar structure [78]. The films cast by slow evaporation of $CHCl_3$
or CH_2Cl_2 were smoother with no indication of texture and when
these solvents were evaporated faster, or a thicker film was grown,
the films appeared uniformly porous [78]. We have observed [79]
that fibers of poly(3-decylthiophene) (65) pulled from solution show
a rather porous surface. However, fracturing a liquid-nitrogen-cooled
fiber and examining the cross section by SEM showed the fiber
was quite dense and the pores were only on the surface, indicating
that the pitting was probably due to bubbles caused by escaping
solvent [79].

For poly(3-dodecylthiophene) (66) freestanding films prepared
electrochemically, surface projections have been observed by SEM
and these continue to grow as the electrochemical polymerization
proceeds [87]. A 0.1-µm-thick film (on a substrate) had a smooth
looking surface, but at high magnification many irregularities, due
to nucleated growth, were observed [87].

The first report of a melt-spun polythiophene fiber was in 1987
where poly(3-dodecylthiophene) (66) produced a tough, flexible
fiber which, after doping with I_2, showed a conductivity up to
55 S cm^{-1} [102]. It showed a somewhat higher conductivity than
a hot-pressed film doped under the same conditions with iodine.
It was suggested that the higher conductivity was due to alignment
of the polymer chains along the fiber [102]. We have produced
both melt-spun and solution-drawn fibers [79]. Indeed, solution-
drawn fibers (ca. 10 µm) of poly(3-decylthiophene) (65) examined
by transmission optical microscopy were red-orange in color. Polarized
optical microscopy showed these fibers to be extensively birefringent
and suggests that the polymer chains are highly oriented with chain
alignment along the fiber axis [79]. Melt-spun fibers of 210 µm
diameter were quite flexible and elastic, had a smooth surface,
as shown by SEM, and showed no cracks or breaks. TGA, under
nitrogen, of the processed fibers was virtually identical to that
of the as prepared 65, and by 420°C had only lost about 10% of
its weight. Doping of these melt-spun fibers with $FeCl_3$ in CH_3CN
occurred extremely rapidly and EDAX studies of a highly conducting
fiber which had been fractured under liquid nitrogen showed quite
uniform Fe and Cl distributions across the cross section. Even
after rather short doping times the doping was quite uniform, demon-
strating that the solvent and $FeCl_3$ are able to penetrate the polymer
matrix quite rapidly and dope the polymer uniformly [79]. Molded
electrically conductive articles can also be prepared from fusible
poly(3-alkylthiophenes) [112].

Recently there have been reports of gels being produced from
poly(3-alkylthiophenes) where the side chain is shorter than 12

carbon atoms, with the maximum amount of gel being obtained from poly(3-butylthiophene) (62) [113]. These gels expanded in chloroform but shrank by about 100-fold on addition of ethanol and showed drastic color changes. Thermochromism was also observed in these gels, the color going from red when contracted to yellow when expanded. The thermochromism was interpreted as due to changes in conjugation length as discussed previously. When doped (I_2) the conductivity of the expanded state was lower than that of the contracted state and the polymer shrunk upon doping. Dedoping restored the original volume. The nature of the cross-linking in these gels is not clear, but it has been suggested there are enhanced intermolecular interactions of the alkyl side chains in the shrunken state compared to the expanded state [113]. Drawn poly(3-alkylthiophene) gels were prepared by either drawing gel films or irradiating soluble drawn films with x-rays. The drawn films showed very anisotropic changes of length with solvent and temperature. The length perpendicular to the draw direction showed drastic changes, but that parallel to the draw direction was fairly insensitive to these changes. A similar anisotropic length change was observed upon doping. The tentative explanation given was that there was substantial alignment of polymer main chains and that cross-linking occurred between these main chains [114].

A recent paper reports on the photoconductivity of poly(3-alkylthiophenes) both in the solid and liquid states [115]. The photoconduction decreased with increasing temperature in the solid and was interpreted in terms of p-type conduction and increasing twisting of polymer chain at higher temperatures [115].

B. Alkyl-Substituted Poly(3-Alkylthiophenes)

A number of alkyl-substituted poly(3-alkylthiophenes) (70) have been prepared and studied. Substituents, R (70) include R = phenyl [81,84,86,116,117], OCH_3 [118], $OCH_2CH_2OCH_3$ [118],

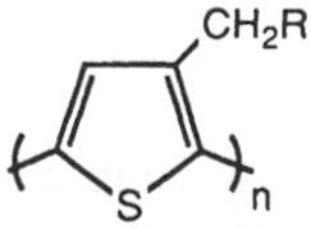

70

$O(CH_2CH_2O)_2CH_3$ [118], $NHC(O)(CH_2)_{10}CH_3$ [118], $O(CH_2)_6CH_3$ [117], CH_2OCH_3 [117], $CH_2O(CH_2)_5CH_3$ [117], $CH_2O(CH_2)_2OCH_3$ [117], $CH_2(OCH_2CH_2)_2OCH_3$ [117], CH_2OCH_2Ph [117], and $CH_2OSO_2(CH_2)_3CH_3$ [119]. All of these polymers were prepared by electrochemical polymerization. All showed reasonable conductivity, between 0.3 and 1050 S cm^{-1}, with the highest (1050 S cm^{-1}) observed for 70, R = O(CH$_2$CH$_2$O)$_2$CH$_3$ and when doped this polymer is quite soluble [118]. Other studies have shown the similarities of these compounds with the poly(3-alkylthiophenes) discussed above.

Interesting substituted polythiophenes with chiral side groups have also been prepared. These include polymers 71 (m = 2,3) and 72 which have been prepared from optically active monomers

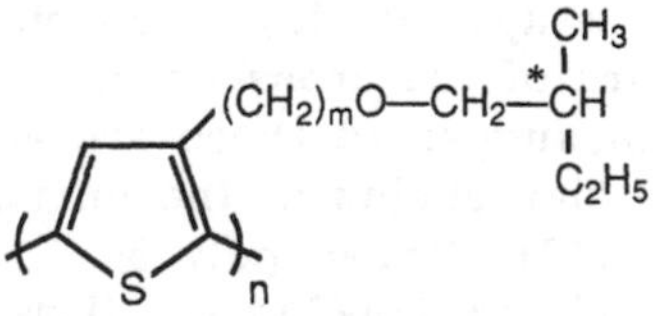

71 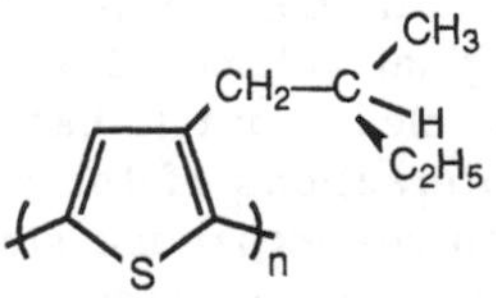   **72**

by electropolymerization [117,120]. The former (71) show high specific rotation (α_D^{22} = ± 3000°), reasonable conductivity, and, most significantly, can recognize and distinguish between chiral anions used as dopants [120a]. The latter (72) showed a two-probe conductivity of ca. 0.2 S cm^{-1}. In addition, a clay-polymer composite was prepared but not well characterized [120b].

C. Poly(3-Alkoxy)- and (3-Alkylthiothiophenes)

Poly(3-alkoxythiophenes) (73) and poly(3-alkylthiothiophenes) (74) have been prepared both by chemical and electrochemical methods. Poly(3-methoxythiophene) (73, R = CH$_3$) has been prepared by electrochemical oxidation of the monomer [121]. Cyclic voltammetry showed one oxidative wave and two broad reduction waves. It is

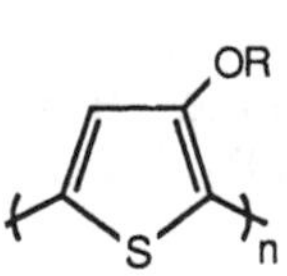

73 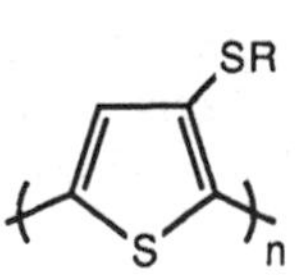   **74**

soluble in a number of organic solvents and produces a deep red solution [121a]. The oxidized form, which is gold in reflected light, is soluble in dimethylformamide and gives a deep blue solution. The low oxidation potential of this polymer relative to other polythiophenes has allowed electrochemical cycling and thus delivery of anions which might be physiologically active. More recent measurements of this system has suggested that it is actually an oligomeric mixture of 5-10 monomer units as determined by GPC [122]. It has been characterized by ^{1}H and ^{13}C NMR, IR and UV-Vis spectroscopy,

as well as by electrochemistry and spectroelectrochemistry. The
neutral material showed λ_{max} near 470 nm and was solvatochromic
[123]. The electrochemically or I_2-doped material showed four bands
at about 630, 705, 760, and 870 nm, and was solvent-dependent.
Further studies including electron paramagnetic resonance (EPR)
spectroscopy led to the conclusion that there were at least two
different types of species and that the main doped species were
bipolarons (dications) in solution [123]. More recent work on higher-
molecular-weight material (DP = 79) showed three peaks for the
polymer at 550, 820, and 1250 nm, all attributed to bipolaron formation [124]. Conductivity studies have shown that this polymer has
a conductivity of 1×10^{-2} S cm^{-1}, which is substantially lower
than other poly(3-substituted thiophenes) [125]. An earlier report,
however, gives conductivities of 0.33 S cm^{-1} for a PF_6^--doped
film and 15 S cm^{-1} after washing with hexane [121b]. Other electrochemically prepared derivatives of 73, where R = C_2H_5 through
$C_{15}H_{31}$, all show conductivities for electrochemically prepared, BF_4^--
doped material, between 10^{-2} and 10^{-4} S cm^{-1} [125]. With the polyether
side chain 73, R = $(CH_2CH_2O)_2CH_3$, the conductivity of PF_6^--doped
material is very slightly better, 5×10^{-2} S cm^{-1} [118]. GPC molecular-
weight studies of the alkoxy-substituted material gave degrees of
polymerization of 5-8 for R = CH_3 through $C_{10}H_{21}$ using a carbon
felt electrode and 10-50 for R = $C_{10}H_{21}$ through $C_{15}H_{31}$ using a
platinum electrode. All were soluble in the doped state.

Poly(3-methoxythiophene) (73, R = CH_3) has also been reported
to have been prepared by $FeCl_3$ polymerization, and a pressed
pellet conductivity of 3 S cm^{-1} was given for the $FeCl_4^-$-doped
species [124].

The sulfur-substituted systems 74 with R = CH_3 and C_2H_5 have
been prepared and studied. The chemical synthesis (Grignard coupling) shown in Eq. (7) provided a polymer 74, R = CH_3, whose
conductivity of I_2-doped material was reported as 0.5 S cm^{-1} [126].
Electrochemical polymerization of 3-(methylthio)thiophene gave a
soluble polymer with a degree of polymerization of 7.4 [124]. UV-
Vis-near-IR spectroscopy gave peaks at 390, 655, and 1020 nm
for the doped (PF_6^-) polymer (74, R = CH_3) in propylene carbonate
solution due to bipolarons. Interestingly, after four days in air
the blue solution reverted to orange, characteristic of the undoped
material.

We have recently reported on the synthesis and properties
of the soluble 3-ethylthio derivative of 74 (R = C_2H_5) prepared
by the Grignard coupling reaction shown in Eq. (7) [127]. The
polymer film exhibited a band gap of 2.0 eV, similar to other polythiophenes, and a solution bandgap of 2.2 eV. Doping with $NOPF_6$
gave two new UV-Vis-near-IR absorptions at about 800 and 1460 nm,

consistent with bipolaron formation. It is interesting that in this
case the UV-Vis spectrum is not particularly sensitive to changes
in solvent [127]. Unfortunately this polymer shows conductivities
of only 2×10^{-5} to 1×10^{-3} S cm^{-1}, depending on dopant, and
the degree of polymerization is only about 15-20. It is interesting
that attempts at electrochemical polymerization of 3-(ethylthio)thiophene
failed. Quantum mechanical calculations were in accord with these
observations and showed that the spin densities at the two α-carbons
of the radical cation, formed by removal of an electron from this
monomer, were not high enough to allow polymerization. An extended
Hückel calculation gave a theoretical band gap of 1.74 eV, in reasonable agreement with the 2.0 eV observed experimentally [127].

D. Poly[3,4-(Disubstituted)Thiophenes]

Poly(3,4-dimethylthiophene) (75) has been prepared by the Grignard
coupling reaction [Eq. (7)] and doping of a film with NOSbF$_6$ gave

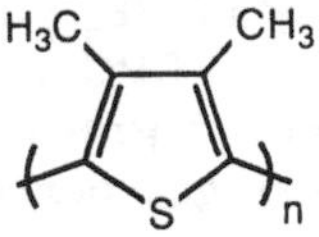

75

a conductivity of 0.5 S cm^{-1} [73]. Iodine, however, did not dope
the polymer. The molecular weight was 26,000, and the yellow color
of the undoped polymer suggested less conjugation than in mono-
alkyl substituted polymers [73]. It has been reported that poly(3,4-
dialkylthiophenes) are not fusible up to 300°C, the band gaps are
much larger (due to greater steric hindrance), and the conductivity
of doped polymers are much less than the mono-alkyl derivatives
[95]. Interestingly, the color is reported to change in an unspecified
way upon heating to 150°C [95].

The electrochemically prepared polymer 76 showed, by cyclic
voltammetry, that steric hindrance was intermediate between that
in poly(3-methylthiophene) (60) and poly(3,4-dimethylthiophene)
(75). It gave a conductivity, for electrochemically doped material,
of 10-20 S cm^{-1} [128].

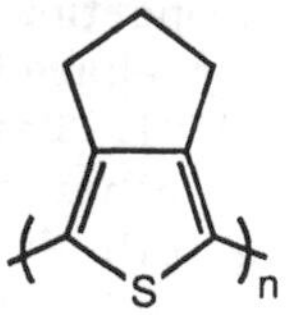

76

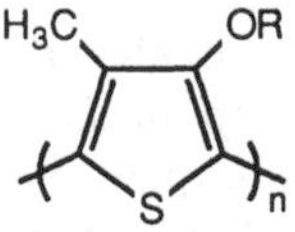

77

Other disubstituted systems include monomethyl-monoalkoxy derivatives 77. These systems, prepared and doped electrochemically, showed good conductivity, 5 S cm^{-1} for 77, R = $C_{12}H_{25}$, 30 S cm^{-1} for 77, R = $CH_2CH_2OCH_3$, and 220 S cm^{-1} for 77, R = CH_3. While the undoped polymers are soluble in common organic solvents, the doped polymers are not [124].

Poly[3,4-bis(ethylthio)thiophene] [127] prepared by the Grignard coupling reaction was an ochre-colored material with a GPC number average molecular weight of about 2600 (M_W = 9000). It was soluble in common organic solvents, but when doped with NOPF$_6$ the conductivity was only 2 × 10^{-7} S cm^{-1}, similar to other 3,4-disubstituted polythiophenes [127].

E. Polythiophene Copolymers

There are a number of soluble copolymers of thiophene derivatives that have been prepared. The copolymer of 3-methyl- and 3-butylthiophene where the monomer ratio was 1:1 and of 3-methyl- and 3-octylthiophene where the ratio was 1.5:1 gave conductivities of NOSbF$_6$-doped films of 5 and 20 S cm^{-1}, respectively, while FeCl$_3$ doping of the former gave 50 S cm^{-1} [124].

A report of a copolymer formed electrochemically from 3-benzyl- and 3-hexylthiophene units has recently appeared [129]. Thermal analysis showed it to be quite stable to 300°C. UV-Vis spectra showed the as-synthesized film had λ_{max} = 450 nm while the solution-cast film showed λ_{max} = 490 nm with an increase in the shoulder at 605 nm. This suggests that the precipitated material has more order than the as-synthesized material. There is also a hyposochromic shift, to 440 nm, on dissolution of the copolymer [129].

The copolymer of 3-hexyl- and 3-methylthiophene units (using a 2:1 feed ratio) prepared chemically either by dehydrohalogenation of the 2-halogeno-3-substituted thiophenes or by metal halide oxidation of the 3-substituted thiophenes had a GPC weight average molecular weight of 76,000 with a polydispersity of 9.2 and a peak maximum at 84,000. Doping with I$_2$ gave a film whose conductivity was 55 S cm^{-1} and upon stretching gave a conductivity of 150 S cm^{-1} for a 2.5 draw ratio and 180 S cm^{-1} for a 3.0 draw ratio. These values were higher than those for the homopolymer, poly(3-hexylthiophene) (63), under comparable conditions [88].

Preliminary molecular weight studies in our laboratories on the copolymer of 3-hexyl- and 3-octylthiophene (using a 1:1 feed ratio and FeCl$_3$ polymerization) has shown a GPC weight average molecular weight of 114,000 and the peak maximum at 28,000, while a MALLS detector gave M_n = 159,000, M_W = 442,000 (polydispersity = 2.8), and the peak maximum at 112,000 [107]. Once again, the absolute molecular weights obtained from light scattering are considerably greater than those obtained by GPC using polystyrene standards. The molecular weight of this copolymer is quite high, as reflected by M_W obtained by the MALLS technique.

F. Polypyrroles

In pyrrole there are two positions which can be substituted with groups which can render the polymer soluble and/or fusible, namely the 3 and nitrogen positions. Chemically it is easier to put groups on the nitrogen, but, because of steric hindrance, this often causes adjacent rings to twist relative to one another and thus reduce the conjugation.

A number of poly(3-substituted pyrroles) have been prepared. The long-chain ketopyrrole polymers 78 (n = 10 and 16) have been electrochemically prepared from the respective monomers and give

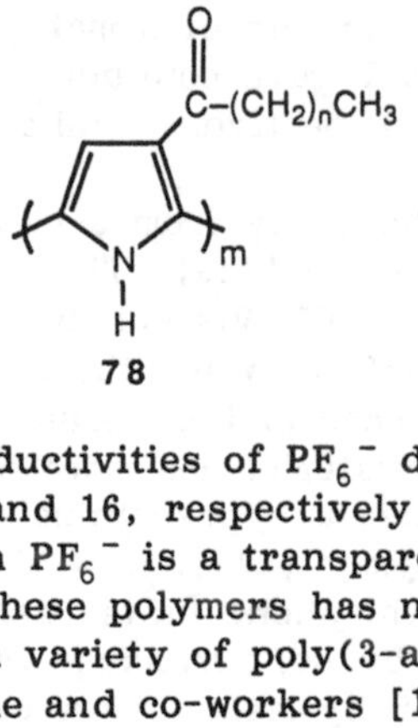

78

conductivities of PF_6^- doped material of 360 and 10 S cm^{-1} for n = 10 and 16, respectively [118]. It is reported that 78, n = 16, doped with PF_6^- is a transparent conductive polymer [130]. The solubility of these polymers has not been reported. The preparation and study of a variety of poly(3-alkylpyrroles) (79) has been carried out by Rühe and co-workers [131]. They were prepared from the respective

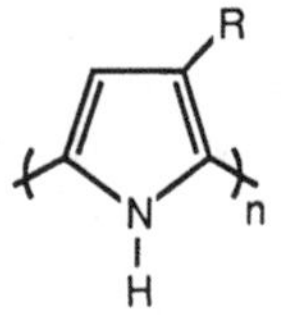

79

monomers electrochemically. Substituents included R = C_2H_5, C_6H_{13}, C_8H_{17}, $C_{10}H_{21}$, $C_{12}H_{25}$, $C_{15}H_{31}$, and $C_{18}H_{37}$, and the neutral polymers were partly soluble in common organic solvents such as $CHCl_3$, THF, and o-dichlorobenzene but insoluble in acetonitrile. The solutions were not stable and, upon standing, a black precipitate formed. GPC (polystyrene standards) indicated molecular weights in the range 5000–10,000. The UV-Vis spectra of the dissolved polymers are reported to be similar to those obtained on solid films [131].

The reduction potential of the polymeric films was shifted about 300 mV toward negative values compared to polypyrrole, and the length of the substituent had little effect on the redox potentials up to 10 carbon atoms. However, with 12 or more carbons the redox potential increases significantly [131]. The ClO_4^- doped as-prepared films had conductivities of 10^{-3} to about 10^{-1} S cm^{-1}. However, after pressing the films at 10 kbar/cm^2 the conductivities increased about 10-fold up to 0.03 to 10 S cm^{-1}. There seemed to be a slight dependence on the side-chain length with the longer side-chain polymers showing the poorest conductivities. Studies of the temperature dependence of conductivity showed a marked decrease with decreasing temperature down to about 10 K. For the side chains up to 10 carbons the decrease from the room temperature conductivity was about three orders of magnitude, whereas for R = $C_{15}H_{31}$ and $C_{18}H_{37}$ the decrease was five to six orders of magnitude [131]. The poly(3-alkylpyrroles) (79) could also be obtained by FeCl$_3$ polymerization, and the conductivity of films was around 0.01 S cm^{-1}, and they were soluble in CHCl$_3$ and o-dichlorobenzene in the oxidized state [131].

Poly(3-octylpyrrole) (79, R = C_8H_{17}) has also been independently prepared by both electrochemical polymerization and oxidative coupling using Cu(ClO$_4$)$_2$ and Fe(ClO$_4$)$_3$ [132]. The ClO$_4^-$-doped film showed UV-Vis-near-IR absorptions at 344, 496, and 1240 nm similar to poly(3-methylpyrrole) but different from polypyrrole, while in pyridine solution, the material prepared by oxidative coupling and doped with Fe(ClO$_4$)$_3$ exhibited a spectrum which consisted of two peaks at 388 and 620 nm. A solution of the electrochemically prepared material as well as solutions of the polymer in CHCl$_3$ gave spectra which resembled those from the pyridine solution. This behavior is in marked contrast to that of poly(3-octylthiophene) (64) [132].

A large number of N-substituted polypyrroles have been prepared and studied, many for modifying electrode surfaces [133]. An N-substitution in polypyrrole usually lowers the electronic conductivity of the doped material, presumably by sterically twisting adjacent rings out of planarity. Unfortunately the properties such as solubility and melting point, which would make these polymers processable have not been reported, most likely due to the fact that their processability is limited. This is true, for example, for the N-alkyl substituents (up to butyl) [134] and for long-chain substituents on the nitrogen containing other functional groups [135]. On the other hand, a benefit of N-substitution is the ability to vary the conductivity of the fully oxidized, and stable, form of the pyrrole copolymer. We have obtained free-standing films of poly(pyrrole-co-N-phenylpyrrole) which have conductivities that range from 10^{+2} S cm^{-1} to 10^{-3} S cm^{-1} as a function of copolymer composition [135a].

In addition, a chiral polypyrrole, 79, R = 2-methylbutyl, as
a co-polymer with pyrrole, has been reported but was not well
characterized [120b]. The lack of processability of these pyrrole-
based polymers precludes further discussion here.

G. Poly(3-Alkylselenophene)

Very recently the preparation and properties of poly(3-alkylseleno-
phenes) (80,81) has been reported [136]. They were prepared by

80: R = C$_6$H$_{13}$
81: R = C$_{12}$H$_{25}$

FeCl$_3$ or electrochemical polymerization of the monomers and were
soluble in solvents such as CHCl$_3$, CCl$_4$, and CH$_2$Cl$_2$. The dodecyl
polymer 81 melted around 250°C, substantially higher than the corresponding thiophene polymer 66. The hexyl derivative 80 did not
melt up to 300°C. The I$_2$-doped polymer films showed conductivities
of about 10^{-3} S cm^{-1}, as did electrochemically prepared films doped
with BF$_4$$^-$ ions, similar to poly(selenophene) itself [136]. Thermochromism was not observed, presumably because of the decreased
coplanarity of the selenophene rings resulting from the steric
hindrance between the alkyl groups and the large selenium atom
of the adjacent ring. This increased torsion of the rings also explains
the relatively high band gap of 2.4 eV [136].

H. Self-Doped Polyheterocycles

Since the conducting state of a polyheterocycle contains positive
charges, there must be compensating anions present. The preparation of charge compensating conducting (self-doped) polyheterocycles,
where anionic sites are incorporated into the polymer, was first
reported in 1987 by the Wudl and Heeger group (82 and 83) [137],
by ourselves (84) [138], and by Pickup (85) [139]. Subsequent
reports have involved preparation and study of 86 and 87 [140a]
and 88 [141]. Polymers 82, 83, and 88 could not be prepared electrochemically from the sulfonic acids or sulfonates directly. The methyl
esters were used for 82 and 83, and the ester groups in the polymers
were then converted to the free sulfonate [137]. The copolymer 84,
containing about 75% of pyrrole rings could be prepared directly
via electrochemical copolymerization [138]. Polymer 88 was prepared

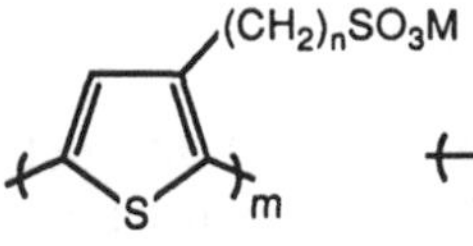

(CH₂)ₙSO₃M structure — **82**: n = 2, **83**: n = 4

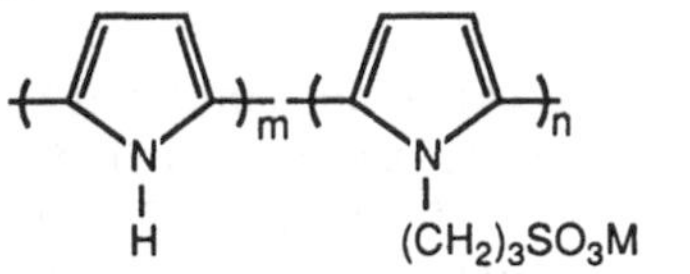

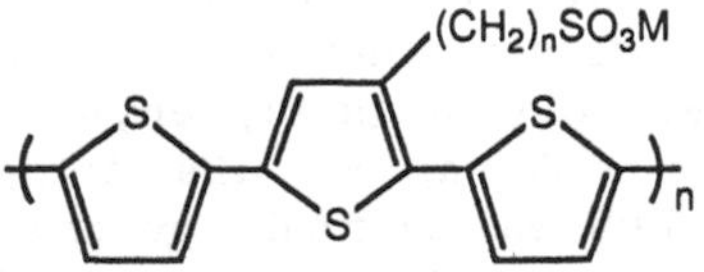

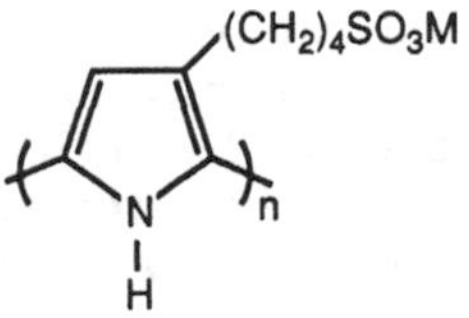

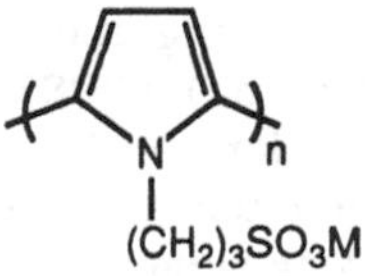

84 — pyrrole copolymer with (CH₂)₃SO₃M

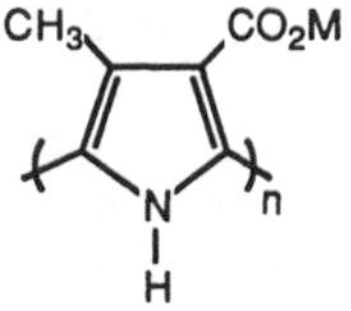

85 — CH₃, CO₂M substituted pyrrole

86 — terthiophene with (CH₂)ₙSO₃M

87 — pyrrole with (CH₂)₄SO₃M

88 — pyrrole with (CH₂)₃SO₃M

by $FeCl_3$ polymerization of the monomer [141]. Polymers 85-87 could be prepared directly by electrochemical polymerization [139,140a]. Polymers 82,83, and 86-88 are reported to be water-soluble in both the undoped and doped states [137,138,140a]. The conductivity of undoped films of 82, M = Na^+, was between 10^{-7} and 10^{-2} S cm^{-1} depending on the relative humidity, while it was ~10 S cm^{-1} after bromine doping. The band gap of this polymer was about 0.5 eV higher than that of polythiophene itself. Unlike the poly(3-alkylthiophenes) the UV-Vis spectra of cast films and aqueous solutions are virtually identical. Upon electrochemical doping, 82, M = Na^+, showed two new UV-Vis bands characteristic of bipolaron formation. Spontaneous doping was also observed at low pH [137]. A series of polymers similar to 84 containing a four-carbon spacer between the pyrrole and the sulfonate has been reported by Bidan et al. [140b].

There have been both electrochemical and microgravimetric studies of films of these self-doped polymers verifying the concept that redox switching of these systems involves cations and/or protons moving in and out [141,142]. Polymer 85, which showed fairly low conductivity when doped (ca. 10^{-5} S cm^{-1}), has been demonstrated to be a cation exchange polymer which can bind large cations reversibly [139]. The high charging and discharging rates indicate that it is also a facile ion conductor. Polymer 84 has also been shown to be capable of binding and transporting cations such as $Ru(NH_3)_6^{3+}$ and $Ru(bpy)_3^{2+}$ (bpy = 2,2'-bipyridyl) [143].

Polymers 86 and 87 showed electrochemically doped conductivities of pressed pellets of 0.01 and 0.1 S cm^{-1}, respectively, but films obtained from aqueous solution by evaporation at 70°C showed conductivities of 10^{-8} S cm^{-1} [140a].

VI. CONDUCTING POLYMER SOLUTIONS

The polyacetylenes, poly(arylene vinylenes), and soluble polyheterocycles are most conveniently processed in their insulating states
and subsequently doped to become conductors. An interesting
advancement has been made in the preparation of solutions of polymers in the conducting form [144]. Thus processing, for example
spin coating, leads directly to conducting polymer films.

   This work was catalyzed by the discovery that AsF_3, a non-
oxidizing inorganic solvent, could swell films of poly(phenylene
sulfide) (PPS) and accelerate its rate of doping with the strong
oxidant AsF_5 [145]. The interest in PPS was originally generated
because it is a commercially available dopable thermoplastic (Phillips
Petroleum, Ryton) which was dopable [146-149]. Typically, treatment
of PPS films with 400 torr of AsF_5 leads to an initial increase in
conductivity to 10^{-5} S cm^{-1} which, over a period of days, ultimately
reaches 10^{-2} S cm^{-1}. Analysis of arsenic distributions in these
films showed that most of the dopant was concentrated at the film
surface. Diffusion into the bulk of the film, and homogeneous doping,
was prevented by an impenetrable skin. If the PPS is exposed to
180 torr of AsF_3, the vapor pressure of AsF_3 at room temperature,
and subsequently, AsF_5 is added to a total pressure of 300 torr,
much faster conductivity increases are observed. Within 10 min
the polymer attains a conductivity of 10^{-2} S cm^{-1} and approaches
1 S cm^{-1} in 2 h. Dopant ion distribution studies indicated the arsenic
atoms to be distributed throughout the entire polymer with little
concentration gradient, thus showing diffusive processes were not
impaired. This is likely due to a solvating effect of the doped sites
on the polymer chain by AsF_3.

   This phenomenon has been exploited in the preparation of solutions of PPS in the conducting form [149a,b]. These solutions are
formed by first dispersing PPS powder in AsF_3 and subsequently
exposing this dispersion to an oxidizing agent (e.g., AsF_5). An
immediate color change occurs as the polymer dopes and dissolves
to yield a deep blue solution. Slow vacuum removal of solvent from
the solution leads to coherent films of conducting PPS with better
electrical properties than films doped using gas-phase techniques.

   Ac measurements (400 Hz) of the conducting PPS solutions yield
a conductivity of 2×10^{-2} S cm^{-1}. Applications of dc techniques,
on the other hand, show the solution to have ionic character since
polarization is observed. Cast films exhibit dc conductivities which
range from 25 to 200 S cm^{-1}, and the conductivity becomes predominantly electronic in the solid state.

   In addition to being electrically conductive, 0.2 *M* conducting
PPS solutions are paramagnetic exhibiting a Dysonian line shape

[150] at room temperature. This is indicative of a highly electrically conductive system and is also observed for metal ammonia solutions.

This concept for forming conducting polymer solutions has not been limited to PPS. For example, poly(*p*-phenylene oxide) (PPO) can be handled in a similar manner, and films of PPO have a conductivity of 100 S cm^{-1} when solution cast.

The AsF$_5$/AsF$_3$ system has also been found to be useful for the polymerization of aromatic molecules to form conducting polymer solutions [144,151]. For example, polythiophene has been prepared from bithiophene, poly(3-methylthiophene) from 3-methylthiophene, polypyrrole from pyrrole and polyphenylene from bi-, ter- and sexiphenyl. This has been studied in detail for the case of poly(3-methylthiophene) (60) [151]. Addition of excess AsF$_5$ (2 moles/monomer) to an AsF$_3$ solution of 3-methylthiophene leads to a deep blue solution. When the AsF$_3$ was removed from this solution, films having conductivities as high as 28 S cm^{-1} were obtained.

VII. POLYANILINE

Although one of the oldest known conducting polymers [152,153], a resurgence of interest has occurred in polyaniline. Its proton dopability has been viewed as a new concept [154] in conducting polymers and techniques have been developed to obtain it as a highly soluble and processable polymer. Polyaniline stands out from the previously mentioned redox dopable polymers in that it can be rendered conductive by partial oxidation of the fully reduced (leucoemeraldine base) form or by partial protonation of the half-oxidized (emeraldine base) form. The resulting emeraldine salt exists as a polysemiquinone radical cation, as shown in Scheme 7, with conductivity attributable to the formation of a polaron band. The proton addition and redox doping are reversible and follow a cyclic pattern as initially suggested by Hjertberg et al. [155]. Thus, the conductivity of the polymer is a function of the pH at which the

Scheme 7

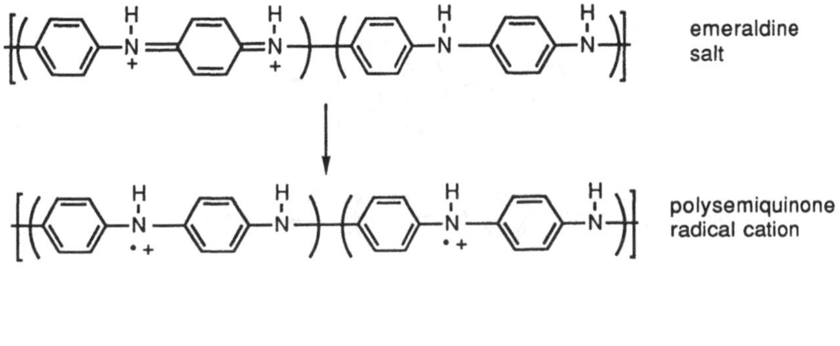

emeraldine salt

polysemiquinone radical cation

polymer is prepared. Initial preparations of polyaniline led to insoluble materials whose structures were difficult to definitively characterize. This was circumvented by both model compound studies [156,157] and a separate synthesis of poly(p-phenylene amineimine) [158]. The specific model included N,N'-diphenyl-p-phenylenediamine (89, x = 1), N,N'-bis[4-(phenylamino)phenyl]-1,4-benzenediamine (89, x = 3) and a phenyl-capped octaaniline (89, x = 7). A series of

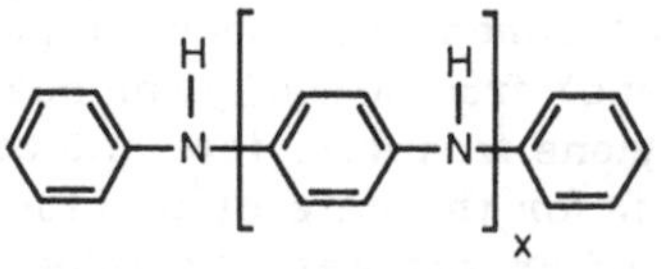

89

structural, electrochemical, spectroscopic, Brønsted acid-doping, and electrical conductivity measurements on these model compounds showed them to be fully representative of the polymer prepared via the oxidation of aniline itself. To elucidate whether or not polyaniline (PANI) is a purely para-linked poly(phenylene amineimine) (PPAI) Wudl et al. [158] synthesized a model polymer via Scheme 8 through which they could "honor the oldest organic polymer with the oldest method of structure proof: synthesis." Their synthetic route was developed to allow linkage through nitrogen atoms only at the para position, would produce high-molecular-weight polymer under mild conditions and would have a known mechanism of polymerization. Comparison of the properties of PPAI with polyaniline showed the PPAI to be an excellent model both structurally and electronically. Of special note here is the fact that polyaniline "is devoid of incorporation of o-phenylenediamine moieties in the backbone" and that the PPAI was DMF soluble.

Scheme 8

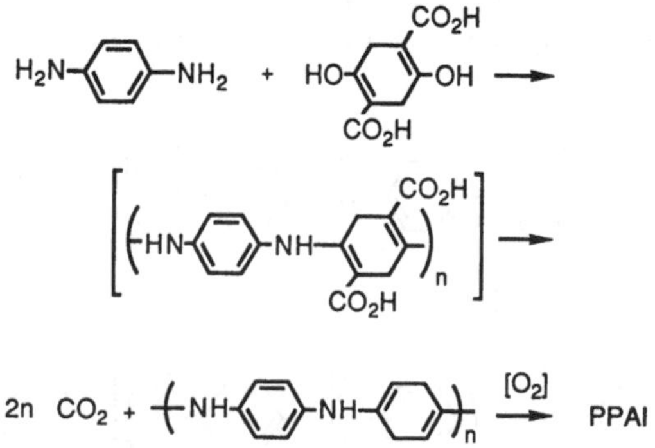

More commonly, polyaniline is prepared by the polymerization of aniline using $(NH_4)_2S_2O_8$ in HCl [159]. It is prepared as the "emeraldine hydrochloride" which can be represented by structure 90 or, a possibly more correct presentation, as 91 in the polaronic form. This form of polyaniline is highly conducting but completely insoluble.

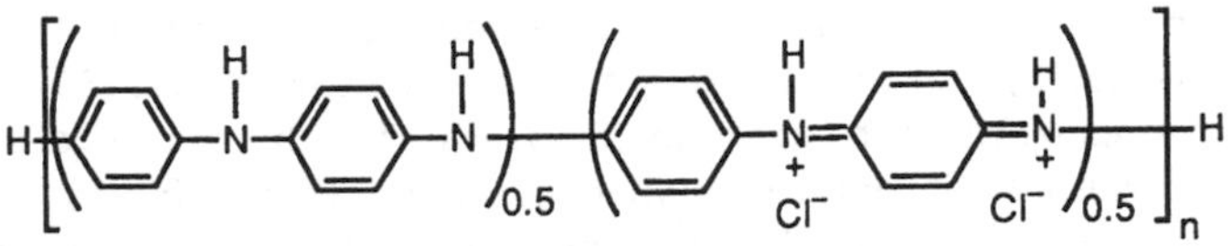

90

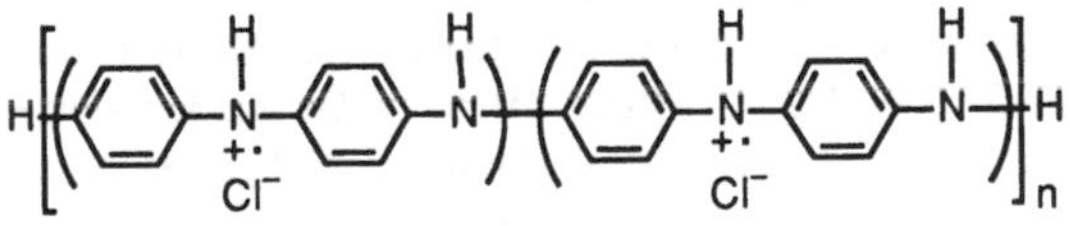

91

A highly soluble and processable form of polyaniline, termed "emeraldine base," (92) can be prepared by the NH_4OH deprotonation

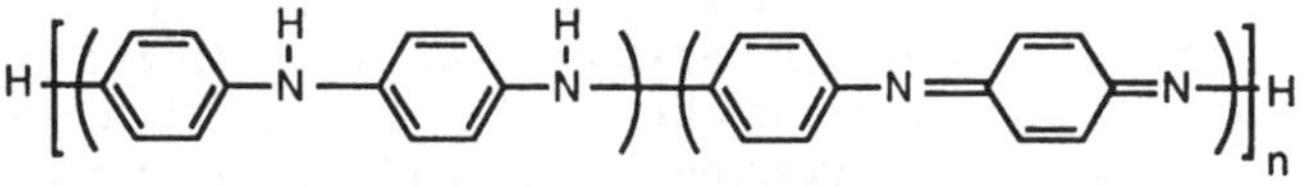

92

of emeraldine hydrochloride [159]. A variety of solvent systems are available for the polymer in this form, including 80% aqueous acetic acid, 60-80% aqueous formic acid, DMSO, and *N*-methylpyrrolidone (NMP). Freestanding films of the base form can be obtained by casting from the organic solvents while films of the doped material can be obtained by casting from acid [160]. Treatment of the neutral insulating base film with HCl subsequently leads to the conducting form as detailed in Scheme 7.

A wide spectrum of molecular weights have been reported for polyaniline. GPC elution times for polyaniline synthesized in DMF/1% acetic acid were compared to a series of structurally similar compounds and a molecular weight of 80,000 was deduced for the polymer [161]. A GPC comparison of polyaniline extracted with a 20% NaOH/THF solution, which had been prepared by electropolymerization of 0.1 *M* aniline in 0.1 *M* H_2SO_4, to polystyrene standards led to a molecular weight of 9000 [162]. Wei and co-workers [163] separated the THF soluble fraction from the THF insoluble/DMF soluble fraction of polyaniline hydrochloride deprotonated with NH_4OH. Due to the inaccuracies inherent in the GPC technique they utilized a tetrameric model compound of polyaniline to correct their molecular weights derived from polystyrene standards. A bimodal molecular weight distribution was observed with the low molecular weight fraction (only THF soluble) having a range of 2200-4800 while the high-molecular-weight fraction had a range of 170,000-200,000.

The above descriptions of soluble polyaniline are limited to the neutral insulating forms. Heeger and co-workers [164,165] have found that the emeraldine hydrochloride form of polyaniline is soluble in 97% sulfuric acid with polymer concentrations up to 20 wt%. Homogeneous viscous, purple black, polymer solutions are obtained. A viscosity molecular weight study of this polyaniline in the base form showed it to have a viscosity molecular weight average between 12,000 and 40,000 dependent upon whether the rigid-chain or flexible-chain limit was used. Semicrystalline films with spherulitic morphologies were reported for materials cast from H_2SO_4. A major benefit of solubility in the doped form is the ability to spin fibers of the polymer as a conductor. Dry-jet wet spinning into cold water from 96% H_2SO_4 led to fibers that dried with a smooth, shiny and metal-like appearance. Although only weak orientation was observed for these fibers, conductivities of 20-60 S cm^{-1} were measured.

As was described earlier for the poly(3-substituted thiophenes), substitution along the polyaniline backbone has been utilized as a means of improving the processability [166-172]. In this case a broad family of polymer derivatives are possible since substitution can be accomplished on the monomers at either the main-chain nitrogen or the aromatic rings as shown in 93. For example, Wei and

93

co-workers [168-171] have investigated polyanilines substituted with alkyl, aryl, sulfonyl, and amino groups on the ring (R_2, R_3) along with alkyl, aryl, benzyl, and $-CH_2CN$ substituents on the nitrogen (R_1). Since aniline preferentially oxidatively polymerizes at the position para to the amine, only aryl substituents (R_4) can be used there. Ring substitution with small methyl and ethyl groups led to a marked decrease in the yield and molecular weights of the polymers when compared to the unsubstituted parent polymer. Increasing the size of the substituent to propyl or larger essentially prohibits polymerization altogether. The position of the substituent on the ring is also important as indicated by an overall yield of

80% for the preparation of poly(2-methylaniline) (94) and only 29%
for poly(3-methylaniline) (95) shown in the base form for simplicity.
Interestingly, the substituents have little influence on the conductivity

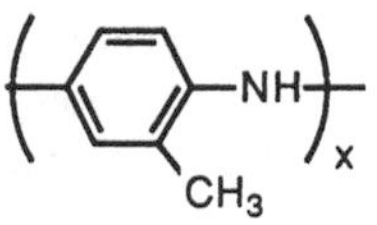

94 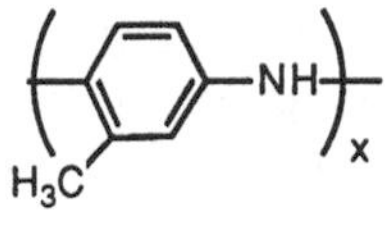   95

of the polymer obtained with both methyl derivatives exhibiting
conductivities of 0.3 S cm^{-1} and the 2-ethyl derivative 1 S cm^{-1},
while the polyaniline itself, prepared under these conditions, has
a conductivity of 5 S cm^{-1}. Although these polyaniline derivatives
are insoluble in the as-prepared, conducting forms, treatment with
base yields polymers that are highly soluble in CHCl$_3$ and THF.

The preparation of *N*-substituted polyanilines also results in
electroactive polymers. In general the polymerization mechanism
is believed to proceed through the typical head-to-tail coupling,
but, in the case of *N*-aryl substituted anilines, an alternative mechanism has been proposed [171], which is outlined in Scheme 9. Both
poly(*N*-phenylaniline) and poly(*N*-naphthylaniline) have been prepared in this manner and exhibit low conductivities of 2 × 10^{-3}
S cm^{-1}.

Scheme 9

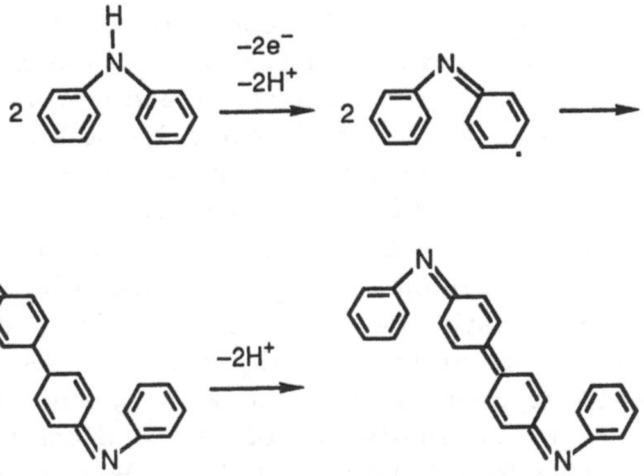

VIII. BLENDS, COMPOSITES, AND LATEXES
OF CONDUCTING POLYMERS

Multicomponent materials, including polymer blends and composites,
are prepared to obtain specific properties which are contributed by

the various components. Blending and composite techniques are
used extensively in polymer science to improve material strength,
thermal stability, gas diffusivity, etc. In the development of electronically conducting polymers, blend and composite formation allows
a conducting polymer to be dispersed in an insulating matrix as
either a continuous, phase-separated, or possibly miscible component.
At sufficiently high compositions of the conductor, electron transport
and electroactivity is possible.

A. Polyacetylene Composites

Again turning to a polyacetylene system as a prototype to illustrate
conducting polymer blend formation, Wnek and Galvin [173,174]
developed methods of in situ polymerization to form polyacetylene
within a polymeric matrix. In their method, low-density polyethylene
(LDPE) was used as a matrix due to its availability as a film, its
mechanical strength and expected inertness to Ziegler-Natta-type
initiator systems. The composite is formed by initially swelling the
LDPE in the initiator solution, impregnating the film throughout.
This swollen matrix is then washed to remove surface catalyst residues
and exposed to acetylene gas. Polymerization occurs as the monomer
diffuses into the membrane with the polymerization rate being maxi-
mized at high temperatures (110°C) where the matrix polymer is
soft. Using this method, $(CH)_x$-LDPE composites with up to 15 wt%
$(CH)_x$ could be obtained. Subsequent doping with I_2 yields strong,
flexible films with conductivities as high as 10 S cm^{-1}. Analysis
of the electrical properties of the doped films as a function of composition shows an interesting percolation-like transition at 3-4 wt %
of $(CH)_x$. Even at such low contents of conducting components,
conductivities as high as 1 S cm^{-1} could be obtained. It has been
proposed [175] that this percolation threshold can be correlated
with the surface tension differences between the two phases.

In extending this idea of in situ polymerization, Rubner et al.
[176] polymerized acetylene in polybutadiene (PB), and this composite
exhibited rubbery elastic properties at low $(CH)_x$ loadings and be-
haved more plasticlike at high $(CH)_x$ loadings. As made, these
composites could be chemically doped with I_2 and $FeCl_3$ or electrochemically doped to conductivities of 10-100 S cm^{-1}. These elevated
conductivities were found down to $(CH)_x$/PB compositions as low as
15/85. Working with these materials in the 40-60% $(CH)_x$ composition,
stretch elongation led to composites with significantly enhanced
conductivities. The most dramatic example of this is illustrated by
a conductivity increase of 15 to 575 S cm^{-1} for a 40/60 $(CH)_x$/PB
composite that was stretched by a factor of 6 prior to I_2 doping.

In addition to in situ polymerization methods, $(CH)_x$ composite materials have been prepared by polymerizing acetylene onto material surfaces. For example, Schoch [177] has developed methods for preparing $(CH)_x$ on both Kevlar and glass fabrics. Polymerization was accomplished by soaking the fabrics with the typical $Ti(OBu)_4$/ $AlEt_3$ initiator system. SEM analyses indicate the $(CH)_x$ forms as a coating on the fabric fibers and does not penetrate into the fibers themselves.

B. Poly(Arylene Vinylene) Composites

The development of the water soluble precursor prepared poly(arylene vinylenes) has led Karasz et al. [178-180] to investigate blending of the precursor polymers with other water-soluble polymers. The observation that polyacrylamide (PAcr) blend films with the PPV-soluble precursor displayed optical clarity and mechanical integrity was suggestive of a compatible polymer blend [178]. With heating, elimination to form the fully conjugated PPV could be accomplished within the PAcr matrix and the films retained their optical transparency. DSC analysis of these blends showed the T_g of PAcr to be unaffected, and thus the blend is actually completely phase separated. This concept was extended to a variety of water-soluble matrix polymers including poly(ethylene oxide) (PEO), poly(vinyl pyrrolidone) (PVP), poly(vinyl methyl ether) (PVME), methyl cellulose (MC), and hydroxypropyl cellulose (HPC) [179]. All of these blends are dopable, after heating to form the PPV, by AsF_5 vapor to conductivities of 1-100 S cm^{-1}. Whereas the in situ polymerization of $(CH)_x$ in LDPE was found to improve the ambient atmospheric stability of conductivity, this is not the case for the PPV/PAcr blends. This is most likely due to the ability of PAcr to absorb atmospheric water which subsequently reacts with, and degrades, the doped complex.

Soluble precursor routes to conjugated polymers typically yield space-filling films. Since both gas-phase and electrochemical doping require diffusion of dopant species through the films, doping rates can be quite slow. The blending of polar matrix polymers, having low glass transitions (i.e., PEO, PVME, and HPC) with PPV was found to greatly enhance the rate of gas-phase AsF_5 doping. On the other hand, use of glassy polymers in these blends (i.e., MC and PVP) depressed the doping rate and ultimate conductivity [179]. For example, a 50:50 wt% unoriented blend of PEO and PPV attained a conductivity of 200 S cm^{-1}, about an order of magnitude higher than that for unoriented PPV films alone.

Electrochemical doping rates are also seen to increase dramatically in certain PPV blends [179,180]. For the case of a PPV/PEO with a

50:50 wt% composition, the elimination process was optimized to yield the most fully conjugated PPV while limiting thermal degradation of the matrix polymer. When the PEO is present at $\geq 40\%$ in the solid state, it is obviously phase separated with the formation of 100-500 μm diameter spherulites. During electrochemical doping this PEO phase can complex Li^+ ion from electrolyte and assist in rapid ion transport through the membrane. Electrochemical doping studies showed that, in addition to accelerated doping rates with increased PEO content, doping homogeneity improved also.

C. Poly(3-Alkylthiophene) Blends

The advent of truly thermoplastic poly(3-alkylthiophenes) has opened up significant opportunities for blend and composite formation using compounding methods commonly used in the plastics industry. As pointed out earlier, Heeger et al. [129] prepared poly(3-hexylthiophene-co-3-benzylthiophene) (P3HBzT), by electropolymerization of monomer mixtures, which was highly soluble in chloroform, toluene, dichloro-methane, and tetrahydrofuran. Although bulk yields via electro-chemical polymerization are typically quite low, the copolymer could be prepared in significantly higher yields than either homopolymer. In these initial studies, miscible solutions of P3HBzT and polystyrene (PS) were prepared in $CHCl_3$ and films cast from these solutions. Spectroscopic results suggest that, upon film formation, the P3HBzT and PS phase separate. This has been observed in our labs with optical microscopy for blends of PS with electrochemically prepared poly(3-hexylthiophene) (63) [181]. At low 63 concentrations, spheres of poly(3-hexylthiophene) are dispersed in the PS matrix while at high 63 volume fractions the polythiophene is observed as a continuous phase. This is consistent with the observation [129] that P3HBzT/PS blends with >30% volume fraction of P3HBzT could be doped with $NOPF_6$ to conductivities greater than 1 S cm^{-1} while the conductivity drops precipitously to ca. 10^{-8} S cm^{-1} at 15% volume fraction P3HBzT, indicating a conductor-to-insulator transition due to percolation.

In addition to solution processing, the fusibility of poly(3-alkylthiophenes) at relatively low temperatures allows melt processing and blending with a wide variety of thermoplastics. This has been exploited by researchers at Neste Oy (Kulloo, Finland) and collabo-rators in the preparation of blends with poly(ethylene-co-vinyl acetate) (EVA), poly(ethylene-co-butyl acrylate), PS, and poly-ethylene [77,182-184]. Most of this work has centered on poly(3-octylthiophene) (64) since it exhibits good melt flow properties at processing temperatures below 200°C with little or no degradation. Films and sheets of 64/EVA blends were made by compression mold-ing at various compositions, along with film blowing extrusion which allowed preparation of 10-cm-wide films over 100 m long. As was observed with the P3HBzT/PS blends, these 64/EVA blends undergo a percolation transition and can be $FeCl_3$ doped to conductivities > 1 S cm^{-1}. In these blends the reported transition is between 5-10% weight fraction although conductivities of 10^{-4} S cm^{-1} have been observed at compositions as low as 3% of 64. This latter point suggests low percolation thresholds may be obtainable, ultimately yielding low-cost conducting polymer blends with very high mechanical integrity. As discussed earlier, the optical absorption behavior of these polymers undergo significant changes with temperature which are related to the chain conformation. This is also the case in blends of poly(3-alkylthiophenes) as demonstrated by Yoshino et al. [95] in polystyrene, poly(methyl methacrylate), and ethylenepropylene copolymer (EPR) matrices. Interestingly, this conformational change can also be induced mechanically. Stretching of poly(3-docosylthiophene) (69)/EPR blends up to draw ratios of 12 led to a significant blue shift in the absorption maximum [104]. It is suggested that increased torsion about the bonds between thiophene rings takes place in elongated samples making the effective conjugation length shorter.

D. Electrochemical Composite Formation

In addition to the formation of multicomponent materials containing conducting polymers by chemical and physical methods, electrochemical techniques have also proven to be useful. The anodic electropolymerization of a variety of heterocycles has been used to produce free-standing films. In the case of pyrrole, electropolymerized in the presence of aromatic sulfonates (e.g., tosylate, benzene sulfonate), flexible films with tensile strengths approximately equal to polystyrene were obtained [185]. By combining the electrical properties of electropolymerized polyheterocycles with the physical properties of a variety of matrix polymers, a new family of multicomponent conducting and electroactive polymers have been obtained.

The electropolymerization of heterocycles within swollen polymers previously deposited on electrode surfaces were the first systems investigated [186-191]. Poly(vinyl alcohol) (PVA) films were cast on Pt electrodes, lightly cross-linked, and subsequently swollen with an aqueous solution of 0.1 M $CuSO_4$ and 0.1 M pyrrole [186]. Non-cross-linked PVA was also used by changing the solvent system to 50:50 $CH_3CN:H_2O$. Application of an anodic potential to the electrode causes pyrrole electropolymerization to begin at the electrode surface and polymer grows through the swollen membrane. The growth of the polypyrrole can be stopped at any point, yielding a film conducting on one side and insulating on the other. On the

other hand, growth of the polypyrrole through the entire PVA
matrix yields films having conductivities of 0.1-10 S cm^{-1} on both
sides. These films were found to exhibit superior mechanical properties to poly(pyrrole sulfate) with a doubling of the ultimate tensile
strength and an order of magnitude increase in the elongation to
break. The generality of this concept was shown by the preparation
of both polypyrrole [187-189] and polyazulene [189] composites
with poly(vinyl chloride) prepared in acetonitrile electrolytes which
exhibited conductivities of 50 S cm^{-1} and 10^{-2} S cm^{-1}, respectively.

The ability to control the physical properties of these composites
has been demonstrated. Utilizing a vinylidene fluoride/trifluoroethylene
copolymer as a matrix, Niwa et al. [190] prepared polypyrrole composites with an elongation to break of greater than 120%. Changing
the matrix to a more flexible polyurethane and utilizing a tosylate
electrolyte, Bi et al. [192] obtained polypyrrole composites with
elongation to break of up to 640%.

Another benefit of electrochemical composite film formation is
the ability to control the optical density of the films by the amount
of conducting polymer incorporated [191]. The changes in optical
absorption during electrosynthesis of films of poly(pyrrole tosylate)
and poly(pyrrole tosylate)/poly(vinyl chloride) composites have
been compared. Homopolymer and composite films prepared with
a charge density of 0.032 C cm^{-2} both exhibit a transmittance to
visible light in the range of 40-70%, indicating approximately the
same amount of pyrrole was polymerized. The mechanical properties,
though, are completely different. The composite was obtained as
a 12-μm-thick freestanding film, while the homopolymer formed a
nonuniform film of 0.05-0.07 μm which could not be removed intact
from the electrode surface. Since the conducting polymer forms
initially at the electrode/matrix polymer interface, a highly conducting surface can be obtained quickly. Conductivities of up to 5 S
cm^{-1} with an optical transmittance of 60% were measured.

A method for overcoming the inhomogeneities across the film
thickness caused by in situ electropolymerization was developed
by Wang et al. [193]. They recognized that the amount of electrolyte
that could diffuse into a membrane was quite low and found that,
by blending electrolyte with host polymer (in this case PVC) prior
to coating, conducting film uniformity was greatly improved. Examination of the conductivity across the film thickness, after electropolymerization of pyrrole through an electrolyte blended and unblended
PVC for 20 min, showed over a 10 order of magnitude difference.

The use of ionomers as ion-containing matrix polymers on electrode surfaces has allowed the formation of conducting polymer
composites where the matrix contributes the dopant anion and acts
as an effective medium for ion transport. Due to its well-known

stability during electrochemical processes, the perfluorosulfonated
ionomer, Nafion, has been used in a number of studies [194-198].
Although free electrolyte is used during synthesis, it is most likely
unable to penetrate into the membrane and thus charge balance is
retained during electrosynthesis by the expulsion of counter cations.
The electrochemical charge transport properties of polypyrrole/Nafion
membranes during switching is highly dependent on the nature of
the cation [197]. Polypyrrole itself, on the other hand, often has
electrochemical charge transport properties which are a strong function of the electrolyte anion employed.

In experiments directed to open up the morphology of these
conducting polymer composites and obtain higher current densities
during electrochemical switching, Martin and co-workers [195,198]
developed a composite electrode based on a fibrous form of polytetrafluoroethylene known as Gore-tex (GT). Although GT is mechanically
strong, chemically stable and has an open structure its fluorinated
surface causes it to be extremely hydrophobic and not easily wetted.
To circumvent this, the GT membrane was impregnated with a coating
of Nafion. In fact, this Nafion-impregnated Gore-tex (NIGT) was
found to be a better ionic conductor than Nafion itself. Pyrrole
was found to polymerize easily in the NIGT membranes to form conducting composites. In addition, polymerization currents were actually
found to be higher at Pt coated with NIGT than at Pt alone due
to a negative shift in the oxidation potential of the monomer.

Conducting polymer/matrix polymer composites have also been
formed by initially solubilizing the matrix polymer in the monomer/
electrolyte solution and allowing the matrix to deposit during electropolymerization. This has been investigated for both neutral thermoplastics [199], such as PVC, and a variety of polyelectrolytes
[200-211]. A benefit here is that, since the matrix is depositing
directly with the conducting polymer, a more homogeneous film is
obtained. In the case of polyelectrolytes, where the matrix polymer
is serving as the dopant species also, there is necessarily an intimate
mixing of polymer chains and the system can be viewed as a molecular
composite.

The one-step deposition of poly(3-methylthiophene) (60)/PVC
composites described by Roncali and Garnier [199] shows how the
morphology of the film obtained is dependent on the mode of growth
of the conducting polymer when the polymer is grown within a pre-
existing matrix. This is not the case in these composites, as the
morphology of the matrix affects the electropolymerization process.
The composition and electrical properties of these polymers can
be controlled by varying either the deposition charge or reagent
concentration. For example, by changing the deposition charge
from 0 to 256 mC cm^{-2} at a current density of 5 mA cm^{-2} films

Table 1. Electrical Conductivities of Various Polypyrrole/Polyelectrolyte Composites

Polyelectrolyte	Solvent	Conductivity ($S\ cm^{-1}$)	Ref.
Poly(2-acrylamido-2-methyl propanesulfonic acid)	CH_3CN/H_2O H_2O	4×10^{-2} 2.9	200 202
Poly(methacrylic acid)	CH_3OH/H_2O	1.7	200
Chlorosulfonate polyethylene (hydrolyzed)	THF/H_2O	1×10^{-3}	200
Poly(acrylamide-co-2-acrylamido-2-methyl propanesulfonic acid)	CH_3OH	1×10^{-1}	200
Sulfonated poly(vinyl alcohol)	DMF/H_2O	2.6	200
Sulfonated styrene/(hydrogenated) butadiene triblock copolymer	$THF/C_6H_5NO_2$	9×10^{-3}	200
Poly(4-styrenesulfonic acid)	H_2O	9.4	202
Sodium poly(4-styrenesulfonate)	$H_2O/$dioxane H_2O	3×10^{-2}–5.0 2×10^{-1}	203 204
Sodium poly(vinylsulfonate)	H_2O	7.4	202
Potassium poly(vinylsulfate)	H_2O	10.0	204
Poly(styrene-co-maleic acid) (VERSA-TL4, National Starch)	H_2O	6.2	202
Sodium sulfonate of naphthalene/formaldehyde copolymer (Stepantan-A, Stepan Co.)	H_2O	6.2	202
Sodium poly(phenyleneterephthalamide propanesulfonate) (sulfonated Kevlar)	H_2O	1.0	211

of 1.5-3.0 μm were obtained that contained 0-42% of poly(3-methylthiophene) (60).

The electropolymerization of heterocycles in solutions of polyelectrolytes leads directly to doped conducting composites films and is an extremely versatile technique. For the case of polypyrrole composites, Table 1 lists a number of the polyelectrolytes that have been used as the ionic matrix. Two major benefits stand out from using these polyelectrolytes. The first is the ability to use aqueous electrolytes making disposal of waste electrolyte significantly less costly. Second, the polyelectrolyte imports its mechanical and physical properties to the overall composites properties. These films are flexible and may be heat-processable. For example, a film of polypyrrole/poly(styrene sulfonate) (PSS) is slightly flexible, but when heated to above the glass transition (T_g) of the PSS it becomes highly flexible and can be easily fit to a nonplanar surface. In some cases where the T_g of the polyelectrolyte is lower than room temperature [200] the conducting composites are actually melt-processable and elastic.

The controllable physical, mechanical, electrical, and ion transport properties of these polyheterocycle/polyelectrolyte molecular composites have also allowed them to be used for some specific applications, including electrochemical deionization [201,208] and electrodes in polymer/metal and all polymeric batteries [205,209].

E. Conducting Polymer Latexes

Colloidally dispersed latexes of polypyrrole [212-217] and polyaniline [218] have been used to prepare processable conducting organic polymers. Initial polypyrrole latexes were prepared by the electropolymerization of pyrrole in the presence of latex particles having a high concentration of bound sulfonate or sulfate groups on their surface [212]. Using a number of commercially available latexes encompassing acrylates, methacrylates, vinylidene chlorides, EVAs and styrenics, thick (up to 1/4 inch) films were obtained having conductivities from 10^{-3} to 5 S cm^{-1}. These materials could be ground, solvent swollen and sheared to yield dispersions from which air-stable electrically conducting films could be cast or spin coated.

The chemical (FeCl$_3$) polymerization of pyrrole in aqueous solutions of methylcellulose was investigated by Bjorklund et al. [213] in the preparation of colloidal polypyrrole which could be cast to form films having conductivities up to 2×10^{-1} S cm^{-1}. This method, originally employed by Edwards et al. [214] with other conducting polymers, was extended by Armes et al. [215], who carefully examined the conditions necessary to reproducibly prepare colloidal polypyrrole with a high degree of control of particle morphology.

Highly monodisperse particles were prepared having mean diameters on the order of 75-150 nm which, when pressed into pellets, yielded electrical conductivities up to 5 S cm^{-1}. These authors conclude that "steric stabilization is required to produce aqueous latex dispersions" and that "charge stabilization is ruled out under the high ionic strength conditions present" in discussing the properties of these colloids. When using poly(4-vinylpyridine) steric stabilizer, it has been found that the polypyrrole latex exhibits reversible acid/base-induced flocculation-stabilization behavior [217]. In strongly acid solution the vinyl pyridine units of the stabilizer are believed to be protonated and thus provides an electrostatic contribution to stabilization which helps disperse the colloid. Addition of base to give a pH of >3.6 ± 0.1 leads to flocculation as the vinyl pyridine units are deprotonated.

IX. SUMMARY

It is evident now that processability and high mechanical integrity can be attained for essentially any conducting polymer system. The full range of techniques available to the polymer chemist—copolymerization, processable precursors, substituent effects, blend formation, composite formation, latex formation, and even electrochemical film formation—can be applied to the development of useful materials. These principles and techniques, coupled with the atmospheric stability of a number of conducting polymer systems (especially the polyheterocycles and polyaniline), indicate these materials are prime candidates for use in specific applications.

ACKNOWLEDGMENTS

Financial assistance through grants from the Defense Advanced Research Projects Agency (University Research Initiative Program), monitored by the Office of Naval Research and the Robert A. Welch Foundation, is acknowledged with gratitude.

REFERENCES

1. T. Ito, H. Shirakawa, and S. Ikeda, *J. Polym. Sci.*, *Polym. Chem. Ed.*, *12*:11 (1974).
2. C. K. Chiang, M. A. Druy, S. C. Gau, A. J. Heeger, E. J. Louis, A. G. MacDiarmid, Y. W. Park, and H. Shirakawa, *J. Am. Chem. Soc.*, *100*:1013 (1978).

3a. A. G. MacDiarmid and A. J. Heeger, *Synth. Met.*, *1*:1013 (1978).

3b. G. B. Street and T. C. Clarke, *IBM J. Res. Develop.*, 25:51 (1981).

3c. G. Wegner, *Angew. Chem. Int. Ed. Engl.*, 20:361 (1981).

3d. K. J. Wynne and G. B. Street, *I&EC Prod. Res. Develop.*, *21*:23 (1982).

3e. R. H. Baughman, *Contemp. Topics in Polym. Sci.*, 5:321 (1984).

3f. R. L. Greene and G. B. Street, *Science*, *226*:651 (1984).

3g. J. L. Brédas and G. B. Street, *Acc. Chem. Res.* *18*:309 (1985).

3h. J. R. Reynolds, *J. Molec. Elec.*, 2:1 (1986).

3i. A. J. Epstein, *Handbook of Conducting Polymers*, Vol. 2, T. A. Skotheim, ed., Dekker, New York, 1986, p. 1041.

3j. R. S. Potember, R. C. Hoffman, H. S. Hu, J. E. Cocchiaro, C. A. Viands, R. A. Murphy, and T. O. Poehler, *Polymer*, *28*:574 (1987).

3k. A. O. Patil, A. J. Heeger, and F. Wudl, *Chem. Rev.*, *88*:183 (1988).

3l. J. R. Reynolds, *Chemtech*, *18*:440 (1988).

4a. N. Basescu, Z.-X. Liu, D. Moses, A. J. Heeger, H. Naarmann, and N. Theophilou, *Nature*, *327*:403 (1987).

4b. H. Naarmann and N. Theophilou, *Synth. Met.*, 22:1 (1987).

5. J. Roncali, A. Yassar, and F. Garnier, *J. Chem. Soc., Chem. Commun.*, 581 (1988).

6a. G. Wegner, *Z. Naturforsch*, *246*:824 (1969).

6b. G. Wegner, *Molecular Metals*, W. E. Hatfield, ed., Plenum, New York, 1979, p. 225.

6c. G. Wegner, *Pure Appl. Chem.*, 49:443 (1977).

6d. B. Tieke, G. Wegner, D. Naegele, H. Ringsdorff, *Angew Chem. Int. Ed. Engl.*, *15*:764 (1976).

6e. D. J. Ando and D. Bloor, *Polymer*, 20:976 (1979).

6f. J. L. Brédas, R. R. Chance, R. Silbey, G. Nicolas, and P. Durand, *J. Chem. Phys.*, 75:255 (1981).

6g. H. Eckhardt, R. H. Baughman, J. L. Brédas, R. R. Chance, R. L. Elsenbaumer, and L. W. Shacklette, *Mater. Sci.*, 7:21 (1981).

6h. D. Bloor, C. L. Hubble, and D. J. Ando, *Molecular Metals*, W. E. Hatfield, ed., Plenum, New York, 1979, p. 243.

7a. J. H. Edwards, W. J. Feast, and D. C. Bott, *Polymer*, 25:395 (1984).

7b. W. J. Feast and J. N. Winter, *J. Chem. Soc., Chem. Commun.*, 202 (1985).

8a. D. White and D. C. Bott, *Polym. Commun.*, 25:98 (1984).

8b. G. Leising, *Polym. Commun.*, 25:201 (1984).

8c. P. D. Townsend, C. M. Pereira, D. D. C. Bradley, M. E. Horton, and R. H. Friend, *J. Phys. C.*, *18*:L283 (1985).

9. Y. B. Moon, M. Winokur, A. J. Heeger, J. Barker, and D. C. Bott, *Macromolecules*, *20*:2457 (1987).
10. T. M. Swager, D. A. Dougherty, and R. H. Grubbs, *J. Am. Chem. Soc.*, *110*:2973 (1988).
11. T. M. Swager and R. H. Grubbs, *J. Am. Chem. Soc.*, *111*:4413 (1989).
12. J. C. W. Chien, G. E. Wnek, F. E. Karasz, and J. A. Hirsch, *Macromolecules*, *14*:479 (1981).
13. G. L. Baker and F. S. Bates, *J. Phys. (Les Ulis, Fr.)*, *C3*:11 (1983).
14. G. F. Dandreaux, M. E. Galvin, and G. E. Wnek, *J. Phys. (Les Ulis, Fr.)*, *C3*:135 (1983).
15. M. E. Galvin and G. E. Wnek, *Polym. Prepr. (Am. Chem. Soc., Div. Polym. Chem.)*, *24*(2):14 (1983).
16. F. S. Bates and G. L. Baker, *Macromolecules*, *16*:704 (1983).
17. G. L. Baker and F. S. Bates, *Macromolecules*, *17*:2619 (1984).
18. M. J. Aldissi, *J. Chem. Soc., Chem. Commun.*, 1347 (1984).
19. S. Destri, M. Catellani, and A. Bolognesi, *Makromol. Chem., Rapid Commun.*, *5*:353 (1984).
20. M. E. Galvin and G. E. Wnek, *Polym. Bull.*, *13*:109 (1985).
21. M. Aldissi and A. R. Bishop, *Polymer*, *26*:622 (1985).
22. J. A. Stowell, A. J. Amass, M. S. Beevers, and T. R. Farren, *Makromol. Chem.*, *188*:1635 (1987).
23. T. R. Farren, A. J. Amass, M. S. Beevers, and J. A. Stowell, *Makromol. Chem.*, *188*:2535 (1987).
24. P. Kovacic and C. Wu, *J. Polym. Sci.*, *47*:448 (1960).
25. P. Kovacic and M. B. Jones, *Chem. Rev.*, *87*:357 (1987).
26. C. E. Brown, P. Kovacic, C. A. Wilkie, J. A. Kinsinger, R. E. Hein, S. I. Yaniger, and R. B. Cody, *J. Polym. Sci., Polym. Chem. Ed.*, *24*:255 (1986).
27. T. Yamamoto, Y. Hayashi, and Y. Yamamoto, *Bull. Chem. Soc. Jpn.*, *51*:2091 (1978).
28. L. W. Shacklette, R. R. Chance, D. M. Ivory, G. G. Miller, and R. H. Baughman, *Synth. Met.*, *1*:307 (1980).
29. D. G. H. Ballard, A. Courtis, I. M. Shirley, and S. C. Taylor, *J. Chem. Soc., Chem. Commun.*, 954 (1983).
30. D. G. H. Ballard, A. Courtis, I. M. Shirley, and S. C. Taylor, *Macromolecules*, *21*:294 (1988).
31. P. R. McKean and J. K. Stille, *Macromolecules*, *20*:1787 (1987).
32. M. Rehahn, A.-D. Schlüter, G. Wegner, and W. J. Feast, *Polymer*, *30*:1054 (1989).
33. M. Rehahn, A.-D. Schlüter, G. Wegner, and W. J. Feast, *Polymer*, *30*:1060 (1989).
34. M. Fukuda, K. Sawada, and K. Yoshino, *Jpn. J. Appl. Phys.*, *28*:L1433 (1989).

35. K. D. Gourley, C. P. Lillya, J. R. Reynolds, and J. C. W. Chien, *Macromolecules*, 17:1025 (1984).
36. J. D. Capistran, D. R. Gagnon, S. Antoun, R. W. Lenz, and F. E. Karasz, *Polym. Prepr. (Am. Chem. Soc., Div. Polym. Chem.)*, 25:282 (1984).
37. D. R. Gagnon, J. D. Capistran, F. E. Karasz, and R. W. Lenz, *Polym. Bull.*, 12:293 (1984).
38a. I. Murase, T. Ohnishi, T. Noguchi, and M. Hirooka, *Polym. Commun.*, 25:327 (1984).
38b. I. Murase, T. Ohnishi, T. Noguchi, M. Hirooka, and S. Murakami, *Mol. Cryst. Liq. Cryst.*, 118:333 (1985).
39. M. Kanbe and M. Okawara, *J. Polym. Sci., Part A-1*, 6:1058 (1968).
40. R. A. Wessling and R. G. Zimmerman, U.S. Patents 3,401,152, 1968 and 3,706,677, 1972, *Chem. Abstr.*, 69:87735q (1968) and 78:85306n (1973).
41. D. R. Gagnon, J. D. Capistran, F. E. Karasz, R. W. Lenz, and S. Antoun, *Polymer*, 28:567 (1987).
42. J. M. Machado, R. R. Denton, III, J. B. Schlenoff, F. E. Karasz, and P. M. Lahti, *J. Polym. Sci., Polym. Phys. Ed.*, 27:199 (1989).
43. D. R. Gagnon, F. E. Karasz, E. L. Thomas, and R. W. Lenz, *Synth. Met.*, 20:85 (1987).
44. I. Murase, T. Ohnishi, T. Noguchi, and M. Hirooka, *Synth. Met.*, 17:639 (1987).
45a. K. Yoshino, T. Takiguchi, S. Hayashi, D. H. Park, and R. Sugimoto, *Jpn. J. Appl. Phys.*, 25:881 (1986).
45b. J. Obzrut and F. E. Karasz, *J. Chem. Phys.*, 87:2349 (1987).
46. D. D. C. Bradley, R. H. Friend, H. Lindenberger, and S. Roth, *Polymer*, 27:1709 (1986). T. Granier, E. L. Thomas, D. R. Gagnon, F. E. Karasz, and R. W. Lenz, *J. Polym. Sci., Polym. Phys. Ed.*, 24:2793 (1986).
47. Y. B. Moon, S. D. D. V. Rughooputh, A. J. Heeger, A. O. Patil, and F. Wudl, *Synth. Met.*, 29:E79 (1989).
48. A. O. Patil, S. D. D. V. Rughooputh, and F. Wudl, *Synth. Met.*, 29:E115 (1989).
49. R. W. Lenz, C.-C. Han, J. Stenger-Smith, and F. E. Karasz, *J. Polym. Sci., Polym. Chem. Ed.*, 26:3241 (1988).
50. J. D. Stenger-Smith, R. W. Lenz, and G. Wegner, *Polymer*, 30:1048 (1989).
51. J. Obzrut and F. E. Karasz, *J. Chem. Phys.*, 87:6178 (1987).
52. T. Murase, T. Ohnishi, and T. Noguchi, German Patent 3,522,720, 1986; *Chem. Abstr.*, 104:169124v (1986).
53. S. Antoun, D. R. Gagnon, F. E. Karasz, and R. W. Lenz, *J. Polym. Sci., Polym. Lett. Ed.*, 24:503 (1986).

54. S. Antoun, D. R. Gagnon, F. E. Karasz, and R. W. Lenz, *Polym. Bull.*, *15*:181 (1986).
55. H.-H. Hörhold, J. Gottschaldt, and J. Opfermann, *J. Prakt. Chem.*, *319*:611 (1977).
56. M. Helbig, H.-H. Hörhold, and A.-K. Gyra, *Makromol. Chem.*, *Rapid Commun.*, *6*:643 (1985).
57. K.-Y. Jen, L. W. Shacklette, and R. L. Elsenbaumer, *Synth. Met.*, *22*:179 (1987).
58. C.-C. Han and R. L. Elsenbaumer, *Synth. Met.*, *30*:123 (1989).
59. S. H. Askari, S. D. Rughooputh, and F. Wudl, *Synth. Met.*, *29*:E129 (1989). *Polym. Mater. Sci. Eng.*, *59*:1068 (1988). *Chem. Abstr.*, *110*:154960q (1989).
60. R. W. Lenz, C.-C. Han, and M. Lux, *Polymer*, *30*:1041 (1989). *Polym. Commun.*, *28*:261 (1987).
61. J.-I. Jin, V.-H. Lee, K.-S. Lee, S.-K. Kim, and Y.-W. Park, *Synth. Met.*, *29*:E47 (1989).
62. R. D. Gooding, C. P. Lillya, and J. C. W. Chien, *J. Chem. Soc., Chem. Commun.*, 151 (1983).
63a. K.-Y. Jen, M. Maxfield, L. W. Shacklette, and R. L. Elsenbaumer, *J. Chem. Soc., Chem. Commun.*, 309 (1987).
63b. R. L. Elsenbaumer, K.-Y. Jen, G. G. Miller, H. Echkardt, L. W. Shacklette, and R. Jow, *Electronic Properties of Conjugated Polymers*, J. Kazmany, M. Mehring, and S. Roth, eds., Springer-Verlag, Berlin, 1987, p. 400.
63c. K.-Y. Jen, R. L. Elsenbaumer, and L. W. Shacklette, PCT Int. Patent WO 88 00,954, 1988; *Chem. Abstr.*, *109*:191098q (1988).
64. I. Murase, T. Ohnishi, T. Noguchi, and M. Hirooka, *Polym. Commun.*, *28*:229 (1987). I. Murase, T. Ohnishi, and T. Noguchi, German Patent DE 3,704,411, 1987; *Chem. Abstr.*, *109*:22489s (1988).
65. S. Yamada, S. Tokito, T. Tsutsui, and S. Saito, *J. Chem. Soc., Chem. Commun.*, 1448 (1987).
66. S. Tokito, H. Murata, T. Tsutsui, and S. Saito, *Jpn. J. Appl. Phys.*, *29*:L1726 (1988).
67. H. Eckhardt, L. W. Shacklette, K.-Y. Jen, and R. L. Elsenbaumer, *J. Chem. Phys.*, *91*:1303 (1989).
68. K.-Y. Jen, H. Eckhardt, T. R. Jow, L. W. Shacklette, and R. L. Elsenbaumer, *J. Chem. Soc., Chem. Commun.*, 215 (1988).
69. K.-Y. Jen, R. Jow, L. W. Shacklette, M. Maxfield, H. Eckhardt, and R. L. Elsenbaumer, *Mol. Cryst. Liq. Cryst.*, *160*:69 (1988).
70. K.-Y. Jen, T. R. Jow, and R. L. Elsenbaumer, *J. Chem. Soc., Chem. Commun.*, 1113 (1987).
71. J.-I. Jin, H.-K. Shim, and R. W. Lenz, *Synth. Met.*, *29*:E53 (1989). H.-K. Shim, R. W. Lenz, and J.-I. Jin, *Makromol. Chem.*, *190*:389 (1989).

72. M. Sato, S. Tanaka, and K. Kaeriyama, *J. Chem. Soc., Chem. Commun.*, 873 (1986).
73. R. L. Elsenbaumer, K.-Y. Jen, and R. Oboodi, *Synth. Met.*, 15:169 (1986).
74. K.-Y. Jen, G. G. Miller, and R. L. Elsenbaumer, *J. Chem. Soc., Chem. Commun.*, 1346 (1986).
75. R. Sugimoto, S. Takeda, H. B. Gu, and K. Yoshino, *Chem. Express*, 1:635 (1986).
76. K. Yoshino, S. Nakajima, and R. Sugimoto, *Jpn. J. Appl. Phys.*, 26:L1038 (1987).
77. J.-E. Österholm, J. Laakso, P. Nyholm, H. Isotalo, H. Stubb, O. Inganäs, and W. R. Salaneck, *Synth. Met.*, 28:C435 (1989).
78. I. Kulszewicz-Bajer, A. Pawlicka, J. Plenkiewicz, A. Pron, and S. Lefrant, *Synth. Met.*, 30:335 (1989).
79. J. J. Tseng, R. Uitz, J. R. Reynolds, M. Pomerantz, H. J. Arnott, and I. Haider. Paper presented at the 1989 International Chemical Congress of Pacific Basin Societies, Honolulu, HI, Dec., 1989.
80. K. Yoshino, S. Hayashi, and R. Sugimoto, *Jpn. J. Appl. Phys.*, 23:L899 (1984).
81. K. Kaeriyama, M. Sato, and S. Tanaka, *Synth. Met.*, 18:233 (1987).
82. M. Sato, S. Tanaka, and K. Kaeriyama, *Synth. Met.*, 14:279 (1986).
83. S. Hotta, T. Hosaka, M. Soga, and W. Shimotsuma, *Synth. Met.*, 9:381 (1984).
84. S. Hotta, *Synth. Met.*, 22:103 (1987).
85. B. Thémans, J. M. André, and J. L. Brédas, *Synth. Met.*, 21:149 (1987).
86. M. Lemaire, R. Garreau, F. Garnier, and J. Roncali, *New J. Chem.*, 11:703 (1987).
87. H. Masuda, S. Tanaka, and K. Kaeriyama, *Synth. Met.*, 31:29 (1989).
88. S. Hotta, M. Soga, and N. Sonoda, *Synth. Met.*, 26:267 (1988).
89. S. Hotta, S. D. D. V. Rughooputh, A. J. Heeger, and F. Wudl, *Macromolecules*, 20:212 (1987).
90a. M. J. Nowack, S. D. D. V. Rughooputh, S. Hotta, and A. J. Heeger, *Macromolecules*, 20:965 (1987).
90b. M. J. Nowack, D. Spiegel, S. Hotta, A. J. Heeger, and P. A. Pincus, *Synth. Met.*, 28:399 (1989).
91. H. B. Gu, S. Nakajima, R. Sugimoto, and K. Yoshino, *Jpn. J. Appl. Phys.*, 27:311 (1988).
92. S. D. D. V. Rughooputh, S. Hotta, A. J. Heeger, and F. Wudl, *J. Polym. Sci., Polym. Phys. Ed.*, 25:1071 (1987).
93a. K. Yoshino, S. Nakajima, H. B. Gu, and R. Sugimoto, *Jpn. J. Appl. Phys.*, 26:L2046 (1987).

93b. K. Yoshino, P. Love, M. Onoda, and R. Sugimoto, *Jpn. J. Appl. Phys.*, *27*:L2034 (1988).

94. O. Inganäs, W. R. Salaneck, J.-E. Österholm, and J. Laakso, *Synth. Met.*, *22*:395 (1988).

95. K. Yoshino, S. Nakajima, M. Onoda, and R. Sugimoto, *Synth. Met.*, *28*:C349 (1989).

96. O. Inganäs, G. Gastafsson, and W. R. Salaneck, *Synth. Met.*, *28*:C377 (1989).

97. B. Thémans, W. R. Salaneck, and J.-L. Brédas, *Synth. Met.*, *28*:C359 (1989). W. R. Salaneck, B, Inganäs, B. Thémans, J. O. Nilsson, B. Sjögren, J.-E. Österholm, J.-L. Brédas, and S. Svenson, *J. Chem. Phys.*, *89*:4613 (1988).

98. W. R. Salaneck, O. Inganäs, J.-O. Nilsson, J.-E. Österholm, B. Thémans, and J.-L. Brédas, *Synth. Met.*, *28*:C451 (1989).

99. M. J. Winokur, D. Spiegel, Y. Kim, S. Hotta, and A. J. Heeger, *Synth. Met.*, *28*:C419 (1989).

100a. K. Yoshino, K. Nakao, M. Onoda, and R. Sugimoto, *Solid State Commun.*, *68*:513 (1988).

100b. K. Yoshino, K. Nakao, and M. Onoda, *Jpn. J. Appl. Phys.*, *28*:L323 (1989).

101. K. Yoshino, D. H. Park, B. K. Park, M. Onoda, and R.Sugimoto, *Jpn. J. Appl. Phys.*, *27*:L1612 (1988). *Solid State Commun.*, *67*:1119 (1988).

102. K. Yoshino, S. Nakajima, M. Fujii, and R. Sugimoto, *Polym. Commun.*, *28*:309 (1987).

103. G. Gustafsson, O. Inganäs, and J. O. Nilsson, *Synth. Met.*, *28*:C427 (1989).

104. K. Yoshino, M. Onoda, and R. Sugimoto, *Jpn. J. Appl. Phys.*, *27*:L2034 (1988).

105. J. R. Linton, C. W. Frank, and S. D. D. V. Rughooputh, *Synth. Met.*, *28*:C393 (1989).

106. K. Yoshino, S. Nakajima, D. H. Park, and R. Sugimoto, *Jpn. J. Appl. Phys.*, *27*:L716 (1988). K. Yoshino, Y. Manda, K. Sawada, M. Onoda, and R. Sugimoto, *Solid State Commun.*, *69*:143 (1989).

107. M. Pomerantz, J. J. Tseng, H. Zhu, and S. J. Sproull, Unpublished observations. We wish to thank C. Jackson and L. M. Nilsson of Wyatt Technology Corp. for several of the MALLS runs.

108. G. Gustafsson, O. Inganäs, J. O. Nilsson, and B. Liedberg, *Synth. Met.*, *26*:297 (1988).

109. P. Love, R. Sugimoto, and K. Yoshino, *Jpn. J. Appl. Phys.*, *27*:L1562 (1988).

110. M. Sato, S. Tanaka, and K. Kaeriyama, *Makromol. Chem.*, *188*:1763 (1987).

111. J. P. Aimé, F. Bargain, M. Schott, H. Eckhardt, G. G. Miller, and R. L. Elsenbaumer, *Phys. Rev. Lett.*, *62*:55 (1989). J. P. Aimé, F. Bargain, M. Schott, H. Eckhardt, R. L. Elsenbaumer, G. G. Miller, M. E. McDonnell, and K. Zero, *Synth. Met.*, 28:C407 (1989).

112. K. Yoshino and R. Sugimoto, *Jpn. Kokai Tokkyo Koho JP* 63,314,234, Dec., 1988; *Chem. Abstr.*, *110*:223876f (1989). R. Sugimoto, *Jpn. Kokai Tokkyo Koho JP* 63,314,245, Dec., 1988; *Chem. Abstr.*, *110*:233877g (1989). K. Yoshino and R. Sugimoto, *Jpn. Kokai Tokkyo Koho JP* 63,264,642, Apr., 1987; *Chem. Abstr.*, *110*:146169y (1989).

113. K. Yoshino, K. Nakao, and R.Sugimoto, *Jpn. J. Appl. Phys.*, *28*:L490 (1989). K. Yoshino, K. Nakao, M. Onoda, and R. Sugimoto, *Jpn. J. Appl. Phys.*, *28*:L682 (1989).

114. K. Yoshino, K. Nakao, and R.Sugimoto, *Jpn. J. Appl. Phys.*, *28*:L1032 (1989).

115. K. Yoshino, K. Sawada, and M. Onoda, *Jpn. J. Appl. Phys.*, *28*:L1029 (1989).

116. M. Sato, S.Tanaka, K. Kaeriyama, and F. Tomonaga, *Polymer*, *28*:107 (1987).

117. J. Roncali, R. Garreau, D. Delabouglise, F. Garnier, and M. Lemaire, *Synth. Met.*, *28*:C341 (1989).

118. M. R. Bryce, A. Chissel, P. Kathirgamanathan, D. Parker, and N. R. M. Smith, *J. Chem. Soc., Chem. Commun.*, 466 (1987).

119. A. O. Patil, *Synth. Met.*, *28*:C495 (1989).

120a. M. Lemaire, D. Delabouglise, R. Garreau, A. Guy, and J. Roncali, *J. Chem. Soc., Chem. Commun.*, 658 (1988).

120b. D. Kotkar, V. Joshi, and P. H. Ghosh, *J. Chem. Soc., Chem. Commun.*, 917 (1988).

121a. R. L. Blankespoor and L. L. Miller, *J. Chem. Soc., Chem. Commun.*, 90 (1985).

121b. S. Tanaka, M. Sato, and K. Kaeriyama, *Polym. Commun.*, *26*:303 (1985).

122. A.-C. Chang, R. L. Blankespoor, and L. L. Miller, *J. Electroanal. Chem.*, *236*:239 (1987).

123. A.-C. Chang and L. L. Miller, *Synth. Met.*, *22*:71 (1987).

124. K. Kaeriyama, S. Tanaka, M. Sato, and K. Hamada, *Synth. Met.*, *28*:C611 (1989).

125. M. Feldhues, G. Kämpf, H. Litterer, T. Mecklenburg, and P. Wegener, *Synth. Met.*, *28*:C487 (1989).

126. R. L. Elsenbaumer, K. Y. Jen, G. G. Miller, and L. W. Shacklette, *Synth. Met.*, *18*:277 (1987).

127. J. P. Ruiz, K. Nayak, D. S. Marynick, and J. R. Reynolds, *Macromolecules*, *22*:1231 (1989). J. R. Reynolds, J. P. Ruiz,

F. Wang, C. A. Jolly, K. Nayak, and D. S. Marynick, *Synth. Met.*, *28*:C621 (1989).

128. J. Roncali, F. Garnier, R. Garreau, and M. Lemaire, *J. Chem. Soc.*, *Chem. Commun.*, 1500 (1987).

129. S. Hotta, S. D. D. V. Rughooputh, and A. J. Heeger, *Synth. Met.*, *22*:79 (1987).

130. K. Poopathy and N. R. M. Smith, *Eur. Pat. Appl.* EP253,595, Jan., 1988; *Chem. Abstr.*, *109*:23522e (1988).

131. J. Rühe, T. Ezquerra, and G. Wegner, *Makromol. Chem., Rapid Commun.*, *10*:103 (1989). *Synth. Met.*, *28*:C177 (1989).

132. H. Masuda, S. Tanaka, and K. Kaeriyama, *J. Chem. Soc., Chem. Commun.*, 725 (1989).

133. A. Deronzier and J.-C. Moutet, *Acc. Chem. Res.*, *22*:249 (1989), and references cited therein.

134. A. F. Diaz, J. Castillo, K. K. Kanazawa, J. A. Logan, M. Salmon, and O. Fajardo, *J. Electroanal. Chem.*, *133*:233 (1982).

135. G. Bidan and M. Guglielmi, *Synth. Met.*, *15*:49 (1986).

135a. J. R. Reynolds, P. A. Poropatic, and R. L. Toyooka, *Macromolecules*, *20*:958 (1987).

136. K. Yoshino, M. Onoda, Y. Manda, K. Sawada, R. Sugimoto, and S. Inoue, *Jpn. J. Appl. Phys.*, *28*:138 (1989).

137. A. O. Patil, Y. Ikenoue, F. Wudl, and A. J. Heeger, *J. Am. Chem. Soc.*, *109*:1858 (1987). A. O. Patil, Y. Ikinoue, N. Basescu, N. Colaneri, J. Chen, F. Wudl, and A. J. Heeger, *Synth. Met.*, *20*:151 (1987).

138. N. S. Sundaresan, S. Basak, M. Pomerantz, and J. R. Reynolds, *J. Chem. Soc., Chem. Commun.*, 621 (1987).

139. P. G. Pickup, *J. Electroanal. Chem.*, *225*:273 (1987).

140a. E. E. Havinga, L. W. van Horssen, W. ten Hoeve, H. Wynberg, and E. W. Meijer, *Polym. Bull.*, *18*:277 (1987).

140b. G. Bidan, B. Ehui, and M. Lapkowski, *J. Phys. D., Appl. Phys.*, *21*:1043 (1988).

141. J. R. Reynolds, N. S. Sundaresan, M. Pomerantz, S. Basak, and C. K. Baker, *J. Electroanal. Chem.*, *250*:355 (1988).

142. Y. Ikenoue, J. Chiang, A. O. Patil, F. Wudl, and A. J. Heeger, *J. Am. Chem. Soc.*, *110*:2983 (1988). Y. Ikenoue, N. Uotani, A. O. Patil, F. Wudl, and A. J. Heeger, *Synth. Met.*, *30*:305 (1989).

143. S. Basak, K. Rajeshwar, and M. Kaneko, *Anal. Chem.*, *62*: 1407 (1990).

144a. J. E. Frommer, *Acc. Chem. Res.*, *19*:2 (1986).

144b. J. E. Frommer, R. L. Elsenbaumer, and R. R. Chance, *Org. Coat. Appl. Polym. Sci. Proc.*, *48*:552 (1983).

145. J. E. Frommer, *J. Polym. Sci., Polym. Lett. Ed.*, *21*:39 (1983).

146. R. R. Chance, L. W. Shacklette, G. G. Miller, D. M. Ivory, J. M. Sowa, R. L. Elsenbaumer, and R. H. Baughman, *J. Chem. Soc., Chem. Commun.*, 348 (1980).
147. J. F. Rabolt, T. C. Clarke, K. K. Kanazawa, J. R. Reynolds, and G. B. Street, *J. Chem. Soc., Chem. Commun.*, 347 (1980).
148. T. C. Clarke, K. K. Kanazawa, V. Y. Lee, J. F. Rabolt, J. R. Reynolds, and G. B. Street, *J. Polym. Sci., Polym. Phys. Ed.*, 20:117 (1982).
149. L. Levy, L. W. Shacklette, R. L. Elsenbaumer, R. R. Chance, H. Eckhardt, J. E. Frommer, and R. H. Baughman, *J. Chem. Phys.*, 75:1919 (1981).
150. F. Dyson, *Phys. Rev.*, 98:349 (1955).
151. D. P. Murray, L. D. Kispert, S. Petrovic, and J. E. Frommer, *Macromolecules*, 22:2244 (1989). *Synth. Met.*, 28:C269 (1989).
152. A. G. Green and A. E. Woodhead, *J. Chem. Soc.*, 2388 (1910).
153. R. deSurville, M. Josefowicz, L. T. Yu, J. Perichon, and R. Buvet, *Electrochim. Acta.*, 13:1451 (1968).
154. A. G. MacDiarmid, J. C. Chiang, A. F. Richter, and A. J. Epstein, *Synth. Met.*, 18:285 (1987).
155. T. Hjertberg, W. R. Salaneck, L. Lundstrom, N. L. D. Somasiri, and A. G. MacDiarmid, *J. Polym. Sci., Polym. Lett. Ed.*, 23:503 (1985).
156. F.-L. Lu, F. Wudl, M. Nowak, and A. J. Heeger, *J. Am. Chem. Soc.*, 108:8311 (1986).
157. L. W. Shacklette, J. F. Wolf, S. Gould, and R. H. Baughman, *J. Chem. Phys.*, 88:3955 (1988).
158. F. Wudl, R. O. Angus, Jr., F. L. Lu, P. M. Allemand, D. J. Vachon, M. Nowak, Z. X. Liu, and A. J. Heeger, *J. Am. Chem. Soc.*, 109:3677 (1987).
159. A. G. MacDiarmid, J.-C. Chiang, A. E. Richter, N. L. D. Somasiri, and A. J. Epstein, *Conducting Polymers*, L. Alcacer, ed., Reidel, Dordrecht, Holland, 1987, p. 105.
160. M. Angelopoulos, A. Ray, and A. G. MacDiarmid, *Synth. Met.*, 21:21 (1987).
161. E. M. Genies, A. A. Syed, and C. Tsintaris, *Mol. Cryst. Liq. Cryst.*, 121:191 (1985).
162. A. Watanabe, K. Mori, Y. Iwasaki, and Y. Nakamura, *J. Chem. Soc., Chem. Commun.*, 3 (1987).
163. X. Tang, Y. Sun, and Y. Wei, *Makromol. Chem., Rapid Commun.*, 9:829 (1988).
164. A. Andreatta, Y. Cao, J. C. Chiang, A. J. Heeger, and P. Smith, *Synth. Met.*, 26:383 (1988).
165. A. Andreatta, Y. Cao, J. C. Chiang, A. J. Heeger, and P. Smith, *Poly. Prepr. (Am. Chem. Soc., Div. Polym. Chem.)*, 30(1):149 (1989).

166. E. M.Genies, M. Lapkowski, and J. F. Penneau, *J. Electroanal. Chem.*, *249*:97 (1988).
167. Y. Wei, W. W. Focke, G. E. Wnek, A. Ray, and A. G. MacDiarmid, *J. Phys. Chem.*, *93*:495 (1989).
168a. J. Guay, M. Leclerc, and L. H. Dao, *J. Electroanal. Chem.*, *251*:31 (1988).
168b. M. Leclerc, J. Guay, and L. H. Dao, *J. Electroanal. Chem.*, *251*:21 (1988).
169. M. Leclerc, J. Guay, and L. H. Dao, *Macromolecules*, *22*:641 (1989).
170. L. H. Dao, J. Guay, M. Leclerc, and J. Chevalier, *Synth. Met.*, *29*:E377 (1989).
171. L. H. Dao, J. Guay, and M. Leclerc, *Synth. Met.*, *29*:E383 (1989).
172. J.-W. Chevalier, J.-Y. Bergeron, and L. H. Dao, *Polym. Commun.*, *30*:308 (1989).
173. M. E. Galvin and G. E. Wnek, *Polym. Commun.*, *23*:795 (1982).
174. M. E. Galvin and G. E. Wnek, *J. Polym. Sci.*, *Polym. Chem. Ed.*, *21*:2727 (1983).
175. G. E. Wnek, *Handbook of Conducting Polymers*, Vol. 1, T. A. Skotheim, ed., Dekker, New York, 1986, pp. 205-212.
176. M. F. Rubner, S. K. Tripathy, J. Georger, and P. Cholewa, *Macromolecules*, *16*:870 (1983).
177. K. E. Schoch and R. J. Sadhir, *Ind. Eng. Chem. Res.*, *26*:678 (1987).
178. J. M. Machado, F. E. Karasz, and R. W. Lenz, *Polymer*, *29*:1412 (1988).
179. J. B. Schlenoff, J. M. Machado, P. J. Glatkowski, and F. E. Karasz, *J. Polym. Sci.*, *Polym. Phys. Ed.*, *26*:2247 (1988).
180. J. M. Machado, J. B. Schlenoff, and F. E. Karasz, *Macromolecules*, *22*:1964 (1989).
181. M. B. Gieselman and J. R. Reynolds, Unpublished results.
182. H. Isotalo, H. Stubb, P. Yli-Lahti, P. Kuivalainen, J.-E. Österholm, and J. Laakso, *Synth. Met.*, *28*:C461 (1989).
183. J. Laakso, J.-E. Österholm, and P. Nyholm, *Synth. Met.*, *28*:C467 (1989).
184. J.-O. Nilsson, G. Gustafsson, O. Inganäs, K. Uvdal, W. R. Salaneck, J.-E. Österholm, and J. Laakso, *Synth. Met.*, *28*:C445 (1989).
185. K. J. Wynne and G. B. Street, *Macromolecules*, *18*:2631 (1985).
186. S. E. Lindsey and G. B. Street, *Synth. Met.*, *10*:67 (1984).
187. M.-A. DePaoli, R. J. Waltman, A. F. Diaz, and J. Bargon, *J. Chem. Soc.*, *Chem. Commun.*, 1015 (1984).
188. O. Niwa and T. Tamamura, *J. Chem. Soc.*, *Chem. Commun.*, 817 (1984).

189. M.-A. DePaoli, R. J. Waltman, R. J. Diaz, and J. Bargon, *J. Polym.Sci.*, *Polym. Chem. Ed.*, *23*:1687 (1985).
190. O. Niwa, M. Kakuchi, and T. Tamamura, *Polym. J.*, *19*:1293 (1987).
191. O. Niwa, M. Hikita, and T. Tamamura, *Appl. Phys. Lett.*, *46*: 444 (1985).
192. X. Bi and Q. Pei, *Synth. Met.*, *22*:145 (1987).
193. T. T. Wang, S. Tasaka, R. S. Hutton, and P. Y. Lu, *J. Chem. Soc.*, *Chem. Commun.*, 1343 (1985).
194. F.-R. F. Fan and A. J. Bard, *J. Electrochem. Soc.*, *133*:301 (1986).
195. R. M. Penner and C. R. Martin, *J. Electrochem. Soc.*, *133*:310 (1986).
196. H. Yoneyama, T. Hirai, S. Kuwabata, and O. Ikeda, *Chem. Lett.*, 1243 (1986).
197. G. Nagasubramanian, S. Di Stefano, and J. Moacanin, *J. Phys. Chem.*, *90*:4447 (1986).
198. R. M. Penner and C. R. Martin, *J. Electrochem. Soc.*, *132*:514 (1985).
199. J. Roncali and F. Garnier, *J. Phys. Chem.*, *92*:833 (1988).
200. N. Bates, M. Cross, R. Lines, and D. Walton, *J. Chem. Soc.*, *Chem. Commun.*, 871 (1985).
201. T. Shimidzu, A. Ohtani, T. Iyoda, and K. Honda, *J. Chem. Soc.*, *Chem. Commun.*, 1415 (1986).
202. L. F. Warren and D. P. Anderson, *J. Electrochem. Soc.*, *134*:101 (1987).
203. D. T. Glatzhofer, J. Ulanski, and G. Wegner, *Polymer*, *28*:449 (1987).
204. T. Shimidzu, A. Ohtani, T. Iyoda, and K. Honda, *J. Electroanal. Chem.*, *224*:123 (1987).
205. T. Shimidzu, A. Ohtani, T. Iyoda, and K. Honda, *J. Chem. Soc.*, *Chem. Commun.*, 327 (1987).
206. R. E. Noftle and D. Pletcher, *J. Electroanal. Chem.*, *227*:229 (1987).
207. L. L. Miller and Q.-X Zhou, *Macromolecules*, *20*:1594 (1987).
208. T. Shimidzu, A. Ohtani, and K. Honda, *J. Electroanal. Chem.*, *251*:323 (1988).
209. A. Ohtani, M. Abe, H. Higuchi, and T. Shimidzu, *J. Chem. Soc.*, *Chem. Commun.*, 1545 (1988).
210. K. Konda, Y. Kishigami, and K. Takemoto, *Makromol. Chem.*, *Rapid Commun*, *9*:101 (1988).
211. M. B. Gieselman and J. R. Reynolds, *Macromolecules*, *23*:3118 (1990).
212. S. J. Jasne and C. K. Chiklis, *Synth. Met.*, *15*:175 (1986).
213. R. B. Bjorklund and B. Liedberg, *J. Chem. Soc.*, *Chem. Commun.*, 1293 (1986).

214. J. Edwards, R. Fisher, and B. Vincent, *Makromol. Chem.*, *Rapid Commun.*, 4:393 (1983).
215a. S. P. Armes and B. Vincent, *J. Chem. Soc.*, *Chem. Commun.*, 288 (1987).
215b. S. P. Armes, J. F. Miller, and B. Vincent, *J. Colloid Interfac. Sci.*, *118*:410 (1987).
216. A. Yassar, J. Roncali, and F. Garnier, *Polym. Commun.*, *28*:103 (1987).
217. S. P. Armes, M. Aldissi, and S. F. Agnew, *Synth. Met.*, *28*:C837 (1989).
218. S. P. Armes and M. Aldissi, *J. Chem. Soc.*, *Chem. Commun.*, 88 (1989).

5

Promising Applications of
Conducting Polymers

Anuntasin Techagumpuch,* Hari Singh Nalwa,† and Seizo Miyata
Tokyo University of Agriculture and Technology, Tokyo, Japan

I. INTRODUCTION

The term *conducting polymer* originated over a decade ago with
the discovery of polyacetylene. The most important structural feature
of polyacetylene is its highly conjugated carbon backbone composed
of an array of alternating single and double bonds. However, pristine
polyacetylene is an insulating or semiconducting material, but its
electrical conductivity increases to the metallic regime simply by
electron-acceptor or electron-donor doping [1-5]. It is the doping
reactivity of the π-electron conjugated system which induces such
unique electronic conductivity in polyacetylene. Since the conjugated

Current affiliations:
*Chulalongkorn University, Bangkok, Thailand
†Hitachi Ltd., Ibaraki, Japan

backbone holds the key to higher conductivity, therefore, based
on similar fundamentals, a large number of organic conducting polymers have been generated in the 1980s. Organic polymers which
have been looked upon for decades as insulators in the electronic
industries have emerged as a new class of electronic material. Some
of the important polymers of the conducting family are polyacetylene,
polyaniline, polypyrrole, poly(p-phenylene), poly(p-phenylene sulfide), poly(metallophthalocyanines), poly(p-phenylene vinylene),
and polythiophene [6,7]. The electrical conductivity of these conjugated organic polymers can be varied over a very wide range
by chemical or electrochemical doping, thus making it possible to
consider them for a variety of electronic applications.

    Conducting polymers are not only comparable to inorganics
from the point of view of metallic conduction, but, in addition,
they also impart a blend of interesting optical and mechanical properties. One of the greater advantages of organic polymers over inorganic materials is their architectural flexibility, since they can
be chemically modified and easily shaped according to the requirements of a particular device. Their versatility and compatibility
coupled with durability, environmental stability, ease of fabrication,
and light weight make them most fascinating materials for electronic
devices. Recently synthetic capabilities as well as the instrumental
techniques have developed to such an extent that electronic devices
based on conducting polymers can be designed and fabricated down
to their molecular levels; thereby an evolution of even more sophisti-
cated technology in the field of microelectronics can be foreseen.
Conducting polymers have attracted the attention of the scientific
community in various fields and are being considered as a new
generation of materials for wide-ranging technological applications.
Since considerable efforts to develop electronic devices from conducting polymers in various research laboratories are on the verge
of satisfying several of the diversified demands of electronic con-
sumer products, the list of their applications is enlarging with
time. Many of these applications are of considerable interest in
solid-state technology.

II. POSSIBLE APPLICATIONS OF
CONDUCTING POLYMERS

Within the last few years, various electronic devices based on conducting polymers have been proposed. These electroactive materials
cover a broad spectrum of applications from solid-state technology
to biotechnology. The electrical conductivities, chemical structures,
and possible applications of some of the important conducting polymers

which are frequently mentioned in the text are listed in Table 1. The first major area of application includes solid-state rechargeable polymer batteries [12]. Because of the versatility of polymeric systems, they can be easily shaped according to the requirement of a device. Furthermore, the electrochemical studies have shown that their rechargeability should be higher than inorganic batteries. In addition to other advantages, these two factors make them competitive to inorganic electrode-active materials. Details of the rechargeable polymer battery are discussed later in the text.

Polyheterocyclic conducting polymers show wide variations in their color with the applied electric voltage when switched between oxidized and reduced states [13,14]. Similarly, they also exhibit photoelectrochromic phenomenon as a result of optical density change [15,16]. Therefore, by employing these physical effects, display devices can be developed by coupling a conducting polymer with a semiconductor. A solid-state electrochromic device from polyaniline has already been fabricated [17].

It has been observed that the widely used n-type inorganic semiconductors in solar-energy devices face an acute photodecomposition problem. The life-span of the semiconductor surfaces can be tremendously enhanced by protecting them against photodegradation. Electrically conducting thin layers of polypyrrole coated on silicon, cadmium sulfide (CdS), gallium arsenide (GaAs), and other surfaces reduce the photodecomposition. Thus they have already found applications in the electrochemical photovoltaic cells and photoelectrolysis cells [18-20].

The conductivity of the conjugated polymers can easily be maintained in the semiconducting regime by controlling the dopant concentration. In this regard, Schottky barrier-type diode [21,22] and field-effect transistors [23,24] from conducting polymers such as polyacetylene polypyrrole, and polythiophene have been developed. In the various fields of industries such as textile, paper technology, photography, printing, etc., an immense danger of electrical shocks, fire, and explosion exists due to the generation of high static electricity. Conducting polymers can easily dissipate electrostatic charges at a much faster rate and can serve as antielectrostatic agents in these industries [25]. Similarly, their variable conductivity can be affectively used in electromagnetic interference shielding [26].

In some applications, conducting polymers could be used as fillers instead of carbon black, graphite, and metals. They can also replace other materials as conductive adhesives [27].

It is well known that by electrochemical polymerization, conducting adherent films of various polyheterocyclics can be deposited on metallic surfaces. Similarly, the composites of conducting polymers with conventional polymers such as poly(methyl methacrylate),

Table 1. Some Important Conducting Polymers and Their Possible Applications

Polymer	Chemical structure	Conductivity (S/cm)	Possible applications
Polyacetylene [3]	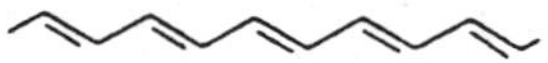 *trans*  *cis*	10^5 a	Rechargeable battery, photovoltaics, gas sensors, chemical indicators, radiation detectors. Schottky diode, antielectrostatic, encapsulation, biotechnology, optoelectronics, solar cells.
Polypyrrole [8]	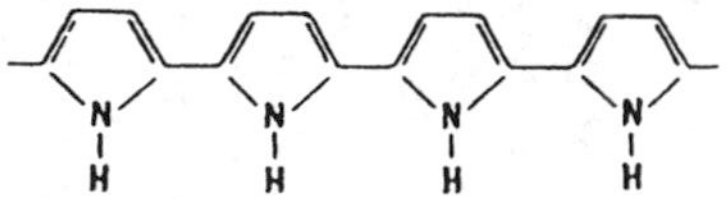	1000^b	Rechargeable battery, condenser, printed circuit boards, gas sensors, potentiometric glucose sensor, electroplating, Schottky diode, electroacoustic device, fillers, adhesive, transparent coating, electromagnetic shielding, electrophotochemical cells, field-effect transistor, photocatalysts, physiological implantations, optoelectronics, conductive textiles.

Polymer	Structure		Applications
Polythiophene [9]	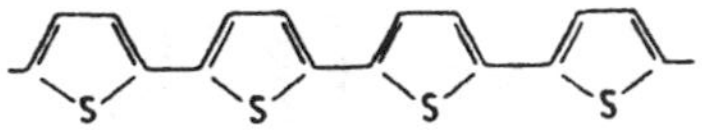	150[c]	Rechargeable battery, display device, fillers, field-effect transistor, optoelectronics, Schottky diode, gas sensor, photocatalysts.
Polyaniline [10]	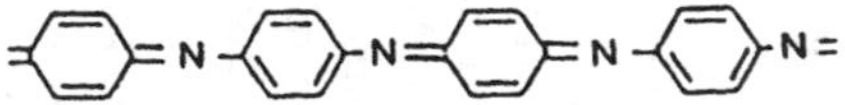	5[d]	Rechargeable battery, electro-chromic devices, indicator devices, biosensors.
Poly(*p*-phenylene) [11]	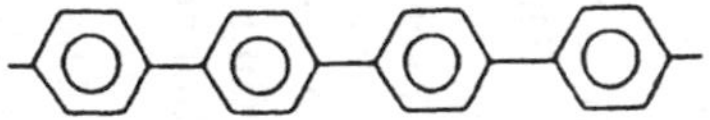	500[e]	Rechargeable battery, fillers, photocatalysts.

[a]Naarmann polyacetylene doped with iodine.
[b]BF_4^- doped.
[c]AsF_5 doped.
[d]Aqueous HCl doped.
[e]AsF_5 doped.

poly(vinyl alcohol), etc., exhibit excellent coatability [28]. Therefore, using their adherent property, a large variety of polymer-
coated substrates can be developed. The coatability of conducting
polymers coupled with their environmental stability provides dual
benefits, namely, controlling the magnitude of electrophysical properties and in improving surface characteristics. In particular, polyheterocyclics have shown their usefulness in electrodeposition which
provides hardness, and solderability and protection against friction,
corrosion, and wear. Some of the applications of conducting polymer
coatings will be discussed, particularly in reference to transparent
coatings. The electrodeposition of conducting polymers is also applicable for decorative as well as for protective purposes.

Conducting polymers have been often pointed out to possess
a combination of undesirable properties as well. Although polyacetylene
led to the discovery of conducting polymers, its poor environmental
stability restricted applications in a number of devices [5]. Recently,
the unwanted instability of polyacetylene proved to be a boon in
chemical sensor technology. Small-scale indicator devices from conducting polymers to detect moisture, radiation, chemicals, and
mechanical abuse have been designed [29]. In addition, their use
as biosensors and gas sensors has been suggested earlier [30-32].
In the field of biotechnology, the use of conducting polymers has
been suggested for an in vivo drug delivery system and for physio-
logical implantation.

The author's laboratory is extensively involved in the designing
of new devices from conducting polymers. The applicability to various
electronic areas was examined. The synthesis of highly conducting
and transparent films of polypyrrole led to several new possibilities
[33,34]. Application of piezoelectric polymers in electroacoustic devices has been known for over a decade. A transparent speaker was
developed by coating transparent conducting films of polypyrrole
on the surface of piezoelectric poly(vinylidenecyanide-vinylacetate).
The speaker generates very good quality sound from CD players,
especially in the high-frequency region. The transparency of conducting polymer is also desirable in other applications. A transparent
coating of polypyrrole improves appearance as well as protects the
substrate surfaces. Furthermore, electroplating on a plastic surface
was also accomplished by first coating a conducting polypyrrole
prepared by the chemical method. The details of electroplating are
discussed later. Among many others, the automobile industry is
one of the targets for this application.

Another application includes the development of a polypyrrole-
aluminum solid electrolyte condenser by Marcon Company. This
condenser exhibits good stability at higher temperatures and shows
application in multilayer printed circuits, audiovisual instruments,
etc.

Recently research activities on conducting polymers have been diverted toward the study of nonlinear optical effects. In addition to the architectural flexibility, mechanical strength, and high damage threshold, the ultra-fast response in subpicoseconds of conjugated polymers is of much significance. There are high expectations that the conducting polymers may find applications in the field of opto-electronics, in such technologies as signal processing, optical communication, data storage, phase conjugation, and bistability devices [35,36]. In conducting polymers, an insulator-to-metal transition occurs upon doping and a similar phenomenon is observed in the optically induced doping process, particularly in polymers which contain photoresponsive dopants. The applicability of such interesting electrooptic effects can be used in designing erasable optical recording and memory devices [37,38]. Polyphenylacetylene and poly(3-alkylthiophene) exhibit transformational and conformational changes, respectively, which can also be used for optical memory and recording [36]. It has been reported that in poly(3-alkylthiophene) spectral changes take place with temperature and pressure; therefore these conducting polymers can be used as mechano-optical elements [36].

A large number of papers and patents has appeared on the prospective applications of conducting polymers and also several reviews have been published in this field from time to time [39]. In this chapter, the emphasis is mainly focused on the newly emerging technologies. Particularly, the details of the electronic devices already placed in the market for commercialization by different companies are discussed. Also, some of the prototypical devices designed and fabricated in the authors' laboratory are described. Details of the promising applications of conducting polymers in connection with rechargeable polymer batteries, transparent loudspeakers, condensers, electroplating, and indicator devices are discussed.

III. PROMISING APPLICATIONS OF CONDUCTING POLYMERS

A. Rechargeable Polymer Battery

A battery is an energy storage device which performs the conversion of chemical energy into electrical energy. Electrical energy is generated as a result of electrochemical reactions operating in a battery. The electrochemical processes involved in primary and secondary batteries are similar, but in a secondary battery a reversible process of chemical-electrochemical reactions occurs; therefore, it has the major advantage of rechargeability over a primary battery. In the charging and discharging processes, a secondary battery acts as an electrolytic and voltage device, respectively. A rechargeable

or secondary battery is a cell consisting of three main components:
positive and negative electrodes, ionically conductive electrolyte,
and a separator. In the design and fabrication of a rechargeable
battery, a careful consideration of these components is necessary.
Some of the practical requirements to be matched are as follows.
The electrodes are the backbone of a battery: therefore, the materials
for the electrodes should possess high electrical conductivity, good
mechanical strength, and inertness to the electrolyte. According
to the applicability of the device, the dimensions of the electrodes
and their porosity have to be defined. Second, the electrolyte plays
an important role, since it governs the process of energy generation.
Therefore, the electrolyte that provides the medium for the electrochemical reactions should impart high ionic conductivity and low
viscosity. In order to facilitate the electrochemical reactions, an
appropriate combination of the electroactive material and electrolyte
is the most important requirement in designing a secondary battery.
Third, the separator which prevents contact and shorting between
electrodes should be made of a nonconducting material. Therefore,
taking these factors into account, a rechargeable battery can be
designed.

The most common rechargeable batteries made of inorganic
electroactive materials are of nickel-cadmium (Ni-Cd) and lead-acid
type. A broad spectrum of applications is available to secondary
batteries. Considering the use of electronic devices, potential of
an organic conducting polymer as an electroactive material was
realized. In 1979 Nigrey et al. [40] first found that the electrochemical doping of polyacetylene is a reversible process. The importance of this reversible electrochemical phenomenon was immediately
materialized into designing a rechargeable polymer battery. In
polyacetylene, both p- and n-type doping can be accomplished
electrochemically; therefore, it can be used as a cathode as well
as an anode in a rechargeable battery.

The electrochemistry of polyacetylene has been extensively
investigated by many research groups. The details of polyacetylene
electrochemistry and its potential as an electroactive material have
been summarized in the literature [12]. The fundamental electrochemistry being similar, this discovery opened avenues to explore
the possibility of other conducting polymers. In addition to polyacetylene, the electrochemistry of polypyrrole, polythiophene, polyaniline, poly(p-phenylenes), etc., has been thoroughly investigated.
In the conventional rechargeable batteries, only very selective
inorganics can be used as electroactive material due to their dissolution in an electrolyte solution. On the other hand, some of
the highly conjugated polymers possess absolute insolubility; thus,
they overcome the problem of decomposition in an electrolyte.

Table 2. Potential Advantages of Polymer Rechargeable Battery

Conventional rechargeable battery (lead-acid, nickel-cadmium)	Polymer rechargeable battery
1. Low cell voltage (about 1 V): material for electrodes are limited due to the electrochemical decomposition in aqueous electrolyte.	Higher cell voltages (about 2-4 V) are possible.
2. Low energy-to-weight ratio leads to relatively heavy batteries.	Higher energy-to-weight ratio means lighter batteries can be expected.
3. Shorter lifetime due to the dissolution and redeposition of electrode material that occurs during the charge-discharge cycles (chemical-electrochemical reactions).	Longer lifetime since the ions involved in the delivery and storage of charge come from the solution rather than from the electrodes themselves.
4. Contains toxic material.	Potentially contains no toxic material.
5. Cannot be easily shaped to fit the device.	Can be easily shaped to fit the device.

Source: Refs. 41 and 42.

Theoretically, these polymer electrode batteries have greater advantage over the conventional inorganic rechargeable batteries as summarized in Table 2.

For the commercial success of a rechargeable battery, technical factors such as cyclability, energy density, and stability are important from the practical point of view. Another important consideration is financial aspects, particularly their inexpensiveness, availability, and competitiveness with existing materials in the market. Technically speaking, in the development of a polymer battery, a polymer electrode should meet the following requirements [43];

1. Self-discharging activity must be low (presumably less than 20% per month).
2. The reversibility of the electrode reaction should be high (99.9% per cycle).

3. In order to obtain a battery with high energy-to-weight
   ratio, the doping level is expected to be about 0.1 per C
   atom or higher.
4. The rate of kinetics has to be fast.

The electrochemical properties of various conducting polymer
electrodes compared with the conventional inorganic electrodes are
listed in Table 3. For practical application, polyacetylene has several
disadvantages; mainly its poor charge retention capability and thermal
stability discouraged its use as an electrode active material. Furthermore, it lacks the processability and environmental stability necessarily required in fabricating an electrochemical device. Some of
these problems to some extent have been overcome by other conducting polymers. Considering the suitability of the electroactive material
in battery applications, polyaniline, polypyrrole, and poly(p-phenylene)
were found to be more promising. These polymers are also interesting
due to their ease of synthesis and fabrication, environmental stability,
and, at least to some extent, cost.

In the development of polymer batteries, low manufacturing
cost and safety must be carefully considered in order to compete
with other kinds of rechargeable batteries. The prospect of developing a rechargeable battery from conducting polymers was explored
in several laboratories. Recently it has become clear that some of
the advantages of conducting polymers in rechargeable batteries
mentioned in Table 2 cannot be fully materialized. In spite of this,
the development of polymer batteries was finally put into reality
by two companies: BASF/VARTA of West Germany and Bridgeston-
Seiko of Japan. Allied Corporation of the United States has also
developed a polymer battery using a poly(p-phenylene) conducting
polymer, but has not commercialized it yet. The authors were provided significant information on polymer batteries developed by BASF
and Bridgestone and polymer batteries developed by BASF/VARTA
and Bridgeston-Seiko are discussed.

BASF/VARTA Polypyrrole Battery

A polypyrrole battery was developed and tested by BASF and VARTA
Batterie A.G. of West Germany [42]. Electrochemically synthesized
polypyrrole doped with tetrafluoroborate ($BF_4{}^-$) was used as a positive electrode. Elemental analysis showed that three pyrrole monomer
units carry one unit of charge. This concludes that the maximum
doping level of polypyrrole is 1/12 per carbon atom. When polypyrrole
is used with a lithium counter electrode at a cell voltage of 3.5 V,
a theoretical energy density of 360 W-h/kg is estimated for the
polypyrrole electrode but it decreases in a practical cell due to
a number of reasons.

Table 3. Capacities of Some Electroactive Materials

Materials	Potential VS M/M (V)	Gravimetric capacity (mA-h/g)	Volumetric capacity (mA-h/cm^3)	Cycle life to 80% capacity
Lithium	0	3830	2040	—
Polyacetylene-BF$_4$ [CH(BF$_4$)$_{0.09}$]$_x$	3.5	116	140	10
Polyacetylene-Li [CHLi$_{0.18}$]^a	1.0	339	423	90
Polyaniline-BF$_4$ [C$_6$H$_5$N(BF$_4$)$_{0.5}$]$_x$	3.4	100	129	200
Polypyrrole-BF$_4$ [C$_4$H$_3$N(BF$_4$)$_{0.33}$]	3.2	94	124	200
Poly(p-phenylene)-Li (C$_6$H$_4$Li$_{0.5}$)^a	0.7	169	220	200
Li$_x$MoS$_2$ (ΔX=0.8)	1.8	128	600	200
Li$_x$CoO$_2$ (ΔX=0.5)	4.0	137	640	80
Li$_x$Al (0<X <1)^a	0.4	788	1320	3
Na$_x$Pb (1<X <3.75)^a	0.2	251	1300	2

^{a}Capacities as an anode material.
Source: Refs. 44 and 45.

In the electrochemical cell, the starting electrolyte is a 0.5 molar solution of lithium perchlorate (LiClO$_4$) in propylenecarbonate. The polypyrrole electrode is used with lithium as a counterelectrode. During the discharging process (Fig. 1), the electrochemical reactions proceed as follows:

$$Li \longrightarrow Li^+ + e^- \tag{1}$$

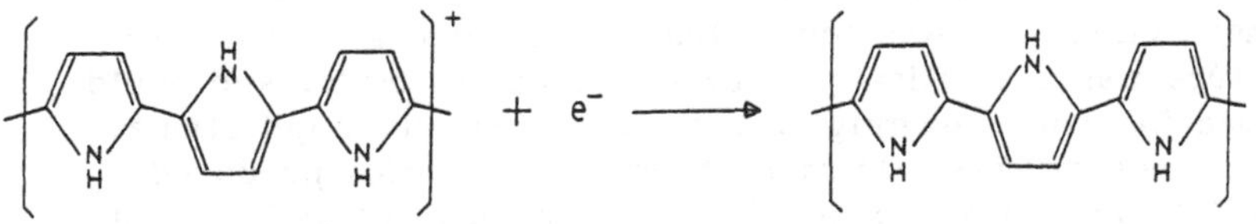

$$\qquad\qquad (2)$$

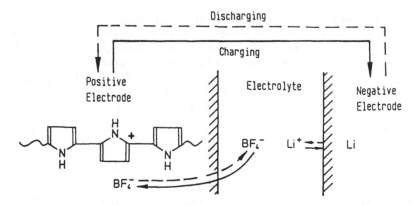

Figure 1. Electrode reactions of a polypyrrole-lithium battery (Ref. 42).

The electron from the negative lithium electrode moves along a circuit to replace BF_4^- in the polypyrrole positive electrode, and the counterion BF_4^- is released from polypyrrole to the electrolyte. During the charging process, a reversible electrochemical reaction operates.

The half-cell charge and discharge curves of polypyrrole electrodes are shown in Fig. 2. The upper and lower voltage limits of the cell were 2 and 4 V versus Li/Li^+, respectively. The cyclability and self-discharge of polypyrrole electrodes are represented in Fig. 3. The lower limit shows that for a deep discharged process, the yield is approximately 100% after more than 400 cycles. Polypyrrole electrodes exhibit better stability compared to the polyacetylene electrode.

In a practical cell, the energy density is lower than the theoretical value. There are many factors to be considered: (i) in a practical cell, excess solvent must be maintained to avoid the increase of viscosity due to the increase of ion concentration in the electrolyte during discharging process. (ii) Further it has been found that an amount of solvent is contained in the polypyrrole electrode, and (iii) the weight of the cell is increased by the weight of the Li electrode, current collector, housing, etc. Moreover, considering the kinetic behavior of the battery, the energy decreases even further. Finally, it was found that polypyrrole of porous surface structure was a kinetically more favorable electrode, and could be used in short-charging-time or in high-power applications.

The polypyrrole-lithium rechargeable batteries produced by BASF/VARTA are shown in Fig. 4a (Mignon-Size) and Fig. 4b (sandwich type). The energy densities of the cells at 20 h discharge time are 15 W-h/kg and 30 W-h/kg, respectively. In the latter case,

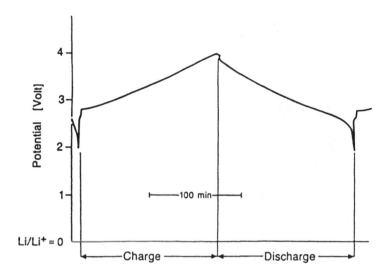

Figure 2. Charge/discharge curve of a polypyrrole electrode. Electrolyte 0.5 M LiClO$_4$ in propylencarbonate (Ref. 42).

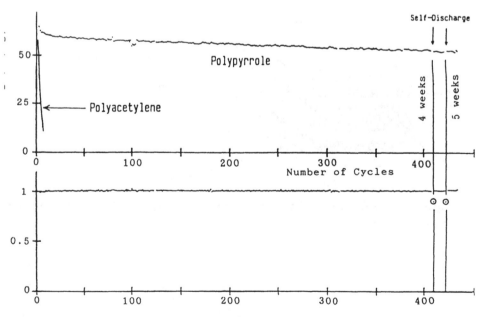

Figure 3. Cyclability and self-discharge of polypyrrole and poly-acetylene electrodes (deep discharge, 0.3 mA/cm^2, 0.5 M LiClO$_4$/ propylencarbonate, capacity values relate to solvent containing electrodes, cycled charge corresponds to approximately 3 monomer units/unit charge). (From Ref. 42.)

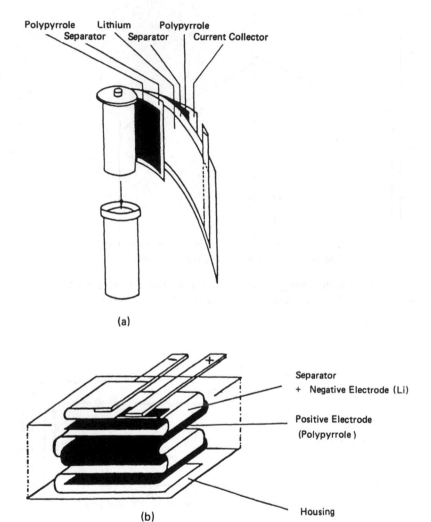

Figure 4. Polypyrrole-lithium battery: (a) mignon-size; (b) sandwich type (Ref. 42).

the energy density is equal to that of the Ni-Cd cell. These
polypyrrole-lithium batteries can be used in dictaphones, pocket
radios, etc.

Bridgestone-Seiko Polyaniline Battery

Aniline black has been known in the literature since 1862 as a low-
molecular-weight oligomer obtained in a powdery form [46,47]. A
major development in the synthesis occurred in 1980 when polyaniline
thin films were prepared by an electrochemical method [6,48]. This
polyaniline is conducting in the oxidized state. Furthermore, a
very interesting electrochemistry operates in polyaniline in that
it can be switched between an oxidized state and a reduced state
similar to polyacetylene and polypyrrole. Therefore, in polyaniline,
a transition takes place from an insulating to a conducting state.
This process is electrochemically reversible and showed that polyaniline can also be used as an electroactive material. Moreover,
color changes in polyaniline were also observed with the switching
behavior indicating that it can be used in electrochromic devices
[17]. Polyaniline also exhibits good environmental stability and
processibility. This blend of electrophysical properties made polyaniline an interesting electroactive polymer. As an electronic device,
it has already found applications in rechargeable polymer batteries.
Compared to other conducting polymers, polyaniline can be inex-
pensively produced in large quantities and thus shows increased
competitiveness over inorganic batteries.

Recently Bridgestone Corporation and Seiko Electronic Components
Ltd. have succeeded in developing rechargeable polyaniline lithium
(PAn-Li) batteries [49]. These polymer batteries have been on
the market since September of 1987. The structure of a coin-type
PAn-lithium battery is shown in Fig. 5. Polyaniline doped with
tetrafluoroborate (BF_4^-) is used as a positive electrode, while the
lithium-aluminum alloy is used as a negative electrode. The electrolyte solution contains lithium tetrafluoroborate ($LiBF_4$) in a mixture
of propylenecarbonate (PC) and 1,2-dimethoxyethane (DME). In
charging and discharging processes, the chemical reaction is similar
to the polypyrrole-lithium cell. The specifications of three types
of polyaniline batteries are listed in Table 4. The self-discharge
of these batteries is about 15% in 90 days. The discharge characteristics for the AL 2016 battery is shown in Fig. 6.

Since this kind of rechargeable battery offers a very small
amount of power, it is expected to be used as a backup power
source of electronic equipment which consumes a relatively small
amount of power, such as electronic memory units in computers,
calculators, multimeters, etc. It is an ideal battery to be used as
a power source in combination with solar cells (Fig. 7). Although

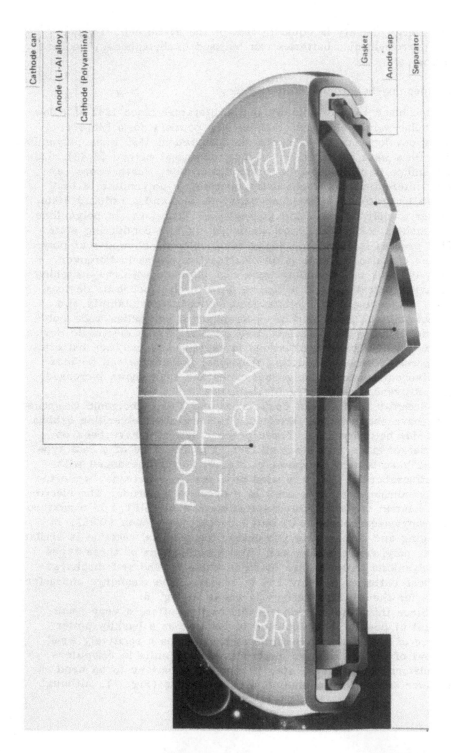

Cathode can

Anode (Li-Al alloy)

Cathode (Polyaniline)

Gasket

Anode cap

Separator

Figure 5.   Bridgestone–Seiko polymer–lithium rechargeable battery.

Table 4. Specification of PAn-Li Batteries

	AL 2016	AL 2032	AL 920
Dimension			
diameter (mm)	20.0	20.0	9.0
thickness (mm)	1.6	3.2	2.0
Weight (g)	1.7	2.6	0.4
Nominal voltage (V)	3	3	3
Nominal operating voltage (V)	3 ~ 2	3 ~ 2	3 ~ 2
Nominal capacity (mA-h)	3	8	0.5
Standard current (A)	1 μ ~ 5 m	1 μ ~ 5 m	1 μ ~ 1 m
Cycle life			
depth (mA-h)	1	3	0.1
life (cycles)	more than 1000	more than 1000	more than 1000
Operating temperature (°C)	-20 ~ +60	-20 ~ +60	-20 ~ +60
Recommended charging method	Constant Voltage Charge		

Source: Ref. 50.

Discharge Characteristics (AL 2016)

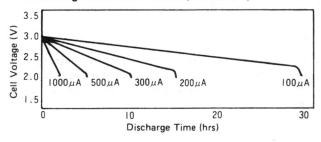

Figure 6. Self-discharge characteristics (at 60°C) of polyaniline-lithium battery. (From Ref. 50.)

Figure 7. The application of polymer-lithium battery as a power source in combination with solar cells. (From Ref. 50.)

the PAn-Li batteries cannot be matched by conventional NiCd batteries in power delivery, they can supply higher voltage, which may suit some devices. The comparison of PAn-Li battery with NiCd battery is shown in Table 5 [50]. Since PAn-Li batteries have very low self-discharge and large cyclability, they should have comparable lifetime to the conventional NiCd batteries.

B. Marcon Polypyrrole Condenser

A simple condenser or capacitor consists of a dielectric material sandwiched between two conducting plates, as schematically shown in Fig. 8a. The capacitance and the impedence of a condenser are given by the equations

$$C = \frac{K \varepsilon_0 A}{d} \tag{3}$$

and

$$X = \frac{1}{2\pi f C} \tag{4}$$

where K, A, and d are dielectric constant, surface area, and thickness of the dielectric material, respectively, f is the frequency, and

Table 5. Comparison of PAn-Li Battery with a Conventional NiCd Battery

Characteristics	PAn-Li	NiCd
1. Open-circuit voltage (volt)	~3.5	1.35
2. Actual voltage (volt)	2.0 ~ 3.5 (2.8 average)	1.20
3. Theoretical energy density (W-h/kg)	370	220
4. Energy density (W-h/kg)	~220	100
5. Self-discharge		
(i) at room temperature	<15%/3 months	<25%/3 months
(ii) high temperature	<30%/2 months (60°C)	<80%/2 months (45°C)
6. Cyclability		
30%	>1000	>1000
100%	> 300	> 500

Source: Ref. 50.

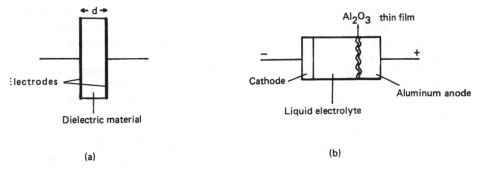

(a)

(b)

Figure 8. (a) A simple condenser. (b) An electrolyte condenser.

ε_0 is the permittivity of free space. From Eq. (3), a condenser of high capacitance can be obtained by using a very thin film dielectric of high dielectric constant and large area. An electrolytic condenser is a commonly used condenser. The diagram of the condenser is represented in Fig. 8b. In this condenser, a very thin film of aluminum oxide (Al_2O_3) is prepared on the rough surface of an aluminum anode. The roughness of the surface is designed to increase the surface area. A liquid electrolyte is used to make good contact between the outer side of the film and the cathode. Due to the ionic nature of conduction, this electrolytic condenser can only be used with a dc voltage source or, in practice, a rippled dc voltage, for which the polarity is shown in Fig. 8b.

Since the conduction process in a liquid electrolyte is of the ionic type, at high frequency, the ions cannot respond fast enough to the applied field. This causes drastic changes in the impedence of the liquid electrolyte condenser at higher frequencies. Moreover, at low temperatures, an increase in the viscosity of the electrolyte liquid retards the motion of ions and thus decreases the conductivity. This activity also increases the impedence of the condenser at low temperatures. There are many ways to overcome this problem. For example, one may use a solid electrolyte rather than a liquid electrolyte such as in the case of a tantalum condenser [51]. Further improvement can be done by using an organic conductor tetracyanoquinodimethane (TCNQ) complex, as a cathode conducting layer [52,53]. This kind of condenser is called a TCNQ complex electrolytic aluminum condenser. This condenser has better frequency stability than the tantalum condenser, but the TCNQ complex condenser cannot resist high temperatures in the automatic soldering and molding processes of the production line.

Recently, the Marcon and Japan Carlit companies jointly developed a polypyrrole aluminum solid electrolyte condenser [54-56]. In this condenser aluminum is used as an anode. The rough side of the aluminum surface is coated with an Al_2O_3 dielectric thin film. The other side of this film is contacted with the cathode via polypyrrole. This condenser requires good contacts between the polypyrrole and Al_2O_3. In a liquid electrolyte condenser, this problem is not encountered. Good contacts can be made by electrochemically synthesized polypyrrole on Al_2O_3. However, since Al_2O_3 is not a conducting material, this process is rather difficult. Finally, this problem can be solved by precoating a thin layer of conducting polypyrrole on the Al_2O_3 surface, using chemical polymerization. This can be accomplished by a coating process similar to that described in Section III.E. Subsequently, polypyrrole can easily be electrochemically synthesized on this precoated surface as required [57].

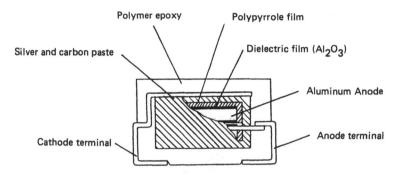

Figure 9. Cross section of polypyrrole condenser.

The diagram of the cross section of a polypyrrole-condenser is shown in Fig. 9. The aluminum anode is surrounded by Al_2O_3 film and then a polypyrrole film. Good electrical contacts between the polypyrrole and cathode are provided by using carbon paste and silver paste. The outer surface is a epoxy polymer which protects the condenser from humidity and chemical degradation.

The impedances of several kind of condensers at various frequencies are represented in Fig. 10. The percentage change of impedances with temperature is shown in Fig. 11. From Eq. (4), for constant capacitance, the plot between ln X and ln f in Fig. 10 should be a straight line of negative slope. From this, it is clear that the capacitance of the liquid electrolyte condenser starts decreasing at about 4×10^3 Hz; therefore, this kind of condenser cannot be used at high frequency, as mentioned before. The tantalum solid electrolyte condenser has better frequency stability, but still it does not perform well at higher frequencies. The TCNQ complex condenser and the polypyrrole condenser are superior since they exhibit good frequency stability up to 1 MHz. As shown in Fig. 11, the polypyrrole condenser also exhibits better temperature stability. This polypyrrole consenser can be produced as a chip type which can stand high temperature in an automatic soldering process.

Polypyrrole condenser also establishes self-recovering process similar to other electrolyte condensers [58]. This process can be described as follows. In preparing a very thin layer, for example 150 nm of Al_2O_3 on an Al surface, in general there are on this layer many defective points through which current can easily pass. This results in big leakage current which damages the condenser. However, in a polypyrrole condenser, the large current density at the defective points quickly degrades the polypyrrole at those points and changes them to insulators. This process immediately stops the

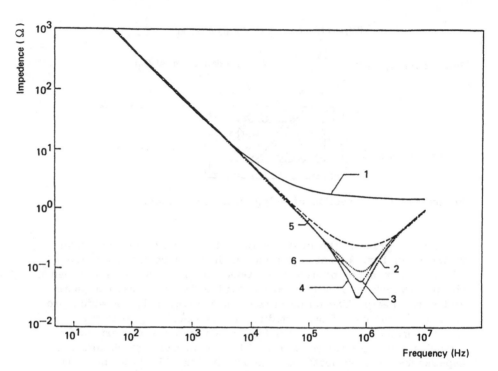

Figure 10.   Impedance of many kinds of condensers at various frequencies (Ref. 54,55): 1. Small-type aluminum electrolyte condenser;
2. metalized film condenser; 3. conducting polymer-aluminum electrolyte condenser (ADC series) 10 V, 3.3 µF; 4. laminated ceramic
condenser; 5. tantalum solid electrolyte condenser; 6. organic semiconductor (TCNQ complex) electrolyte condenser.

leakage current and recovers the condenser. It has been found
that for an applied voltage of 10 V at 105°C, the leakage current
of a polypyrrole condenser reduces from 10 to 0.03 µA in 500 min.
This recovering process is a vital part which shows the reliability
of polypyrrole condensers.

A polypyrrole conducting polymer has advantages over other
conducting polymers because it is more stable and has high conductivity (about 100 S/cm). Although polyacetylene has still higher
conductivity, it possesses almost no resistance to humidity and
other ambient conditions. For polyaniline, the conductivity is relatively lower (about 1 S/cm), and polymerization is carried out in
strong acidic media, which can degrade an aluminum cathode. In
a polypyrrole condenser, the polymerization process is almost the
same as the conventional electrochemical method.

It can be concluded that polypyrrole is chosen as a condenser material because it has relatively high and stable electrical conductivity, it can be easily prepared to attach well with Al_2O_3, and it also provides recovering properties to the condenser. Polypyrrole condensers are under test and manufacturing at Marcon Company. The expected lifetime is about 10,000 h. The maximum working voltage is 16 V, but only lower-voltage types are now produced, and the limit of reverse applied voltage is -3 V. The thermal resistance is 260°C for 5 s. Figure 12 shows two types of polypyrrole condenser which have already been produced. The dimensions of the polypyrrole condensers are listed in Table 6.

Since the height of polypyrrole condensers is less than 3 mm, they can be used in multilayer printed circuits. Polypyrrole condensers are also expected to be used in audiovisual instruments and office automation apparatus. The Marcon Company produces 100 million condensers per month. The company expects to replace 10 million of those condensers by polypyrrole condensers. These polypyrrole condensers are now available in the market by Marcon Company.

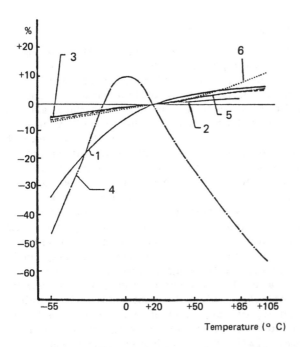

Figure 11. The percentage change of impedance with temperature for various type of condensers. The legends are the same as in Fig. 10 (Ref. 54,55).

Figure 12.   The polypyrrole condensers.

C.   Remotely Readable Indicator Devices

The freestanding films of polyacetylene synthesized in 1971 became
the first example of a conducting polymer. This polyacetylene is
more often called *Shirakawa polyacetylene* [59,60]. Raising the
conductivity of an insulating polymer (polyacetylene) into the metallic
regime by doping was a breakthrough in new electronic materials,
but polyacetylene also suffered from environmental instability and
insolubility. The problem of processability was solved to some extent
by the development of another polyacetylene known as *Durham polyacetylene*, obtained through a precursor route [61]. Attempts to
synthesize polyacetylene by new methods were continued by numerous
groups. After almost a decade, a research team at BASF, West
Germany, succeeded in preparing a new form of polyacetylene,
recently known as *Naarmann polyacetylene*. This form of polyacetylene
possesses an electrical conductivity of the order of 10^5 S/cm and
even shows slightly higher stability [3,4]. Naarmann polyacetylene
exceeds the conductivity of copper by weight, but the problem
of instability remains the same. This inherited undesirable property
always hindered the use of polyacetylene in electronic device applications. Finally the instability of polyacetylene proved to be a boon
in remotely readable indicator devices recently designed and developed
by Allied Signal Inc. (USA) [29]. These indicator devices were
designed based on the changes occurring in the electrical conductivity
of the conjugated polymers due to environmental degradation.

Table 6. Dimensions (in mm) of Polypyrrole Condensers

Case size	L1	L2	W
C	6.0±0.3	5.8±0.1	3.2±0.3
D	7.3±0.3	7.1±0.1	4.3±0.3

Case size	H	a	b
C	2.5±0.3	1.3±0.3	2.2±0.2
D	2.8±0.3	1.3±0.3	2.4±0.2

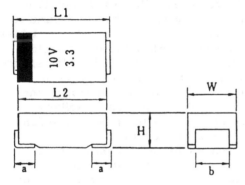

The other specifications are

Temperature range	−55 to +105°C
Voltage	4 to 20 V
Capacitance	±20% at 20°C, 120 Hz
leakage current	1 μA

Source: Ref. 56.

The devices which have been developed are *remotely readable indicator devices* for in-box monitoring of various exposures of products such as time and temperature, temperature limit (freeze or defrost), humidity, radiation dosage, etc. These devices are based on a combination of ambient-responsive conducting polymer and radio-frequency (rf) or microwave (mw) frequency antitheft devices. To understand the function of this indicator device, an antitheft device made of conducting polymer and developed by Allied Signal Inc. is discussed below.

The Antitheft Devices

An antitheft device [29] consists of a target tag, which is a passive resonance circuit of radiowaves, and a detector, which is a transmitter-receiver device. The operation frequencies are either radio frequency or microwave frequencies. For rf antitheft devices developed by Checkpoint System Inc., the target tags are about 4 cm^2 and cost only \$0.03 each (see Fig. 13). This tag consists of a thin strip of aluminum foil configured as a squarish spiral working as an rf resonant circuit. The antitheft device operates at a resonant frequency of about 8.2 MHz. In operation, the transmitter of the detector sweeps over this frequency range in a millisecond. The target tag resonates and emits rf radiation back, which is detected and stored in the memory by the receiver. After many repeating sweeps, the signal at the resonant frequency of the tag accumulates and provides the response of the detector. For practical use, target tags are attached to the items in a store. When the item is sold, this tag is covered with adhesive-backed aluminum foil which shields the tag from the rf signal. This means that the tag will never be detected by the receiver. However, if the item is taken from the store without being paid for, then while the person is walking through an exit, the tag will be detected by the receiver, which in turn will trigger an alarm to reveal the activity.

For microwave antitheft devices produced by Sensormatic Electronics Corp., the targets consist of an encapsulated antenna, diode, and resonant circuit. This tag is different from the rf tag in that it cannot be shielded by aluminum foil but it can be easily shielded by water. In operation, the detector transmits microwaves of two frequencies; the diode of the target generates sum and difference frequencies which are detected by the receiver of the detector, thus providing the response of the target. The microwave antitheft targets are more expensive than rf targets. This microwave system is advantageous because it can transmit as a directional beam and can detect the target up to a distance of 8 ft, which is about twice the range of the rf system.

The antitheft devices mentioned above were modified to act as remotely readable indicator devices [29]. For example, the rf target tag is covered by a thin layer of a doped conducting polymer of high conductivity. The target tag is shielded from the rf signal, and it is always in an Off state. If this tag is exposed to high temperature, the conductivity of the polymer drastically reduces due to the thermal degradation, thus resulting in a decrease of the shielding effect which will finally turn the target to the On state. The change from the Off to the On state of the target reveals the thermal history of place where it was located. This device can be used to check the thermal history of the products whenever required.

(a)

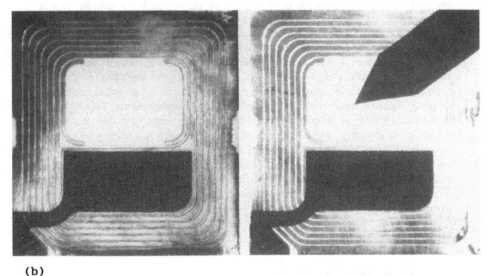

(b)

Figure 13. (a) A hand-held detector, a roll of rf antitheft targets and a roll of deactivating stickers of the antitheft devices from Checkpoint Systems, Inc. (b) Antitheft targets from Checkpoint Systems, Inc., as sold (left) and as modified (right) with a partial covering of an ambient-sensitive conducting polymer (Ref. 29).

It is known that polyacetylene is highly sensitive to chemical
as well as to thermal changes. The indicator devices are based
on the changes observed in the electrical conductivity due to either
an increase by the reactivity of a chemical (dopant) or an enormous
decrease resulting from thermal degradation. Considering the applicability of an indicator device, either possibility can be exploited.
In a simple antitheft device, a layer of doped conducting polymer
is deposited on the surface of an rf antitheft target. On the other
hand, in a time-temperature indicator, the rf target is covered
by an undoped conducting polymer; the solution of a dopant in
encapsulated form is placed on the target as well. Due to the mechanical release of the dopant, the reactivity of undoped polymer leads
to a change in electrical conductivity and provides a response to
the device during thermal exposure. This device is especially useful
in items such as frozen foods. A freeze indicator based on the
principle of a time-temperature indicator can be fabricated. In this
case, the mechanical rupture is caused by the freezing and conductivity sweeps from the insulating state to the metallic state as a
result of doping. This device can be changed into a freeze-activated
defrost indicator with slight modifications.

It is also known that diacetylene goes through a color change
as the solid-state polymerization proceeds. This activity of diacetylene
has also been applied to a design for a time-temperature indicator [29].

Use of a conducting polymer as a radiation sensor is of particular interest. A radiation indicator was developed by depositing a
polyaniline/poly(vinyl chloride) film on the surface of an rf target
[29]. The indicator works on a doping procedure. The γ-ray exposure
degrades the poly(vinyl chloride), which leads to the production of
hydrogen chloride (HCl). The polyaniline film becomes conducting
on reacting with the released HCl and helps in absorbing the radiation. An ultraviolet radiation indicator can also be constructed with
doping agents such as hydrogen hexafluorophosphate (HPF_6).
The sensitivity of a conducting polymer to humidity can also be
used in designing a moisture indicator. Similarly, using the doping
reactivity, an indicator for mechanical abuse can be constructed.
In this indicator, the mechanical rupture releases the dopant, and
the increased conductivity is indicated by lack of rf response.
The conducting-polymer-based indicators also provide flexibility,
since according to the suitability of an indicator appropriate modifications can be made. Polyacetylene, poly(p-phenylene), and polyaniline have been used in these indicator devices.

D. Transparent Loudspeaker

In an electrochemically synthesized polypyrrole, the film has an
absorption band in the visible to near-infrared region. This film

is black because visible light cannot pass through even if its thickness is only 2 μm. For some practical applications, it is desirable to have both a transparent and conducting polymer film. Recently, polypyrrole films which have both conducting and transparent properties have been developed [33,34]. One example is a polypyrrole-poly(vinyl alcohol) composite film, which can be easily prepared by a chemical vapor deposition (CVD) technique. In the preparation of this film [34], poly(vinyl alcohol) (PVA) and ferric chloride ($FeCl_3$) are first dissolved in the water. The films are cast from the solution on a polyethylene terephthalate (PET) surface. The coated film is then exposed to pyrrole and deoxygenated water vapors. This process is performed in a desiccator at -15°C. The composite films are dried under vacuum at room temperature.

It has been observed that the conductivity and transparency of the film depend on the $FeCl_3$ concentration as well as on the polymerization time as shown in Figs. 14 and 15. From these results, it has been observed that

1. The highest conductivity obtained for the film was about 8 S/cm. The higher conductivity of the film leads to lower transmittance.
2. Although the conductivity begins to saturate after a polymerization time of 30 min, the transmittance decreases gradually even before saturation.
3. The conductivity increases with $FeCl_3$ concentration, which tends to saturate for over 30 wt% of $FeCl_3$ concentration while transmittance decreases with the increase of $FeCl_3$ concentration.

SEM studies evidenced that a high concentration of $FeCl_3$ produces a rough surface film which tends to reduce the transmittance. From these results, it was concluded that a high-quality transparent and conducting film can be obtained by controlling the polymerization time and concentration of $FeCl_3$. The best results were obtained with optimum conditions: 5 wt% of $FeCl_3$ with 24 h or 30 wt% of $FeCl_3$ with 30 min. To obtain the composite film of desired properties, other combinations can be selected.

The conducting and transparent films obtained by this technique can be used in many applications. An electroacoustic device (loudspeaker) based on a piezoelectric polymer such as polyvinylidene fluoride (PVDF) and copolymer of vinylidenecyanide-vinylacetate [P(VCN-VAc)], etc., have been reported [62]. In fabricating a simple loudspeaker, metallic aluminum (Al) is evaporated on both the surfaces of a piezoelectric polymer to form electrodes (Fig. 16). The aluminum-coated piezoelectric polymer film is clamped at both ends, and an ac electric signal is applied on the electrodes. A

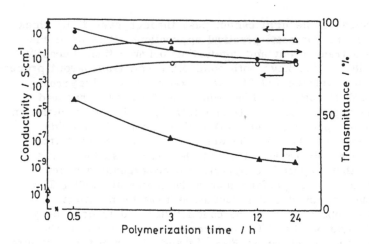

Figure 14. Polymerization time dependence of the conductivity and transmittance of polypyrrole-PVA composite films: (o, ●), PVA/FeCl$_3$ = 95/5 wt%; (△ , ▲), PVA/FeCl$_3$ = 70/30 wt%; composite film thickness, 2 µm; wavelength, 550 nm (Ref. 34).

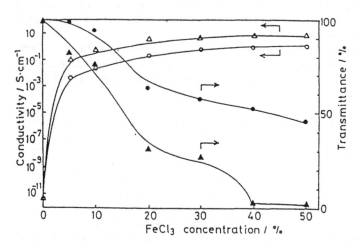

Figure 15. FeCl$_3$ concentration dependence of the conductivity and transmittance of polypyrrole-PVA composite films: (o, ●), 30 min, polymerization time; (△ , ▲), 24 h; composite film thickness, 2 µm; wavelength, 550 nm (Ref. 34).

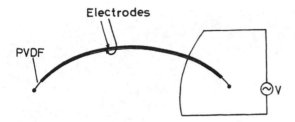

Figure 16.   The piezoelectric loudspeaker.

sinusodal strain is produced due to the piezoelectric effect, which
causes the film to vibrate and emit acoustic waves.

Because of the coated Al electrodes, this piezoelectric loudspeaker
is opaque. In some cases, it is desirable to have a transparent
loudspeaker. Therefore, to develop a transparent loudspeaker transparent electrodes are required. In the authors' laboratory, the
transparent electrodes on the surface of a piezoelectric polymer
were deposited as follows; first, a solution of PVA and $FeCl_3$ was
prepared in water and the films were cast on the surface of the
piezoelectric transparent film of P(VCN-VAc). Then the piezoelectric
films were exposed to pyrrole and water vapors in a desiccator.
The chemically deposited polypyrrole, thin, transparent, composite
film serves as a transparent electrode. This kind of loudspeaker
has for the first time been developed in Miyata's laboratory (see
Fig. 17). It has been tested and produces high-quality sound from
a CD player in the higher-frequency regions.

Generally, the vacuum tube TV is bulky, heavy, and expensive.
Recently, liquid crystal display (LCD) TVs have been developed
and are gradually appearing on the market. The size of the LCD TV
is comparatively small, though larger ones should appear soon.
The large, flat, light-weight LCD TV should attract much attention.
However, loudspeakers for this TV cannot be made as thin as required. This problem can be solved by covering the LCD screen
with this transparent loudspeaker film. This kind of TV is now
under development.

E.   Electroplating

Highly conducting polypyrrole films can be synthesized by electrochemical polymerization. Recently, Yoshikawa et al. [63] showed
that this film can also be easily synthesized by chemical polymerization. The chemical method of preparing highly conducting polypyrrole
is as follows. First polymethyl methacrylate (PMMA) is coated on a
PET surface by solution casting. Then this coated film is dipped

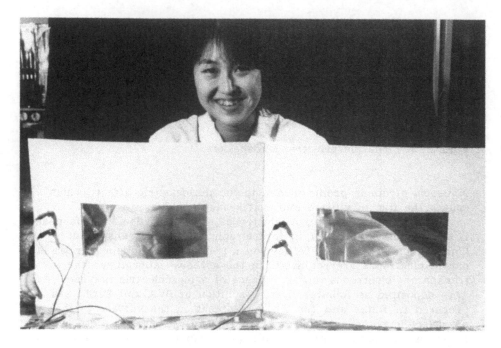

Figure 17. The transparency loudspeaker.

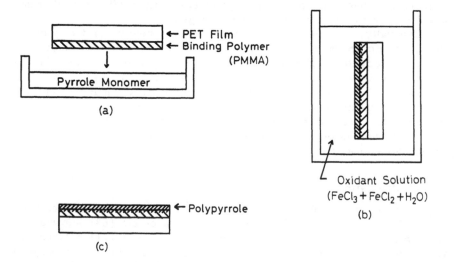

Figure 18. The chemical synthesis of highly conducting polypyrrole film (Ref. 63).

in purified pyrrole for 5 s (Fig. 18). After this, the film is trans-
ferred to 1 M $FeCl_3$ solution in distilled water and maintained at
0°C for 5 min. A suitable amount of ferrous chloride ($FeCl_2$) is
added to this solution to maintain its oxidation potential ($\overline{VS}.SCE$)
at 640 mV at the earlier stage. After polymerization, the film is
dried under vacuum at 40°C for 24 h. The polypyrrole film prepared
by this method has conductivity ca. 110 S/cm. Polypyrrole of very
high conductivity prepared by the chemical method has also been
reported by Machida et al. [64].

In electrochemical synthesis, a black film of polypyrrole is obtained on the positive electrode. If the polarity of the electrode
is reversed, then the dopant is released from the polypyrrole film
to the electrolyte. By this process, the polypyrrole film is undoped
and it becomes an insulator. If the polypyrrole film obtained by
chemical synthesis is used instead of the electrochemically synthesized
film, then the dopant is not released that easily and the film be-
haves like a common resistor. This situation persists for many
days before the dopant gradually starts releasing from the polypyrrole film, and finally this film becomes an insulator. The difference of these undoping times may result from the difference in
morphology of these two kinds of films. From the fact that the
chemically prepared polypyrrole film is very difficult to undope,
one can use this film in electroplating, as discussed below.

In an electroplating technique, it is desired in some cases to
deposit a thin layer of metals like copper on the surface of the

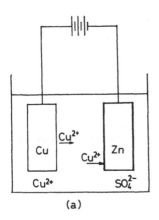
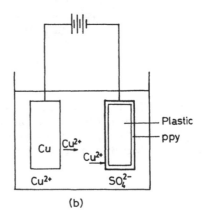

(a)   (b)

Figure 19.   The electroplating of copper on (a) Zn surface, (b)
plastic surface.

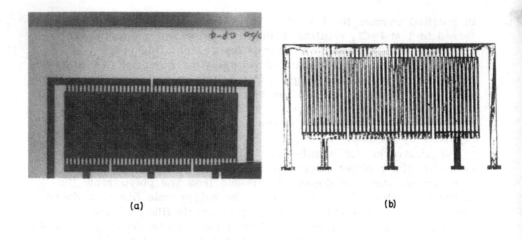

(a)   (b)

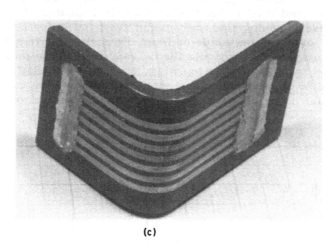

(c)

Figure 20.   The electroplating of copper on insulating surface:
(a) the polypyrrole film coated on insulating surface; (b) the
electroplating of copper on polypyrrole film; (c) the electroplating
of copper on bending polypyrrole film.

plastic. In most cases, the difficulty arises from the insulating behavior of the plastics. Figure 19a schematically shows the electroplating of copper (Cu) on zinc (Zn). As the current flows, Cu will be gradually deposited on the Zn negative electrode. This process is possible because Zn is a conductor and works as the negative electrode. This type of electroplating on a plastic surface can be accomplished by coating the surface of a plastic with conducting polypyrrole prepared by the chemical method, as discussed above. The deposition of Cu on a plastic surface is shown in Fig. 19b. In this case, the coated polypyrrole works as a negative electrode. Since the undoping process of chemically synthesized polypyrrole is very slow, it can work as a negative electrode for a longer time for the electroplating of copper. Polypyrrole film adheres well to the plastic surface, and since the surface of polypyrrole is very rough the copper also adheres well on the polypyrrole surface. This process cannot be carried out by using electrochemically synthesized polypyrrole films because polypyrrole becomes an insulator a short time after starting copper deposition.

Figure 20a,b,c shows the results of electroplating of Cu on plastic specimens by using chemically synthesized polypyrrole. The Cu surface obtained by this method is very smooth, and Cu adheres well to the specimen surface. This electroplating method will be used by Toyota Company to coat metal on the plastic surfaces of automobile parts in the near future.

IV. CONCLUSIONS

Although the newly developed technologies from conducting polymers are still in their early stages, tremendously increased research activities may lead to rapid advancements in this field. The expectation of further breakthroughs in conducting polymer devices lies with the research and development of new materials. Conducting polymers are newly emerged electronic materials, and their replacement of inorganic semiconductor-based technologies is a challenging task for the scientific community. At this stage, a substantial challenge to inorganic semiconductors does not seem feasible. It would be more appropriate to value them as specialty electronic materials. The superiority of these electronic materials still resides in their versatility and architectural flexibility to be tailored for a particular end use. New materials await origination from the chemist's laboratory; therefore, only the future will tell to what extent conducting polymers will reach their promise for applications.

ACKNOWLEDGMENTS

We are greatly indebted to Miss Kazuko Hashimoto for her constant
help and patience in preparing this manuscript. We also sincerely
thank Professors Toshiyuki Watanabe and Sukhant Tripathy for
many stimulating discussions and for critically reading the manuscript.
Literary assistance provided by the companies mentioned in this
chapter is also gratefully acknowledged.

REFERENCES

1. H. Shirakawa, E. J. Louis, A. G. MacDiarmid, C. K. Chiang,
   and A. J. Heeger, *J. Chem. Soc. Chem. Commun.*, 578 (1977).
2. K. Akagi, M. Suezaki, H. Shirakawa, H. Kyotani, M. Shimomura,
   and Y. Tanabe, *Proc. Int. Conf. on Science and Technology
   of Synthetic Metals* (ICSM '88), Santa Fe, New Mexico, 1988;
   Synth. Met. 28:D1 (1989).
3. H. Naarmann and N. Theophilou, *Synth. Met.*, *22*:1 (1987).
4. H. Naarmann, *Synth. Met.*, *17*:223 (1987).
5. J. C. W. Chien, *Polyacetylene: Chemistry, Physics and Material
   Science*, Academic Press, New York, 1984.
6. T. A. Skotheim, ed., *Handbook of Conducting Polymers*, Dekker,
   New York, 1986.
7. *Proc.* ICSM '88, M. Aldissi, ed., *Synth. Met. 28, 29* (1989).
8. M. Ogasawara, K. Funahashi, T. Hagiwara, and K. Iwata,
   Synth. Met., *14*:61 (1986).
9. S. Hotta, *Synth. Met.*, *22*:103 (1987).
10. A. G. MacDiarmid, J. C. Chiang, A. F. Richter, and A. J.
    Epstein, *Synth. Met.*, *18*:285 (1987); J. C. Chiang and A. G.
    MacDiarmid, *Synth. Met.*, *13*:193 (1986).
11. R. H. Baughmann, *Contemporary Topics in Polymer Science*,
    Vol. 5, Plenum, New York, 1984, pp. 321–350.
12. A. G. MacDiarmid and R. B. Kaner, in *Handbook of Conducting
    Polymers*, Vol. 1 (T. A. Skotheim, ed.), 1986, pp. 690–727.
13. A. F. Diaz, J. J. Castillo, J. A. Logan, and W. Y. Lee,
    J. Electroanal. Chem., *129*:115 (1981).
14. F. Garnier, G. Tourillon, M. Gazard, and J. C. Dubois,
    J. Electroanal. Chem., *148*:299 (1983).
15. O. Inganas and I. Lundstrom, *Synth. Met.*, *21*:13 (1987).
16. H. Yoneyama, K. Wakamoto, and H. Tamura, *J. Electrochem.
    Soc.*, *132*:2414 (1985); R. B. Bjorklund and F. Lundstrom,
    J. Electron. Mat., *14*:39 (1985).
17. M. Akhtar, H. A. Weakliem, R. M. Paiste, and K. Gaughan,
    Synth. Met., *26*:203 (1988).

18. A. J. Frank, *Energy Resources Through Photochemistry and Catalysis*, M. Gratzel, ed., Academic, New York, 1983, pp. 467-505.
19. R. Noufi, D. Tench, and I. F. Warren, *J. Electrochem. Soc.*, 128:2596 (1981).
20. T. A. Skotheim, O. Inganas, J. Prejza, and I. Lundstrom, *Mol. Cryst. Liq. Cryst.*, 83:1361 (1982).
21. S. Glenis and A. J. Frank, *Synth. Met.*, 28:C681 (1989).
22. A. E. Hadri, O. Maleysson, and H. Robert, *Synth. Met.*, 28:C697 (1989).
23. H. Koezuka and A. Tsumura, *Synth. Met.*, 28:C753 (1989).
24. M. Josowicz and J. Janata, *Chemical Sensor Technology*, Vol. 1, T. Seiyama, ed., Elsevier, Amsterdam, 1988, pp. 153-177.
25. C. B. Duke and H. W. Gibson, *Encyclopedia of Chemical Technology*, Vol. 18, Wiley, New York, 1982, pp. 755-793.
26. H. Kuzmany, M. Mehring, and S. Roth, eds., *Electronic Properties of Polymers and Related Compounds*, Springer-Verlag, Berlin, 1985, and references therein.
27. T. F. Otero and E. D. Larreta, *Synth. Met.*, 26:79 (1988).
28. O. Niwa and T. Tamamura, *Synth. Met.*, 20:235 (1987).
29. R. H. Baughman, R. L. Elsenbaumer, Z. Iqbal, G. G. Miller, H. Eckhardt, U.S. Patent, 4, 646, 066 (1987); *Electronic Properties of Conjugated Polymers*, Vol. 76, H. Kuzmany, M. Mehring, and S. Roth, eds., Springer-Verlag, Berlin, 1987, pp. 432-439.
30. M. K. Malros, U.S. Patent, 4, 334, 880, 1982.
31. G. Berthet, J. P. Blanc, J. P. Germain, A. Larbi, C. Maleysson, and H. Robert, *Synth. Met.*, 18:715 (1987).
32. K. Yoshino, H. S. Nalwa, W. F. Schmidt, and J. G. Rabe, *Polym. Commun.*, 26:103 (1985).
33. M. Kobayashi, N. Colaneri, M. Boysel, F. Wudl, and A. J. Heeger, *J. Chem. Phys.*, 82:5717 (1985).
34. T. Ojio and S. Miyata, *Polym. J.*, 18:95 (1986).
35. A. J. Heeger, J. Orenstein, and D. R. Ulrich, eds., *Nonlinear Optical Properties of Polymers*, Material Research Society Symp. Proc., Vol. 109, 1988.
36. K. Yoshino, *Synth. Met.*, 28:C669 (1989).
37. K. Yoshino, R. Sugimoto, J. G. Rabe, and W. F. Schmidt, *Jpn. J. Appl. Phys.*, 24:L33 (1985).
38. K. Yoshino, M. Ozaki, and R. Sugimoto, *Jpn. J. Appl. Phys.*, 24:L373 (1985).
39. A. F. Diaz, J. F. Rubinson, and H. B. Mark, Jr., *Adv. Polym. Sci.*, 84:113 (1988); J. E. Frommer and R. R. Chance, *Encyclopedia of Polymer Science and Engineering*, Vol. 5, 1986, pp. 462-507.

40. P. J. Nigrey, A. G. MacDiarmid, and A. J. Heeger, J. Chem. Soc. Chem. Commun., 594 (1979).
41. R. B. Kaner and A. G. MacDiarmid, Scientific American, February 1988, pp. 60-65.
42. D. Naegele and R. Bittihn, Sixth Int. Conf. on Solid State Ionics, Germisch partenkirchen, September 1987.
43. P. Passiniemi and J. E. Osterholm, Synth. Met., 18:637 (1987).
44. R. H. Baughmann, private communication.
45. L. W. Shacklette, T. R. Jow, M. Maxfield, and R. Hatami, Synth. Met., 28:C655 (1989).
46. H. Lethely, J. Chem. Soc., 15:161 (1962).
47. D. M. Mohilner, R. N. Adams, and W. J. Argensinger, Jr., J. Am. Chem. Soc., 84:3618 (1962).
48. A. F. Diaz and J. A. Logan, J. Electroanal. Chem., 111:111 (1980).
49. T. Nakajima and T. Kawagoe, Synth. Met., 28:c629 (1989).
50. Bridgestone Co., private communication, 1989.
51. D. A. McLean and F. S. Power, Proc. IRE, 44:872 (1965).
52. S. Niwa, Synth. Met., 18:665 (1987).
53. Y. Ito and S. Yoshimura, J. Electrochem. Soc., 124:1128 (1987).
54. I. Isa, Nikkei New Material, December 19, 1988.
55. I. Isa, Nikkei New Material, January 30, 1989.
56. T. Maruyama and Shimada, Electronic Packing Technology, 5:63 (1989).
57. M. Fukuda, H. Yamamoto, and I. Isa, U.S. Patent, 4, 780, 796, 1988.
58. I. Isa, private communication.
59. H. Shirakawa and S. Ikeda, Polym. J., 2:231 (1971).
60. T. Ito, H. Shirakawa, and S. Ikeda, J. Polym. Sci. Polym. Chem. Ed., 12:11 (1974).
61. J. H. Edwards and W. J. Feast, Polymer, 21:595 (1980).
62. G. M. Garner, Application of Ferroelectric Polymers, T. T. Wang, J. M. Herbert, and A. M. Glass, eds., 1988, pp. 190-208.
63. T. Yoshikawa, S. Machida, T. Ikegami, A. Techagumpuch, and S. Miyata, Polym. J., 22:1 (1990).
64. S. Machida, A. Techagumpuch, and S. Miyata, Synth. Met., 31:311 (1989).

Index

Milton Keynes UK
Ingram Content Group UK Ltd.
UKHW020020071024
449327UK00032B/2869